Analyzing Spatial Models of Choice and Judgment

Second Edition

Chapman & Hall/CRC
Statistics in the Social and Behavioral Sciences Series

Recently Published Titles

Multilevel Modelling Using Mplus
Holmes Finch and Jocelyn Bolin

Applied Survey Data Analysis, Second Edition
Steven G. Heering, Brady T. West, and Patricia A. Berglund

Adaptive Survey Design
Barry Schouten, Andy Peytchev, and James Wagner

Handbook of Item Response Theory, Volume One: **Models**
Wim J. van der Linden

Handbook of Item Response Theory, Volume Two: **Statistical Tools**
Wim J. van der Linden

Handbook of Item Response Theory, Volume Three: **Applications**
Wim J. van der Linden

Bayesian Demographic Estimation and Forecasting
John Bryant and Junni L. Zhang

Multivariate Analysis in the Behavioral Sciences, Second Edition
Kimmo Vehkalahti and Brian S. Everitt

Analysis of Integrated Data
Li-Chun Zhang and Raymond L. Chambers, Editors

Multilevel Modeling Using R, Second Edition
W. Holmes Finch, Joselyn E. Bolin, and Ken Kelley

Modelling Spatial and Spatial-Temporal Data: A Bayesian Approach
Robert Haining and Guangquan Li

Measurement Models for Psychological Attributes
Klaas Sijtsma and Andries van der Ark

Handbook of Automated Scoring: Theory into Practice
Duanli Yan, André A. Rupp, and Peter W. Foltz

Interviewer Effects from a Total Survey Error Perspective
Kristen Olson, Jolene D. Smyth, Jennifer Dykema, Allyson Holbrook, Frauke Kreuter, Brady T. West

Statistics and Elections: Polling, Prediction, and Testing
Ole J. Forsberg

Big Data and Social Science: Data Science Methods and Tools for Research and Practice, Second Edition
Ian Foster, Rayid Ghani, Ron S. Jarmin, Frauke Kreuter and Julia Lane

Analyzing Spatial Models of Choice and Judgment
Second Edition
David A. Armstrong II, Ryan Bakker, Royce Carroll, Christopher Hare, Keith T. Poole and Howard Rosenthal

For more information about this series, please visit:
https://www.routledge.com/Chapman--HallCRC-Statistics-in-the-Social-and-Behavioral-Sciences/book-series/CHSTSOBESCI

Analyzing Spatial Models of Choice and Judgment

Second Edition

David A. Armstrong II
Ryan Bakker
Royce Carroll
Christopher Hare
Keith T. Poole
Howard Rosenthal

CRC Press
Taylor & Francis Group
Boca Raton London New York

CRC Press is an imprint of the
Taylor & Francis Group, an **informa** business
A CHAPMAN & HALL BOOK

First edition published 2014
by CRC Press
6000 Broken Sound Parkway NW, Suite 300, Boca Raton, FL 33487-2742

and by CRC Press
2 Park Square, Milton Park, Abingdon, Oxon, OX14 4RN

First edition published by CRC Press

CRC Press is an imprint of Taylor & Francis Group, LLC

ISBN: 9781138715332 (hbk)
ISBN: 9780367612542 (pbk)
ISBN: 9781315197609 (ebk)

Typeset in Computer Modern font
by KnowledgeWorks Global Ltd

Contents

Preface

> *[I]t is very common to think and to speak about politics in positional terms. Indeed it is very difficult to analyze real political debates without using positional language and reasoning... Most people—including those who are blissfully unaware of the mysteries of political science, as well as those who are utterly dismissive of them—find it difficult to talk about real politics in tooth and claw without using the notions of position, distance, and movement on the important matters at issue. These notions thus seem to have deep roots in the ways that people, from many different walks of life, think about and describe politics.* – Kenneth Benoit and Michael Laver, *Party Policy in Modern Democracies* (London: Routledge, 2006), p. 12.

The spatial model of voting is the most successful model in the field of political science. It has a theoretically rich history, with roots stretching back to Aristotle (Hinich and Munger, 1997, pp. 21–23). Virtually all political conflict can be (and routinely is) represented spatially, from the enduring left-right division that originated in the seating arrangements of the Jacobins and the Girondins in the National Assembly of revolutionary France to the cross-cutting civil rights division between Northern and Southern Democrats in the United States in the mid-twentieth century.

Spatial models have greatly enriched the study of politics. Indeed, as articulated in the above quote, it seems natural to model political competition in spatial terms. With recent advances in computing power and the widespread availability of political choice data (e.g., legislative roll call and public opinion survey data), the empirical estimation of spatial models has never been easier or more popular.

The aim of this book is to demonstrate how to estimate and interpret spatial models from different types of data sources using a variety of methods. We focus primarily on implementations using the popular, open-source programming language R (Team, 2018). Where appropriate, we also make use of the JAGS (Plummer, 2003a) software for Bayesian inference. This second edition expands our coverage to discussions of several new packages and reflects a number of updates to the methods and software discussed in the first edition. Further, we have assembled an R package that provides all of the data and new functions used in the book. This makes the second edition easier to engage than the first as many of the routines that the user had to run "by hand" in the first edition, now exist in functions. You can install the package with the following:

```
library(devtools)
install_github("davidaarmstrong/asmcjr")
```

In addition to the functionality that the package provides, we have also reconfigured the visualizations to use the ggplot2 package, where possible.

Overview of the Book

We have written this book to be accessible to researchers who will apply the methods to their data as well as to more expert methodologists. In each chapter we explain the basic theory behind a spatial model, show the associated estimation techniques and detail their historical development. In the last section of each chapter we discuss some benefits and drawbacks of the methods. In terms of software, we assume that the reader has a basic familiarity with R. In terms of mathematics, we assume that the reader is familiar with matrix algebra, although most readers will be able to use the applications without following all the parts of the exposition that use matrix expressions.

We demonstrate how each method can be performed, detailing each step of the process with R code used on actual data sets (which we have posted on this book's website located at `https://quantoid.net/asmcjr).()`

In Chapter 2, we discuss how to analyze data from issue scales. Issue scales encompass a large class of data that is commonly analyzed with spatial models. Our focus is on surveys that ask respondents to place themselves and/or stimuli on issue or attribute scales. This is a very common type of data gathered by social scientists. For instance, the American National Election Study has been collecting seven-point issue scale data since 1968. The endpoints of these scales are labeled and the respondent is asked to place herself on the scale (her "ideal point") along with a set of political figures, parties, and, in some cases, current government policy.

Issue scales often include the standard ideological continuum (where the endpoints are "extremely liberal" and "extremely conservative") or major policy issues like the role of government in health care. Issue scale questions frequently appear as Likert-type items in which the respondent is read a political statement (e.g., "homosexual couples should be allowed to marry") and asked to register her opinion on a scale ranging from "strongly agree" to "strongly disagree." Chapter 2 demonstrates how individual issue scales and sets of issue scales can be analyzed using the maximum likelihood implementation of Aldrich-McKelvey scaling (Aldrich and McKelvey, 1977; Hare et al., 2015), and Poole's (1998*a*; 1998*b*) basic space method in R with the basicspace package (Poole et al., 2013). We also discuss the Ordered Optimal Classification (Hare, Liu and Lupton, 2018), semiparametric extension of the nonparametric Optimal Classification procedure for the analysis of issue scale data.

Chapter 3 is devoted to the analysis of similarities and dissimilarities data. This data is relational and organized as a square matrix, where the stimuli comprise both the rows and columns. The entries represent the level of similarity or dissimilarity between objects (the diagonal takes on the highest value of similarity, since objects are, of course, identical to themselves). Frequently, this class of data comes in the form of the familiar correlation matrix. The analysis of this type of data stretches back more than a century to early data reduction techniques: Karl Pearson's (1901) development of a form of principal components analysis using least squares and Charles Spearman's (1904) computation of correlation matrices, which he analyzed with a form of factor analysis. The publication of the Eckart-Young theorem in 1936 (Eckart and Young, 1936) solved the general least squares problem of approximating one matrix by another of a lower rank. Later, Warren Torgerson (1952; 1958), using the Eckart-Young theorem and the results of Young and Householder (1938), developed classical multidimensional scaling where points are calculated directly from a transformation of a symmetric matrix of squared distances (dissimilarities).

In discussing the statistical analysis of similarities data in Chapter 3, we focus primarily on multidimensional scaling (MDS) methods.* The aim of MDS techniques is to extract a configuration of coordinates in low-dimensional space where the inter-point distances are monotonic with the level of dissimilarity between stimuli. MDS methods can be divided into two categories: metric and nonmetric. Metric MDS extracts distance data from an interval-level transformation of the data, while nonmetric MDS uses only the ordinal properties of the data to construct distances. In both cases, we wish to minimize a loss function based on the disparities between the observed and reproduced distances between the stimuli. To do so we introduce the SMACOF (Scaling by Majorizing a Complicated Function) optimization method (de Leeuw, 1977, 1988; de Leeuw and Heiser, 1977) that is implemented in a variety of MDS functions in the `smacof` (de Leeuw and Mair, 2009) package in `R`.

In Chapter 4, we discuss unfolding analysis of rating scale data. This type of data is in the form of a rectangular matrix where the rows are respondents and the columns are stimuli. Examples of this kind of data are feeling thermometers in which respondents place a politician or group on a 0 to 100 point scale labeled "cold and unfavorable feeling" to "warm and favorable feeling" with 50 being neutral. Thermometer data on political figures have been gathered by the American National Election Studies since 1968.† Propensity to vote measures are another kind of rating scale data that are commonly employed in European public opinion surveys like the European Election Study (EES). With this type of item, respondents are asked to rate their propensity to vote

*For a discussion of alternative methods to analyze similarities data, such as factor analysis, we refer the reader to Bartholomew et al. (2008) and Mulaik (2009)

†Some social group data was asked in the 1964 National Election Study.

for a certain party or candidate on a 0 to 10 point scale. In this chapter we return to the `smacof` package (de Leeuw and Mair, 2009).

Chapter 5 is concerned with unfolding binary choice data such as legislative roll calls. There is now a very large literature on methods used to analyze this type of data. We begin the chapter with a short historical overview of the research on roll call voting prior to the 1980s. We then continue with the work of Poole and Rosenthal. In the 1980s Poole and Rosenthal combined the random utility model developed by McFadden (1976), the spatial (geometric) model of voting, and alternating estimation methods developed in psychometrics (Chang and Carroll, 1969; Carroll and Chang, 1970; Young, de Leeuw and Takane, 1976; Takane, Young and de Leeuw, 1977) to develop NOMINATE, an unfolding method for parliamentary roll call data (Poole and Rosenthal, 1985, 1991, 1997; Poole, 2005; McCarty, Poole and Rosenthal, 1997).

The NOMINATE model assumes that legislators have ideal points in an abstract policy space and vote for the policy alternative closest to their ideal point. Each roll call vote has two policy points—one corresponding to Yea and one to Nay. Consistent with the random utility model, each legislator's utility function consists of (1) a *deterministic* component that is a function of the distance between the legislator and a roll call outcome; and (2) a *stochastic* component that represents the idiosyncratic component of utility. The deterministic portion of the utility function is assumed to have a normal distribution and voting is probabilistic. An alternating method is used to estimate the parameters in which legislator ideal points and roll call parameters are each estimated given values of the other. Poole and Rosenthal's W-NOMINATE program is available in the **`wnominate`** package in R (Poole et al., 2011) and includes a parametric bootstrap option to obtain standard errors for all the parameters (Lewis and Poole, 2004).

As advances in computing power popularized simulation methods for the estimation of complex multivariate models, these methods were fused with long-standing psychometric models. Specifically, Markov chain Monte Carlo (MCMC) simulation methods (Metropolis and Ulam, 1949; Hastings, 1970; Geman and Geman, 1984; Gelfand and Smith, 1990; Gelman, 1992) within a Bayesian framework (Gelman et al., 2014; Gill, 2008) provide a class of sampling algorithms that can be used to estimate these models, even with large numbers of parameters.

We then proceed to an exposition of Poole's (2000; 2005) Optimal Classification (OC) estimation procedure. OC is a nonparametric unfolding method for binary choice data. As a nonparametric procedure, OC does not rely on distributional assumptions about the functional form of legislators' deterministic utility or the underlying error process present in roll call voting. OC can be performed in R using the oc package (Poole et al., 2012). We conclude Chapter 5 with a comparison of all the methods discussed in the chapter.

The final chapter, Chapter 6, focuses on Bayesian approaches to the types of data and methods described in the previous chapters. We first discuss the Bayesian implementation of Aldrich-McKelvey scaling (Aldrich and McKelvey,

1977; Hare et al., 2015) which enables applications to data with missing responses and provides a measure of uncertainty for estimates. Next we demonstrate how Bakker-Poole (2013) Bayesian metric unfolding can be performed in `R`.

In addition, we discuss the application of a Bayesian framework to metric MDS (Bakker and Poole, 2013). The advantage of the Bayesian approach is that it allows the researcher to "explore" the posterior densities of the parameters in order to assess and measure uncertainty in the point estimates.

We then examine several Bayesian models designed for binary data. First, a Bayesian implementation of the NOMINATE model known as α-NOMINATE tests which functional form (the quadratic or normal [Gaussian] distribution) of the deterministic utility function best fits the data (Carroll et al., 2013). We discuss the α-NOMINATE model and use of the the α-NOMINATE procedure, which is available in the `anominate` package (Lo et al., 2013).

Next, we have an extensive discussion of Bayesian methods adapted to perform unfolding analysis of binary and ordinal choice data using Item Response Theory (IRT). The Bayesian IRT model has been applied in political science to study binary choices such as roll call data from legislatures and courts (especially due to the work of Martin, Quinn [Schofield et al. 1998; Quinn, Martin and Whitford 1999; Martin and Quinn 2002; Quinn and Martin 2002; Martin 2003; Quinn 2004] and Jackman [Jackman 2000a,b, 2001; Clinton, Jackman and Rivers 2004]) as well as to analyze public opinion data (Treier and Hillygus, 2009; Jessee, 2009). In these methods, the foundation is the spatial theory of voting and the random utility model described above, but estimations are based on sampling from conditional distributions for the legislator and roll call parameters using the Gibbs sampler (Geman and Geman, 1984; Gelfand and Smith, 1990). Although not intrinsic to the estimation method, Bayesian MCMC applications thus far have used a quadratic deterministic utility function. We will show a variety of applications using Simon Jackman's pscl package and Martin and Quinn's `MCMCpack` package (Jackman, 2012; Martin, Quinn and Park, 2011). We also discuss heteroskedastic IRT (Lauderdale, 2010), which allows variation in the degree to which subgroups of respondents deviate from a predicted response pattern.

Finally, Chapter 6 addresses a number of extensions to spatial voting models using Bayesian methods. First, we demonstrate the extension of the Bayesian IRT model to the analysis of polytomous (i.e., ordinal) choice data and dynamic IRT models. We conclude with a discussion of Imai, Lo, and Olmsted's (2016) application of expectation-maximization (EM) algorithm to the IRT models discussed in this and earlier chapters, which provides dramatically improved efficiency in obtaining ideal points comparable to those in computationally intensive IRT models and promising results for "big data" applications. This includes a discussion of text analysis techniques designed for the frequency data associated with word counts.

An important distinction we wish to draw throughout this book is that these methods are intended to model distances between the points, not the locations of the points themselves. As an analogy, one can picture the configurations formed by a double helix or a tinker toy structure; we can move and rotate these objects about, but the configurations themselves are unaffected. The lesson here is that the locations of these points are arbitrary for descriptive and inferential purposes so long as the interpoint distances remain identical. This distinction has important implications for how we understand scaling results, particularly how uncertainty promulgates throughout the entire point configuration and how results are identified.

Further Reading

Readers may find the following works useful in learning more about spatial voting theory, measurement models and estimation methods, and the R programming language:

Bartholomew, David J., Fiona Steele, Irini Moustaki and Jane I. Galbraith. 2008. *Analysis of Multivariate Social Science Data.* 2nd ed. Boca Raton, FL: Chapman & Hall/CRC.

Borg, Ingwer and Patrick J.F. Groenen. 2010. *Modern Multidimensional Scaling: Theory and Applications.* 2nd ed. New York: Springer.

Borg, Ingwer, Patrick J.F. Groenen, and Patrick Mair. 2018. *Applied Multidimensional Scaling and Unfolding.* 2nd ed. New York: Springer.

Enelow, James M. and Melvin J. Hinich. 1984. *The Spatial Theory of Voting.* Cambridge: Cambridge University Press.

Gelman, Andrew, John B. Carlin, Hal S. Stern, David B. Dunson, Aki Vehtari, and Donald B. Rubin. 2014. *Bayesian Data Analysis.* 3rd ed. Boca Raton, FL: CRC Press.

Jackman, Simon. 2009. *Bayesian Analysis for the Social Sciences.* New York: Wiley.

Jacoby, William G. 1991. *Data Theory and Dimensional Analysis.* Thousand Oaks, CA: Sage.

Jones, Owen, Robert Maillardet and Andrew Robinson. 2009. *Introduction to Scientific Programming and Simulation Using R.* Boca Raton, FL: Chapman & Hall/CRC.

Matloff, Norman. 2011. *The Art of R Programming: A Tour of Statistical Software Design.* San Francisco: No Starch Press.

Poole, Keith T. 2005. *Spatial Models of Parliamentary Voting.* Cambridge: Cambridge University Press.

Poole, Keith T. and Howard Rosenthal. 2007. *Ideology and Congress.* New Brunswick, NJ: Transaction.

Teetor, Paul. 2011. *R Cookbook.* Sebastopol, CA: O'Reilly.

Using This Book with R

The methods discussed in this book draw on the resources of a wide range of separate packages in R. Because R is an open-source, collaborative enterprise that undergoes constant revision, some of these packages may become deprecated or removed over time. To ensure compatibility, we have bundled these packages into a single package (asmcjr), which is available at the book github and website described above. The website also stores the R code and data sets used in the book and required by the exercises at the end of each chapter.

Thanks

The authors would like to thank the Alston family and their funding of the Philip H. Alston, Jr. Distinguished Chair in the Department of Political Science at the University of Georgia. The Alston Chair generously provided funding throughout the course of this project. We would also like to extend our gratitude to Jeffrey B. Lewis and James Lo for their invaluable assistance with the interfacing between R and C in the sections on Bayesian metric unfolding and α-NOMINATE. Finally, we wish to acknowledge the contributions of William G. Jacoby, Jan de Leeuw, Walter R. Mebane, Jr. and Kevin M. Quinn. This manuscript was greatly improved by their helpful comments and suggestions at various stages of its development. We also would like to thank the reviewers for the second edition who provided valuable feedback on drafts of the manuscript and David Liao, Laurynas Stankevicius and Charlotte Massetti for feedback and editorial assitance.

Author Biographies

David A. Armstrong II (http://quantoid.net) is Canada Research Chair in Political Methodology and Associate Professor of Political Science at Western University in Ontario, Canada. He received a Ph.D. in Government and Politics from the University of Maryland in 2009 and was a post-doctoral fellow in the Department of Politics and Nuffield College at the University of Oxford. His research interests revolve around measurement and the relationship between Democracy and state repressive action. His research has been published in the *American Political Science Review,* the *American Journal of Political Science,* the *American Sociological Review* and the *R Journal* among others. Dave is an active R user and maintainer of a number of packages. DAMisc has a number of functions that ease interpretation and presentation of GLMs. `factorplot` implements a novel method for visualizing pairwise comparisons optionally with multiple testing corrections. **bsmds**, written with Bill Jacoby, implements a bootstrapping algorithm for multidimensional scaling solutions (see Jacoby and Armstrong, 2014). Dave has taught courses on advanced linear regression, R and measurement at the Inter-university Consortium for Political and Social Research (ICPSR) Summer Program in Quantitative Methods of Social Research since 2006.

Ryan Bakker is Reader in Comparative Politics at the University of Essex. He received his Ph.D. in Political Science from the University of North Carolina at Chapel Hill in 2007. His research and teaching interests include applied Bayesian modeling, measurement, Western European politics, and EU elections and political parties. He is a principal investigator for the Chapel Hill Expert Survey (CHES), which measures political party positions on a variety of policy-specific issues in the European Union. Ryan has taught the Introduction to Applied Bayesian Modeling for the Social Sciences course at the ICPSR Summer Program since 2008. He has also taught classes on data analysis and political methodology at the University of Oxford. His work has appeared in *Political Analysis, Electoral Studies, European Union Politics,* and *Party Politics.*

Royce Carroll is Professor in Comparative Politics at the University of Essex, where he teaches graduate and undergraduate courses on comparative politics and American politics. He received his Ph.D. in Political Science at the University of California, San Diego in 2007. In addition to political methodology, his research focuses on comparative politics of legislatures, coalitions and political parties, as well as measurement of ideology. Carroll's publications have appeared in a number of academic journals, including *American Journal of Political Science, Electoral Studies, Political Analysis*, and *Legislative Studies Quarterly*. Carroll is also Director of the Essex Summer School in Social Science Data Analysis.

Christopher Hare is Assistant Professor in Political Science at the University of California, Davis. He received his Ph.D. in Political Science at the University of Georgia in 2015. His substantive research agenda focuses on ideology and voting behavior in the American electorate, campaign strategy, and political polarization. Methodologically, his work focuses on measurement models, ideal point estimation, and the application of machine learning methods to model political behavior. Since 2017, Hare has taught courses and workshops on the use of machine learning in the social sciences at the ICPSR Summer Program. His research has appeared in the *American Journal of Political Science, Political Science Research and Methods, and Public Choice.* He has also consulted for political campaigns on statistical modeling problems.

Keith T. Poole is Philip H. Alston Jr. Distinguished Professor, Department of Political Science, University of Georgia. He received his Ph.D. in Political Science from the University of Rochester in 1978. His research interests include methodology, political-economic history of American institutions, economic growth and entrepreneurship, and the political-economic history of railroads. He is the author or coauthor of over 50 articles as well as the author of *Spatial Models of Parliamentary Voting* (Cambridge University Press, 2005), a coauthor of *Political Bubbles: Financial Crises and the Failure of American Democracy* (Princeton University Press, 2013), *Polarized America: The Dance of Ideology and Unequal Riches* (MIT Press, 2006), *Ideology and Congress* (Transaction Publishing, 2007), and *Congress: A Political-Economic History of Roll Call Voting* (Oxford University Press, 1997). He was a Fellow of the Center for Advanced Study in Behavioral Sciences 2003–2004 and was elected to the American Academy of Arts and Sciences in 2006.

Howard Rosenthal is Professor of Politics at NYU and Roger Williams Straus Professor of Social Sciences, Emeritus, at Princeton. Rosenthal's coauthored books include *Political Bubbles: Financial Crises and the Failure of American Democracy, Polarized America: The Dance of Ideology and Unequal Riches, Ideology and Congress*, and *Prediction Analysis of Cross Classifications.* He has coedited *What Do We Owe Each Other?* and *Credit Markets for the Poor.* Rosenthal is a member of the American Academy of Arts and Sciences. He has been a Fellow of the Center for Advanced Study in Behavioral Sciences and a Visiting Scholar at the Russell Sage Foundation.

1

Introduction

The purpose of this book is to give a comprehensive introduction to estimating spatial (geometric) models from political choice data with `R` (Team, 2018). These models are geometric in that they use the relative positions of points in an abstract space to represent data that can be interpreted as distances (also known as *relational* data). We will use the term "geometric model" interchangeably with "spatial model" throughout this book. The reason is that the term "spatial" is also used in the field of spatial statistics to refer to the analysis of *physically* proximate units. Unfortunately, this has caused some confusion because the word "spatial" is used for both classes of models.

The aim of a spatial model is to produce a geometric representation or *spatial map* of some quantity. For example, a spreadsheet that tabulates all the distances between pairs of sizable cities in France contains the same information as the corresponding map of France, but the spreadsheet gives you no idea what France looks like.* Embedded within the distance data is the information necessary to construct a map, and the goal of scaling methods is to recover this spatial map. Though a spatial map contains essentially the same information as the table of inter-city distances, the spatial map provides a more readily interpretable visual representation of patterns in the data (Tufte, 1983).

In the above example, the *dimensionality*—understood as "the number of separate and interesting sources of variation among the objects" (Jacoby, 1991, p. 27)—of the data is known *a priori*. That is, we know in advance the number (two) and meaning (north-south and east-west) of the dimensions needed to represent geometrically the inter-city distances data. However, this is not always the case. For instance, we may only have suspicions about the number and meaning of the dimensions necessary to model a data set of legislative roll call votes (in which legislators vote Yea or Nay or abstain on a series of policy proposals). In this case, spatial models can *discover* as well as present patterns in the data.

In political science, spatial models are also used to *measure* latent (unobservable) quantities from observed indicators. For instance, we might measure

*See, for example, Jordan Ellenberg, "Growing Apart: The Mathematical Evidence for Congress' Growing Polarization," *Slate Magazine*, 26 December 2001, available from `http://www.slate.com/articles/life/do_the_math/2001/12/growing_apart.html` for an application of the analogy between spatial and geographic maps to contemporary American politics.

a voter's level of conservatism (i.e., her position on an ideological dimension) using a series of survey questions about her issue positions. This type of information can be used to address a range of important questions in political science: the level of polarization in legislatures and electorates (McCarty, Poole and Rosenthal, 2006; Ansolabehere, Rodden and Snyder, 2006; Shor and McCarty, 2011), the ideological makeup of campaign contributors (Bonica, 2013), and the quality of representation of voter preferences (Gerber and Lewis, 2004; Bafumi and Herron, 2010).

We deal with a broad class of spatial (geometric) models in this book because what is loosely called "spatial models" is actually a collection of theory and estimation methods developed in the fields of psychology, economics, and political science (see Poole, 2017). In psychology, various methods of multidimensional scaling (MDS) have been developed during the past fifty years to analyze similarity and preferential choice data. For example, a set of respondents are asked to judge how similar various colors are to each other (e.g., Ekman 1954). MDS methods model these similarities as distances between points representing the colors in a geometric space. MDS techniques are designed to produce a spatial map that summarizes a large set of data graphically, where the axes of the graph represent the latent, organizing dimensions that account for variation in the data (e.g., light to dark, poor to rich, or liberal to conservative). The research literature has shown that geometric models are not only convenient ways of representing simple patterns in a stream of data but also, (indeed, *because* this is the case), are good models of how humans process information and make decisions (see, e.g., Hare and Poole (2014*a*)).

1.1 The Spatial Theory of Voting

Economists, psychologists and political scientists all contributed to the development of the spatial theory of voting, which states that political preferences can be represented in abstract space and that individuals vote for the candidate or policy alternative closest to them. In this regard, a spatial map is literally a visual representation of where voters and candidates are to be found in a low-dimensional (ideological) space. A spatial map shows locations in a space defined by, for example, a left-right or liberal-conservative dimension. Although Hotelling (1929) and Smithies (1941) are credited with originating the idea, it was the publication of Anthony Downs's *An Economic Theory of Democracy* in 1957 that really established spatial theory as a conceptual tool. Hotelling studied the logic of the location of a grocery store in a "linear" town—that is, a town strung out along a highway, where all the houses face a single road. In Hotelling's famous model, the optimum location for a grocery store is the median of the town (the median minimizes the sum of the walking distances to the store). Hotelling (1929, p. 54) noted the relevance of this result to the political world as he saw it:

> The competition for votes between the Republican and Democratic parties does not lead to a clear drawing of issues, an adoption of two strongly contrasted positions between which the voter may choose. Instead, each party strives to make its platform as much like the other's as possible. Any radical departure would lose many votes, even though it might lead to stronger commendation of the party by some who would vote for it anyhow. Each candidate "pussyfoots," replies ambiguously to questions, refuses to take a definite stand in any controversy for fear of losing votes.

Downs took the Hotelling-Smithies model of spatial competition of stores and applied it to the competition between political parties. He assumed that voters were distributed over a choice dimension—for example, the level of government spending on military defense—and that political parties played the role of the stores. For example, a right-wing, militaristic party would be near one end of the spectrum, while a left-wing, pacifistic party would be on the opposite end. If voters vote for the party closest to them on the dimension, the parties will converge to the median ideal point. Duncan Black (1948; 1958) had earlier derived the median voter theorem for voting in committees.[†]

In the spatial theory of voting, voters are assumed to have *ideal points* that mark their most preferred outcome. For example, a voter whose ideal point is for the government to allocate 10% of the budget to military spending will receive the most utility if this policy is adopted. Voters also have *utility functions* across the range of policy alternatives—in this case, from 0% to 100% of the budget allocated to national defense—that specify the amount of utility voters obtain from each outcome. Utility functions are assumed to be single-peaked (that is, highest at the voter's ideal point) and symmetric (monotonically decreasing in the distance from the ideal point) (Enelow and Hinich, 1984).

Figure 1.1 shows three commonly used utility functions in spatial voting models: the quadratic form, the normal (Gaussian) form, and the linear (absolute distance) form.[‡] Each functional form satisfies the criteria of single-peakedness and symmetry: individuals (in this case, a voter whose ideal point or most preferred outcome is at "0") gain the most utility when the outcome is at their ideal point, and grow increasingly dissatisfied as the proposal moves away from their ideal point. Historically, the quadratic and linear forms were

[†]Black's median voter theorem is central not just to spatial voting models, but the development of the discipline of political science in general. William Riker (1990*b*, p. 178) writes that it was "certainly the greatest step forward in political theory in this century."

[‡]The quadratic is also known as parabolic, the normal as bell-shaped, and the linear as tent.

most commonly used, in large part because they are analytically simpler and more computationally tractable than the normal form (for a voter with ideal point 0 and policy x, the equation for the linear form is: $y = -|x| + a$ and the quadratic form is simply: $y = -x^2 + a$, where a is an arbitrary constant). The quadratic form is more commonly used than the linear form because it is differentiable at the ideal point and because, in models involving uncertainty, it permits a mean variance decomposition. The quadratic form is also attractive in modeling risk-averseness, since losses in utility accelerate as the policy moves away from the voter's ideal point. However, recent work by Carroll et al. (2013) demonstrates that legislators' utility functions are better approximated by the normal (Gaussian) functional form. The normal form is also attractive because it captures the phenomenon of *alienation from indifference* (Riker and Ordeshook, 1973, pp. 324-330). That is, when two policy alternatives are very distant from a voter's ideal point, the voter will be nearly indifferent between the two options.

FIGURE 1.1: Gaussian (solid line), Quadratic (dashed line) and Linear (dotted line) Utility Functions for Voter with Ideal Point of 0

1.1.1 Theoretical Development and Applications of the Spatial Voting Model

The work of Downs and Black led to a surge of theoretical work on spatial models of voting in mass elections and in legislatures. In the decades that followed, formal work blossomed involving the spatial theory of voting. Much of this work involves advances made in the field of social choice theory, which is concerned with the aggregation of individual choices (Davis, Hinich and Ordeshook, 1970). Spatial modeling has produced a number of insights into the role of instability in processes of preference aggregation (see Shepsle (2010, chaps. 3-7) for a thorough exposition of this subject). Scholars have long been aware (since at least Marquis de Condorcet's development of Condorcet's paradox in the late-eighteenth century) that even if individuals hold complete and transitive preferences over a set of three or more alternatives, cycles may nonetheless be present in their *aggregated* preference order. That is, a majority of voters may prefer candidate A over candidate B, candidate B over candidate C, but also candidate C *over* candidate A.

In his seminal work *Social Choice and Individual Values*, Arrow (1951) developed his famous impossibility theorem, which showed that the Condorcet paradox applies to any reasonable method of preference aggregation. More formally, Arrow's impossibility theorem states that the possibility of cycles in the ordering of group preferences is present in any aggregation method that satisfies a set of minimal conditions, often referred to as CUPID: complete and transitive individual preferences, Pareto optimality, independence of irrelevant alternatives, and non-dictatorship.

The result has been obviously troubling for democratic theorists since it rules out the certainty of stable majority rule. Theoretical hopes for stability were partially salvaged by two later works. First, Black (1958) demonstrated that majority rule is guaranteed to produce a stable outcome (at the location of the median voter) if preferences are single-peaked and ordered along a single dimension. Plott (1967) generalized this result to multidimensional space on the very specific condition that ideal points are radially symmetric around the multidimensional equivalent of the median voter. However, McKelvey (1976, 1979) dealt a lethal blow to such hopes with his chaos theorem (see also Schofield 1978). McKelvey's chaos theorem shows that Black's result does not generalize when the policy space is multidimensional and ideal points lack a radially symmetric distribution. In this case, which describes virtually all multidimensional policy spaces, there is an absence of a stable equilibrium.

One implication of this finding is that losers in a single-dimensional situation have an incentive to raise new, advantageous dimensions to upset the existing equilibrium. This type of maneuver was developed and named *heresthetic* by William H. Riker (Riker, 1980, 1982, 1986, 1990*a*, 1996). Unlike rhetoric, which is designed to persuade, heresthetic is designed to strategically manipulate the choice space. Riker (1986) provides several examples of heresthetical maneuvers in the legislative arena (e.g., "killer" amendments), and other scholars (notably Carsey (2000) and Vavreck (2009)) have extended the analysis of heresthetical strategy to issue emphasis in political campaigns.

Of course, as McKelvey (1976, 1979) notes, instability is a *condition* rather than a *description* of majority rule. Simply because stable outcomes cannot be guaranteed does not mean they are not the norm in democratic systems (for a comprehensive exposition of this argument, see Mackie (2003)). Indeed, empirical studies suggest that legislative voting is only rarely thrust into instability by strategic behavior (e.g., "killer" amendments) (Poole and Rosenthal, 1997; Jenkins and Sala, 1998; Finocchiaro and Jenkins, 2008). Moreover, even if stable policy outcomes are not certain, there may exist regions in multidimensional space (e.g., the uncovered set) in which policy is likely to converge (Jeong et al., 2011).

Another strand of theoretical work on the spatial voting model has attempted to explain the factors that inhibit candidate and party convergence to the ideal point of the median voter. Across political systems, it is quite clear that rival candidates and parties do not adopt identical platforms. Hence, the median voter theorem is at best a partial explanation of the behavior of political elites.

While competition is a centripetal force that leads to policy convergence, there are also a number of centrifugal processes that push political elites away from the political center. Aldrich (1983) argues that candidates must also consider the participation decisions of non-centrist party activists (cf. Baron 1993). If opposing candidates adopt identical or near-identical positions, activists will have less incentive to contribute (monetarily or otherwise) to the campaign.

The directional voting model developed by Rabinowitz and Macdonald (1989) offers another resolution to the paradox of non-convergence. In their model, individuals evaluate political stimuli based on *direction* rather than *proximity*; that is, whether or not the stimuli are on the same *side* of an issue as the individual. For example, consider a 7-point scale on defense spending where 1 indicates a preference for greatly reducing the level of defense spending, 7 indicates a preference for greatly increasing the level of defense spending, and 4 is a neutral point indicating a preference for maintaining the current level of defense spending. Suppose a voter's ideal point is at 3: she would like to decrease defense spending, but only somewhat. Now imagine there are two candidates A and B who advocate positions at points 1 and 4.5 on the scale, respectively. The classical proximity voting model predicts that the voter will prefer Candidate B since she is closer to the voter than is Candidate A (1.5 units vs. 2 units). The directional voting model, on the other hand, predicts that the voter will prefer Candidate A since she is on the same side of the issue (i.e., reduce defense spending) as the voter.

Empirical issues have rendered resolution of the proximity vs. directional model debate elusive (Lewis and King, 1999), although some recent experimental work by Tomz and Van Houweling (2008) suggests that voters most frequently employ proximity-based decision rules. In this book we focus on the classical proximity model of spatial voting, which is the dominant model in political science. To our knowledge, there are no R packages that include functions to estimate directional spatial voting models. For readers interested in estimating directional voting models, we recommend the multidimensional preference scaling procedure (Chang and Carroll, 1969) discussed in Weller and Romney (1990) and Jacoby (2013).

The Hotelling-Downs-Black framework assumes two party competition with no entry. Convergence can break down when there are more than two parties and entry is possible. Palfrey (1984) discusses the need to guard against attracting an opponent from the ideological fringes. For instance, in the 2000 presidential election, Democratic nominee Al Gore couldn't move too far to the center without losing votes on the left to Green Party nominee Ralph Nader. A configuration in which the parties or candidates are sharply differentiated but not totally extreme maximizes their share of the vote while preventing third party entry. Particularly in situations in which a candidate must first acquire her party's nomination (e.g., through a primary process) to run in a general election, candidates must actively court support from the left or the right rather than just focusing on the center (Aranson and Ordeshook, 1972). Finally, even in cases in which political elites would prefer to adopt more centrist policies, they are not fully mobile around the policy space. Namely, they are constrained by prior policy commitments and their ideological reputations (Hinich and Munger, 1994; Alesina, 1988; Alesina and Cukierman, 1990; Alesina and Rosenthal, 1995).

1.1.2 The Development of Empirical Estimation Methods for Spatial Models of Voting

The spatial voting models that arose in economics and political science are based on the concept that individual actors have ideal points and that preference declines with distance from the ideal point. This concept was echoed in psychology by the work of Coombs (1964) in his seminal development of unfolding methods for preference data.[§] With specific reference to the estimation of empirical models of spatial voting from roll call and public opinion survey data, the first to apply modern statistical techniques to the study of legislative roll call data were Rice (1928) and Thurstone (1932). Duncan MacRae's (1958; 1967; 1970) path-breaking work used scaling methods (factor and cluster analysis) to evaluate the latent dimensions of roll call voting in the United States Congress and the French Parliament. In 1970, Weisberg and Rusk used data from the newly created feeling thermometer questions (in which survey respondents are asked to rate their affinity for candidates and parties on a $0-100$ scale) to demonstrate that voters relied on essentially two

[§]We return to unfolding in Chapter 4.

dimensions (left-right ideology and partisanship) to evaluate candidates and issues in the 1968 presidential election.

The development of Keith T. Poole and Howard Rosenthal's NOMINATE scaling procedure provided reliable estimates of the ideological positions of members of Congress throughout the span of American history, beginning with the first NOMINATE prototype in 1982 (Poole and Rosenthal, 1983) and concluding in the completion of DW-NOMINATE (McCarty, Poole and Rosenthal, 1997). This work sparked considerable scholarly interest in ideal point estimation and scaling techniques. Standard errors for NOMINATE ideal points can be generated using the parametric bootstrap procedure introduced in Lewis and Poole (2004), based on the variance in random samples drawn from the fitted model.

Following Poole and Rosenthal (1997, 2007) and continued technological leaps in PC computing power, political scientists began to use NOMINATE and alternative procedures (e.g., Clinton, Jackman and Rivers' (2004) IDEAL or Poole's (2005) nonparametric Optimal Classification) to scale political actors in a variety of contexts: international legislatures (Desposato, 2006), the United Nations (Voeten, 2000), the Constitutional Convention (Dougherty and Heckelman, 2006), American state legislatures (Shor and McCarty, 2011), public opinion surveys (Hare and Poole, 2014*b*; Jacoby and Armstrong, 2014), and the European Parliament (Hix, Noury and Roland, 2006, 2007). Advances in computing power have been especially crucial to the growing popularity of Bayesian methods because they allow for simulation-based estimation (via Markov chain Monte Carlo (MCMC) methods) of what were previously intractable models (Gill, 2008). Over the last decade, Bayesian models have been developed to estimate legislative ideal points (Clinton, Jackman and Rivers, 2004), the structure of political actors' utility functions (Carroll et al., 2013), and perceptual data from issue scales (Hare et al., 2015).

1.1.3 The Basic Space Theory

A common thread running through the empirical studies is that political choice behavior can generally be modeled with the use of only a few (usually just one or two) basic dimensions. While legislators and citizens may have preferences across a dizzying array of policy issues—abortion, tax rates, gun control, foreign policy—these attitudes appear to be organized by positions along a small number of latent dimensions. The low-dimensional regularity has a natural connection to the theoretical concepts of ideology and, in particular, Converse's (1964) notion of *constraint*.[¶] Constraint simply refers to the bundling or linkage of many different issue positions as part of a political ideology or

[¶]The theory of ideological constraint squares with the results of psychological applications of multidimensional scaling. These results consistently produce low-dimensional maps, indicating that humans organize information in a geometric manner (Shepard, 1987).

"belief system." For example, political conservatives oppose amnesty for illegal immigrants, support a free-market health care system, and oppose cuts in military spending.

Ideological constraint has a natural geometric interpretation. It means that the complex, high-dimensional issue space (where each attitude is represented by a separate dimension) maps onto an underlying low-dimensional *basic space* (Cahoon, Hinich and Ordeshook, 1976). The dimensions of the basic space are synonymously referred to as *basic* dimensions, *predictive* dimensions, or *ideological* dimensions (Hinich and Munger, 1994). Indeed, the primary latent dimension recovered by most applications of scaling procedures is the classic left-right (or liberal-conservative) ideological continuum.

The central idea of the basic space theory is straightforward: a few fundamental dimensions underlie political preferences, and the goal of scaling procedures is to recover these dimensions and individuals' ideal points in the abstract space. The success of scaling techniques in confirming that an array of voting behavior can be explained in low-dimensional space means that spatial models are not just useful abstractions, but also accurate depictions of decision making processes (see, e.g., Hare and Poole 2014*a*).

Scaling methods project or map multidimensional issue spaces onto a low-dimensional basic space. An issue space has a separate dimension for each issue (e.g., affirmative action, environmental regulation, stem cell research, etc.). We can think of a matrix of hundreds of parliamentary roll call votes or public opinion survey items as giant issue spaces, and legislators or respondents place themselves in the space by voting or answering on each item.

Basic space theory holds that these individual positions in high-dimensional issue spaces can be captured with a few basic dimensions. In many instances, two ideological dimensions—one for economic items such as taxation, the other for cultural items such as abortion—are sufficient to capture variation in the issue space. For example, a bloc of voters may be "libertarians": economically conservative but socially liberal. Figure 1.2 shows a two-dimensional model—with the horizontal axis representing economic attitudes and the vertical axis representing social views—as the basic space that generates the multitude of preferences expressed in a hypothetical issue space. In other contexts (namely, contemporary American politics), a single left-right or liberal-conservative dimension constrains political attitudes. In these cases, the basic space will be unidimensional or nearly unidimensional, with the second dimension representing minor sources of attitudinal variation (e.g., regional differences) or statistical noise.

What are the sources of ideological constraint—the forces that collapse or "map" the complex issue space onto the low-dimensional basic space? As Jost, Federico and Napier (2009) discuss in a recent review essay, psychologists and political scientists often differ on the process by which policy preferences are bundled together. Political scientists tend to emphasize the role of political elites—parties, interest groups and elected officials—in packaging a comprehensive set of issue positions into a coherent ideological structure. The fact

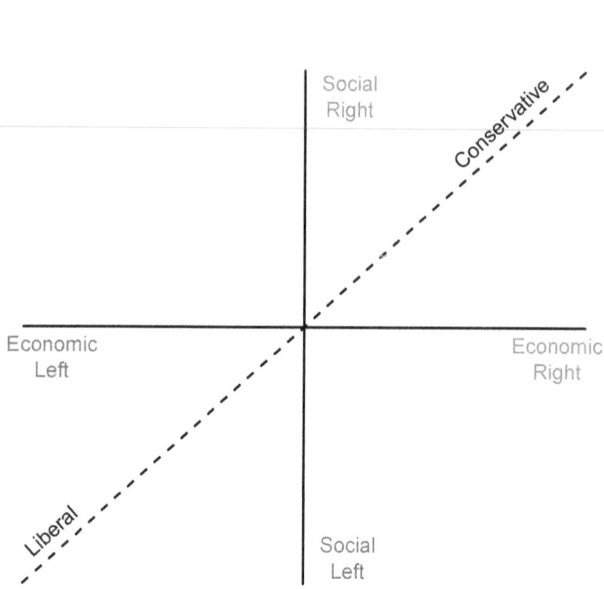

FIGURE 1.2: Conceptualizing the Basic Space: One and Two-Dimensional Models of Political Ideology

that what constitutes the conservative or liberal position on a given issue can change (sometimes dramatically, as in the case of abortion in the 1970s (Stimson, 2004, pp. 58-60)) illustrates the importance of elite political actors in the mapping process. Conversely, psychologists tend to view ideological structure as stemming from a variety of personality traits, needs, motives, and attitudes (Jost, Federico and Napier, 2009). For example, openness to change has a leftward influence, while preferences for order and structure exert a rightward effect. Both forces—elite-driven and psychological—serve to organize the political agenda in simple ideological terms. Hence, whatever the source of constraint, the empirical results from spatial analyses will be observationally equivalent. In this sense, using spatial models of choice need not be tied to any particular approach in political science (e.g., the "rational choice school" or the "Michigan (behavioral) school") (see Poole 2005, chap. 7; Poole 2007).

The basic space theory of ideology formulated by Hinich and colleagues facilitates a richer understanding of the substantive meaning of spatial maps

(Cahoon, Hinich and Ordeshook, 1976; Hinich and Pollard, 1981; Enelow and Hinich, 1984; Hinich and Munger, 1994, 1997). The construction of simple "pictures" of legislative bodies or groups of voters from complex matrices of hundreds or even thousands of votes has a strong theoretical foundation. The empirical results obtained by these methods—for example, the success of techniques like NOMINATE in modeling legislative roll call voting in a variety of contexts—corroborate the applicability of the spatial theory of voting to real-world political behavior.

1.2 Summary of Data Types Analyzed by Spatial Voting Models

In Table 1.1 we provide an outline of the scaling techniques discussed in each chapter with a focus on which types of data they are used to analyze.

It is important to note that with several of these methods, it may be possible and useful to use data from multiple sources. In some cases, the data can be "glued" or "bridged" together across separate units or time using common actors (e.g., legislators or voters) and/or common stimuli (e.g., roll call votes or survey questions). For example, a group of survey respondents may be asked to "vote" on a series of roll calls that have also been voted on by members of Congress. These groups can then be scaled in a common space by merging the two sets of votes together (e.g., Tausanovitch and Warshaw (2013)). This is relevant to several of the methods described in this book, and particularly those discussed in further detail in Chapter 5.

1.3 Conclusion

The application of spatial models of choice and judgment is built upon applying statistical procedures that analyze observed data and extract latent (i.e., abstract) dimensions upon which the objects or subjects can be placed. In political science, scholars seeking to measure the behavior of legislators or survey respondents are generally interested in policy or ideological scales using individuals' judgments or observed choices. The results of these scales, often referred to as "ideal point estimates," can uncover the basic dimensionality of choice behavior and reveal the latent preferences underlying that behavior. We begin first with the analysis of issue scale data in Chapter 2.

Table 1.1: Data Types and Appropriate Methods

Data Type	Example	Method	Chapter
Perceptual Data: Single Issue Scales	Individuals place themselves and/or parties on a liberal-conservative scale.	Maximum Likelihood and and Bayesian Aldrich-McKelvey Scaling, Basic Space Scaling	2 and 6
Perceptual or Preferential Data: Multiple Issue Scales	Survey respondents register their attitudes on a series of ordinal policy scales.	Basic Space Scaling, Ordinal Item Response Theory (IRT), Ordered Optimal Classification	2 and 6
Perceptual or Preferential Data: Single Square Matrices of Similarity Ratings of Objects	An agreement score matrix is created that shows how often each legislator voted on the same side as every other legislator.	Metric, Non-metric and Bayesian Multidimensional Scaling (MDS)	3 and 6
Perceptual or Preferential Data: Multiple Square Matrices of Similarity Ratings of Objects	Individuals or groups rate how similarly they view a series of political objects/stimuli (e.g., taxes and liberals).	Metric and Non-Metric Individual Differences Scaling (INDSCAL) Bootstrapped MDS	3
Preferential Data: Rectangular Matrices with Preferential Ratings using Interval Ratio-Level Scales	Individuals rate parties on a 0-100 scale or rank candidates from most to least-preferred.	Least Squares and Bayesian Unfolding	4 and 6
Preferential Data: Choices between Binary Alternatives	Legislators cast a series of roll call (Yea or Nay) votes.	Parametric Unfolding (NOMINATE and α-NOMINATE), Nonparametric Unfolding (Optimal Classification), Bayesian IRT, and EM IRT	5 and 6

2

Analyzing Issue Scales

Public opinion surveys often ask respondents to place themselves and political parties, candidates, and public figures on issue scales. Some issue scales have labeled endpoints such as "strongly agree" and "strongly disagree". These are also known as *Likert-type questions* or *items*. Likert-type items provide a prompt and a gradated, symmetric scale for respondents to register their opinion. A *Likert scale*, on the other hand, is a simple scaling method in which the responses to several items are added together to create a summated scale (Likert, 1932).

For example, in every presidential election year since 1972, the American National Election Study (ANES) has included a seven-point "guaranteed jobs" scale, in which respondents are read the following prompt: "Some people feel that the government in Washington should see to it that every person has a job and a good standard of living. Others think the government should just let each person get ahead on his/her own." One end of the scale (1) represents the position that the government should guarantee jobs and a good standard of living, and the other end of the scale (7) denotes the attitude that the government let each person get ahead on her own.

Another scale often used in public opinion surveys asks respondents to place themselves, the parties, and candidates on the liberal-conservative spectrum. The ANES version of this question is: "We hear a lot of talk these days about liberals and conservatives. Here is a 7-point scale on which the political views that people might hold are arranged from extremely liberal to extremely conservative." The endpoints are labeled "extremely liberal" (1) and "extremely conservative" (7). Respondents use this scale to locate themselves and political stimuli on the scale. Issue scales, then, can be used to collect both preferential (what is my position?) and perceptual (what are the positions of others?) data. Both types of data are thus tailor-made to create and test spatial voting models. Moreover, issue scale data allows for the estimation of the ideal points of political actors who have not served in a legislature (e.g., Ralph Nader, Ross Perot and Colin Powell) and thus have no roll call voting record.

Issue scales are often combined into Likert scales and Guttman scales. The difference between the two is that Guttman scaling (also known as scalogram analysis) is a method of cumulative scaling. Cumulative methods create a scale by arranging a series of questions in order of how well they discriminate in measuring some underlying attribute (Torgerson, 1958, p. 307). This frequently involves skills-based tests, where, for example, a student who correctly answers a question about differential calculus will correctly answer less difficult math questions. The classical Item Response Theory (IRT) model used in skills-based tests is a cumulative scaling model and inconsistent with Coombs's (1964) unfolding model, which treats individuals as having *ideal points* and *single-peaked preference functions.* The unfolding model is more appropriate to the spatial theory of voting (see van Schuur 1992, pp. 42-43 for an elaboration on the distinction between the probability functions of cumulative scaling and unfolding analysis). However, the application of the IRT model to analyze political choice data—which we discuss in Chapters 6 and 7—is consistent with the unfolding model's assumption of ideal points with single-peaked preferences, and hence can be used to estimate spatial models of voting—particularly as part of a Bayesian approach (e.g., Clinton, Jackman and Rivers, 2004; Jessee, 2009). The latent dimensions recovered from these methods are *policy*—not *ability*—dimensions.

This chapter covers several widely used methods for the analysis of issue scales: Aldrich-McKelvey scaling, Blackbox scaling, and Blackbox transpose scaling. These methods are included in the `basicspace` package in `R` (Poole et al., 2013). We also demonstrate the Ordered Optimal Classification method (Hare, Liu and Lupton, 2018) for the analysis of issue scale data and how *anchoring vignettes* can be used to facilitate the cross-comparability of survey responses.

2.1 Aldrich-McKelvey Scaling

Issue scales are well-suited for the application of spatial models, but they nonetheless present a formidable methodological difficulty. Namely, respondents may interpret the meaning of the scale differently—a problem that has come to be known as "interpersonal incomparability" or "differential item-functioning" (DIF) in the literature (Brady, 1985; King et al., 2004). DIF can stem from several sources.

First, respondents have preferences and affective orientations that can bias their evaluations of the political world. Those with extreme preferences may warp the meaning of the scale. For example, a very liberal survey participant may view President Barack Obama as insufficiently progressive, thus rating him as an ideological moderate or even conservative. Likewise, we should also expect that respondents' affective orientations lead them to exaggerate the policy distances between themselves and stimuli they view unfavorably while understating the policy distance between themselves and stimuli they favor. For instance, a conservative Republican who holds President Obama in very low regard may place him at the extreme leftward end of the scale.

Even if respondents were unbiased, the meaning of the points on an issue scale may be ambiguous. This problem becomes more acute as the number of response categories grows (e.g., 100 point feeling thermometer scales (Wilcox, Sigelman and Cook, 1989)). Finally, citizens tend to be poorly informed on most political matters (Delli Carpini and Keeter, 1996). Many respondents will even reverse the placement of the stimuli on the scale; for example, placing the Democratic Party to the right of the Republican Party. The stimuli placements of low sophistication respondents will be more error-prone since they have less information on which to base their evaluations.

Aldrich and McKelvey's (1977) insight was that even though survey respondents distort their placement of political stimuli, they typically nonetheless perceive and report an accurate *ordering* of the stimuli. For example, a conservative respondent may view the 2008 Republican presidential nominee Senator John McCain (R-AZ) as a moderate, but they will still correctly place him to the right of the Democratic nominee Senator Barack Obama (D-IL). Responses to issue scales, then, can be understood as a linear transformation of the true locations of political stimuli along a latent dimension, plus random noise, u_{ij}, which satisfies the Gauss-Markov assumptions of zero mean, homoscedasticity, and independence (Aldrich and McKelvey, 1977, p. 113).

The goal of Aldrich-McKelvey (A-M) scaling is to estimate the perceptual distortion for each respondent, and from those weights back out estimated locations for stimuli and respondents along the issue dimension. In particular, let z_{ij} be the perceived location of stimulus j $(j = 1,...,q)$ by individual i $(i = 1,...,n)$. The A-M model assumes that the individual reports a noisy linear transformation of the true location of stimulus j (z_j); that is

$$\alpha_i + \beta_i z_j = z_{ij} + u_{ij} \tag{2.1}$$

where u_{ij} satisfies the usual Gauss-Markov assumptions of zero mean, homoscedasticity, and independence (Aldrich and McKelvey, 1977, p. 113).

Let $\hat{z}_j$ be the estimated location of stimulus j, and let $\hat{\alpha}_i$ and $\hat{\beta}_i$ be the estimates of α_i and β_i; define

$$\hat{\alpha}_i + \hat{\beta}_i \hat{z}_j - z_{ij} = e_{ij} \tag{2.2}$$

Aldrich and McKelvey set up the following Lagrangian multiplier* problem

*Lagrangian multipliers are a method for finding function maxima and minima given a set of restrictions. Riker and Ordeshook (1973, p. 242) provide an illustration of a Lagrangian multiplier problem: How can a stick that is r units long be cut into four pieces to form a rectangle with the greatest possible area? Since it is a rectangle, there will be two pieces of length a and two pieces of length b. The constraint can be expressed as $2a + 2b = r$ or $2a + 2b - r = 0$; we wish to maximize ab, the area of the rectangle. The Lagrangian multiplier problem, then, is expressed as $ab - \lambda(2a + 2b - r)$, where λ is the Lagrangian multiplier. Differentiating with respect to a and then to b and setting the equations equal to 0 produces the expressions $b - 2\lambda = 0$ and $a - 2\lambda = 0$, which reduce to $\lambda = \frac{b}{2}$ and $\lambda = \frac{a}{2}$. Hence, the area is maximized by cutting the pieces of equal length $(a = b)$ forming a square.

$$L(\alpha_i,\beta_i,z_j,\lambda_1,\lambda_2)=\sum_{i=1}^{n}\sum_{j=1}^{q}e_{ij}^2+2\lambda_1\sum_{j=1}^{q}\hat{z}_j+\lambda_2\left[\sum_{j=1}^{q}\hat{z}_j^2-1\right] \tag{2.3}$$

that is, minimize the sum of squared error subject to the constraints that the estimated stimuli coordinates have zero mean and sum of squares equal to one. Define the q by 2 matrix X_i as

$$X_i=\begin{bmatrix} 1 & z_{i1} \\ 1 & z_{i2} \\ \cdot & \cdot \\ \cdot & \cdot \\ \cdot & \cdot \\ 1 & z_{iq} \end{bmatrix} \tag{2.4}$$

then the solution for the individual transformations is simply the "least-squares regression of the reported on the actual (unknown) positions of the candidates" (Aldrich and McKelvey, 1977, p. 115). That is

$$\begin{bmatrix} \hat{\alpha}_i \\ \hat{\beta}_i \end{bmatrix}=\left[X'_iX_i\right]^{-1}X_i'\hat{z} \tag{2.5}$$

where $\hat{z}$ is the q by 1 vector of the "true" positions of the candidates. To get the solution for $\hat{z}$, define the q by q matrix A as

$$A=\left[\sum_{i=1}^{n}X_i(X'_iX_i)^{-1}X'_i\right] \tag{2.6}$$

Aldrich and McKelvey show that the partial derivatives for the $\hat{z}_j$ can be rearranged into the linear system

$$[A-nI_q]\hat{z}=\lambda_2\hat{z} \tag{2.7}$$

where I_q is the q by q identity matrix. From the above equation, $\hat{z}$ is simply an eigenvector of the matrix $[A-nI_q]$ and λ_2 is the corresponding eigenvalue, making this a characteristic value problem. To determine which of the q possible eigenvectors is the solution, Aldrich and McKelvey show that:

$$-\lambda_2=\sum_{i=1}^{n}\sum_{j=1}^{q}e_{ij}^2=-\hat{z}'\left[A-nI_q\right]\hat{z} \tag{2.8}$$

Hence, the solution is the eigenvector of $[A-nI_q]$ "with the highest (negative) nonzero" eigenvalue. The solution for $\hat{z}$ from Equation 2.7 can be taken back to Equation 2.5 to solve for the individual transformation parameters.

Aldrich and McKelvey measure model fit as the reduction of the normalized variance of perceptions:

$$\hat{\sigma}^2 = -\frac{n\lambda_2}{q(n+\lambda_2)^2} \tag{2.9}$$

Note that in equation (2.1) the A-M model assumes homoscedastic error. Substantively, this means that respondents are assumed to have an equal likelihood of reporting an incorrect ordering of the stimuli. This assumption is almost certainly unrealistic given variation in individuals' levels of political knowledge and affective orientations towards the stimuli. However, Palfrey and Poole (1987, pp. 514-516) report the results of Monte Carlo simulations that show the A-M estimator remains robust up to massive Gauss-Markov violations, recovering an accurate configuration of the stimuli even in the presence of very high levels of heteroscedastic error. This is important because in cases of high heteroscedasticity, a "naive" method (e.g., using the mean placement of each stimuli as its estimated location) is prone to failure since errors are less likely to cancel out. A-M scaling, though, is resilient in this instance.

As we noted above, most US respondents have low levels of political information and do not think in ideological terms. Some respondents will confuse the standard left-right ideological scale entirely, placing conservative stimuli to the *left* of liberal stimuli. Reversed rankings, in particular, can substantially bias "naive" measures like mean stimuli placements. Counterintuitively, these reversed rankings actually help provide a better fit to the model, since the ordering of the stimuli is corrected after a negative weight is applied by the A-M routine.

This has been a source of some controversy, but Palfrey and Poole (1987) demonstrate that this feature has the advantage of "filtering" respondents by level of political sophistication. Respondents with negative weights will generally have lower levels of political information than respondents with positive weights. Thus, this aspect of the A-M model can be leveraged to study ideological heterogeneity in the mass electorate. The correlation between respondent placements and the recovered configuration of the stimuli can also be utilized as a measure political information (Palfrey and Poole, 1987). One of the important substantive findings of Palfrey and Poole (1987) is that low information respondents in the United States tend to be ideological moderates, while politically sophisticated respondents tend to cluster away from the center towards the liberal and conservative endpoints of the scale (see also Abramowitz (2010) and Lauderdale (2013)).

2.1.1 The basicspace Package in R

The `basicspace` package (Poole et al., 2013) includes several functions for the recovery of latent dimensions of choice and judgment from issue scale data. Indeed, most of the functions covered in this chapter are included in the `basicspace` package. Namely, the A-M scaling routine is implemented in the aldmck() function and Poole's (1998*a*; 1998*b*) basicspace procedure can be executed with the blackbox() and blackbox_transpose() functions.

2.1.2 Example 1: 2009 European Election Study (French Module)

To demonstrate use of the A-M scaling routine with the aldmck() function, we use data from the French module of the 2009 European Election Study (EES). The EES asked 1,000 French citizens to place themselves and eight major political parties on a 0-10 left-right scale (0 representing the most left-wing position, 10 the most right-wing position). Responses are coded 77, 88, or 89 if the respondent refuses to answer or does not know of the party or where to place it.

The data is stored in the matrix franceEES2009. franceEES2009 is arranged such that the respondents are on the rows and the stimuli (parties) are on the columns. The command head() prints the first six rows of an object to allow for a quick inspection of the data.

```
library(asmcjr)
library(ggplot2)
data(franceEES2009)
head(franceEES2009)
```

```
     self Extreme Left Communist Socialist Greens
[1,]   77            0         0         1      5
[2,]   77            0         5         4      5
[3,]   77           89        89        89     89
[4,]    3           89        89        89     89
[5,]   77           77        77        77     77
[6,]    5            0         0         3     89
     UDF (Bayrou) UMP (Sarkozy) National Front Left Party
[1,]            5             9             10          1
[2,]           89             8             10          4
[3,]            6            89             10         89
[4,]           89            89             89         89
[5,]           77            77             77         77
[6,]            0            89             89          5
```

The aldmck() function requires five arguments: the matrix to be analyzed (franceEES2009), the column number for respondent self-placements (1), the column number for a stimuli to be placed on the left side of the dimension (2 for the Extreme Left party), a vector of missing value codes, and a logical argument (TRUE/FALSE) that specifies whether verbose output is desired as the function is executed. The requirement to specify "polarity" via the left stimulus is a function of the recovered space being defined only up to a rotation. By convention, aldmck() places left-leaning stimuli on the left end of the scale by assigning them negative scores.

Note that the `aldmck()` function uses listwise deletion to remove respondents who failed to place all of the stimuli on the issue scale, so only rows that contain no missing values are included in the analysis.

```
#Loading the 'basicspace' package
library(basicspace)
#Running Bayesian Aldrich-Mckelvey scaling on France EES
result.france <- aldmck(franceEES2009, respondent=1, polarity=2,
    missing=c(77,88,89), verbose=FALSE)
```

The `aldmck()` function stores eight objects in the list `result`. The objects can be accessed using the commands `"result.france$respondents"`, and `"result.france$respondents$intercept"`, etc:

stimuli Estimated locations of the stimuli.

respondents Estimates of the respondents:

> **intercept** Perceptual distortion intercept term ($\hat{\alpha}_i$).
>
> **weight** Perceptual distortion weight term ($\hat{\beta}_i$).
>
> **idealpt** Respondent ideal point; missing values are coded `NA`.
>
> **R2** Respondent R^2 of bivariate regression between estimated and reported stimulus locations.
>
> **selfplace** Self-reported placement.
>
> **polinfo** Respondent correlation between "true" and reported stimulus locations; used as a measure of political information.

eigenvalues List of eigenvalues.

AMfit Aldrich and McKelvey's measure of fit from Equation 2.9 (lower values indicate a better fit).

R2 Total R^2.

N Number of respondents included in the analysis.

N.neg Number of respondents with negative weights.

N.pos Number of respondents with positive weights.

The command `summary(result)` provides a useful overview of the A-M scaling output:

```
#Output of the results of Aldrich-Mckelvey scaling
summary(result.france)
```

```
SUMMARY OF ALDRICH-MCKELVEY OBJECT
----------------------------------

Number of Stimuli: 8
Number of Respondents Scaled: 883
```

```
Number of Respondents (Positive Weights): 825
Number of Respondents (Negative Weights): 58
Reduction of normalized variance of perceptions: 0.13

                 Location
Extreme Left       -0.507
Communist          -0.342
Left Party         -0.197
Socialist          -0.098
Greens             -0.024
UDF (Bayrou)        0.122
UMP (Sarkozy)       0.433
National Front      0.612
```

The results have a high degree of face validity. The Extreme Left and National Front Parties are the most ideologically extreme stimuli, and the left-right ordering of the parties squares with a basic understanding of French party politics. For example, Nicolas Sarkozy's UMP (Union for a Popular Movement) Party is center-right, while the Socialist Party is a left-wing party, but to the right of the Extreme Left, Communist, and Left Parties. Moreover, the data fit a one-dimensional spatial model reasonably well: the R^2 value is 0.72 and the AMfit statistic is 0.06. Finally, most of the 617 scaled respondents—583—have positive weights, meaning that they saw the political space "correctly" (i.e., that the Extreme Left Party is to the left of Le Pen's National Front Party). Only 28 respondents confused this ordering and have negative weights.

The `basicspace` package includes a function to do a summary four-panel plot with the command `plot(result)`. These plots show the locations of the stimuli and all respondents in a standard format, as a cumulative density function, isolating respondents with positive weights, and isolating respondents with negative weights. Below we provide code to produce these plots separately.

The following code plots the estimated ideal points of the respondents in Figure 2.1. Only respondents who were included in the scaling are assigned to the matrix `voters` using the command `na.omit()`. The estimated respondent ideal points are stored in the third column of `result$respondents` and `voters`. The `density()` function in `R` is a Kernel density estimator that can be used to plot smoothed histograms, as below.

```
# plot density of ideal points
plot_resphist(result.france, xlab="Left-Right")
```

We next add the estimated locations of the parties and a legend to the plot in Figure 2.1 with the code below to produce Figure 2.2:

```
# plot stimuli locations in addition to ideal point density
plot_resphist(result.france, addStim=TRUE, xlab = "Left-Right") +
```

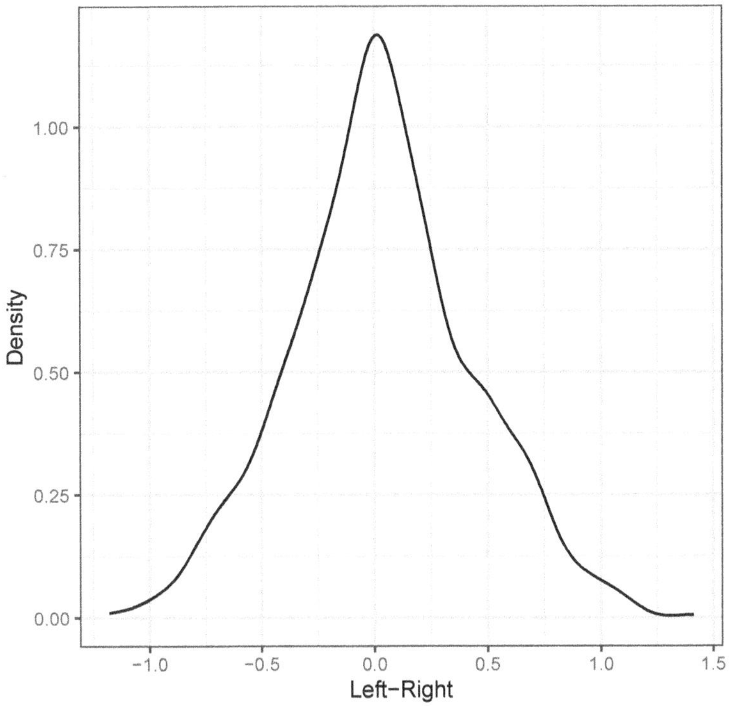

FIGURE 2.1: Aldrich-McKelvey Scaling of Left-Right Self-Placements of French Respondents (2009 European Election Study)

```
  theme(legend.position="bottom", aspect.ratio=1) +
  guides(shape = guide_legend(override.aes = list(size = 4),
                              nrow=3)) +
  labs(shape="Party", colour="Party")
```

To isolate respondents with positive weights, we use the following code to produce the plot in Figure 2.3:

```
# Isolate positive weights
plot_resphist(result.france, addStim=TRUE, weights="positive",
    xlab = "Left-Right")  +
  theme(legend.position="bottom", aspect.ratio=1) +
  guides(shape = guide_legend(override.aes = list(size = 4),
                              nrow=3)) +
  labs(shape="Party", colour="Party")
```

Finally, to isolate respondents with negative weights (those with reversed perceptions of party locations), we use the following code to produce the

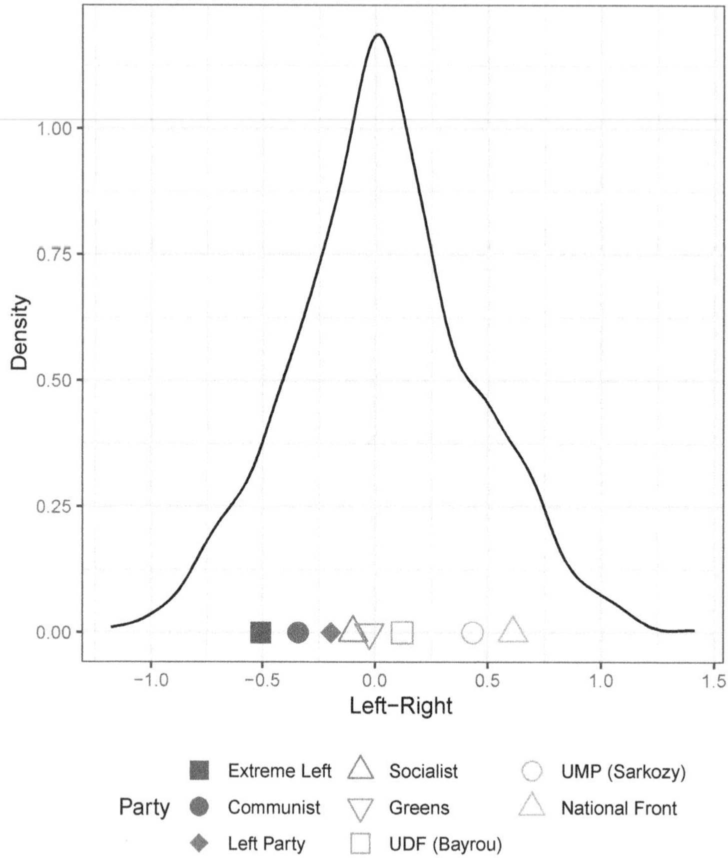

FIGURE 2.2: Aldrich-McKelvey Scaling of Left-Right Placements of French Political Parties (2009 European Election Study)

plot in Figure 2.4. Indeed, we see in Figure 2.4 that those respondents with negative weights cluster near the center of the ideological continuum. This makes intuitive sense, since left-wing and right-wing ideologues should be less likely to confuse the left-right ordering of major political actors.

```
# Isolating negative weights
plot_resphist(result.france, addStim=TRUE, weights="negative",
    xlab = "Left-Right")  +
  theme(legend.position="bottom", aspect.ratio=1) +
  guides(shape = guide_legend(override.aes = list(size = 4),
                              nrow=3)) +
  labs(shape="Party", colour="Party")
```

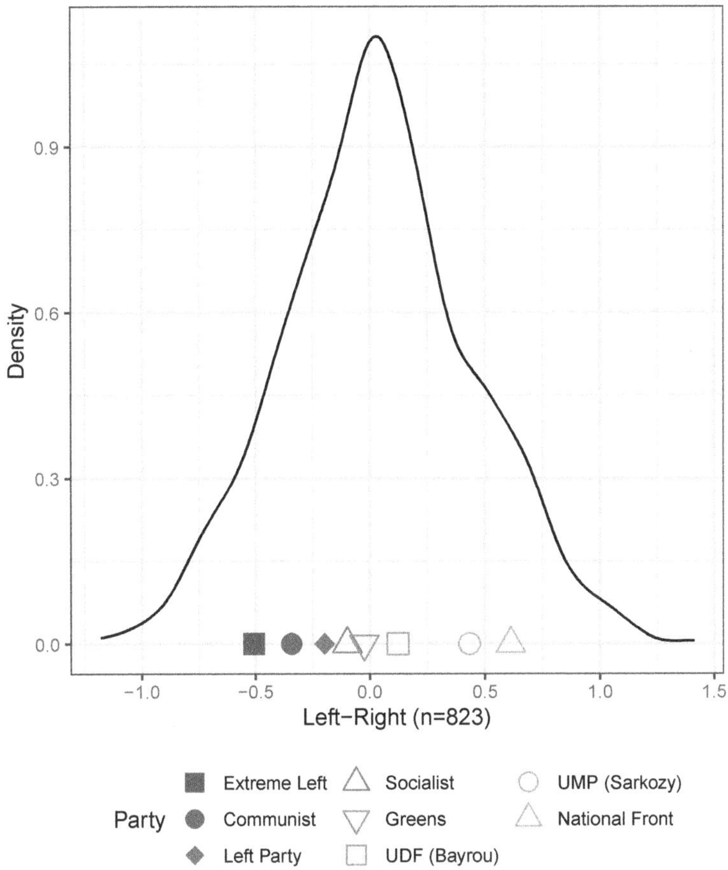

FIGURE 2.3: Aldrich-McKelvey Scaling of Left-Right Placements of French Political Parties: Positive Weights (2009 European Election Study)

2.1.3 Example 2: 1968 American National Election Study Urban Unrest and Vietnam War Scales

We next use the A-M scaling procedure to analyze two issue scales included in the 1968 American National Election Study (stored in the data sets `nes1968_urbanunrest` and `nes1968_vietnam`). Respondents were asked to place themselves, President Lyndon Johnson, and the three major presidential candidates (Democrat Hubert Humphrey, Republican Richard Nixon, and American Independent George Wallace) on two seven-point issue scales. The two were the Urban Unrest and Vietnam War scales. The Urban Unrest scale

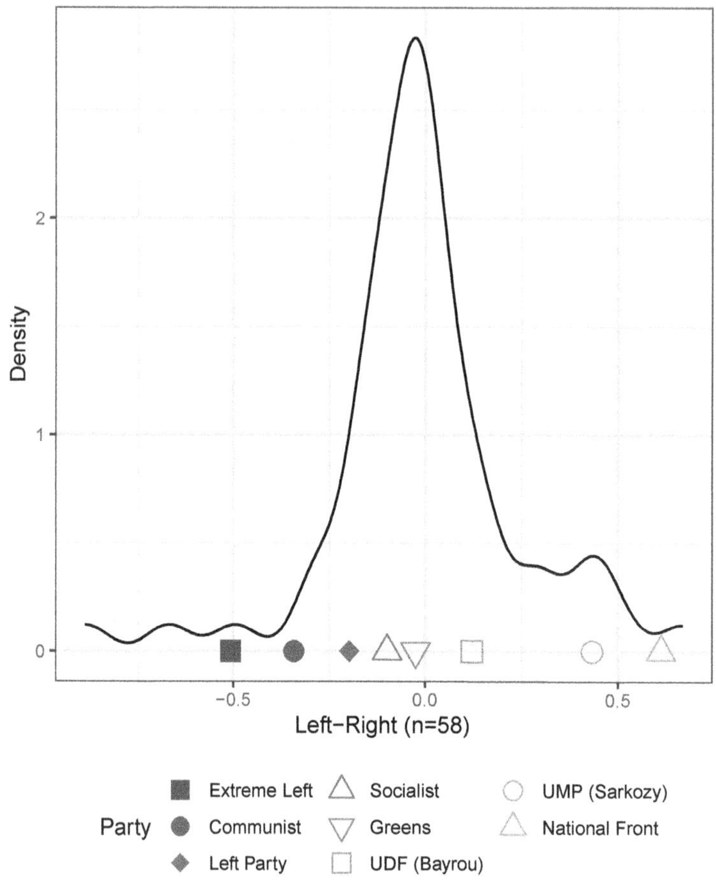

FIGURE 2.4: Aldrich-McKelvey Scaling of Left-Right Placements of French Political Parties: Negative Weights (2009 European Election Study)

uses the endpoints "Solve Problems of Poverty and Unemployment" and "Use All Available Force" in the question prompt:

> There is much discussion about the best way to deal with the problem of urban unrest and rioting. Some say it is more important to use all available force to maintain law and order – no matter what results. Others say it is more important to correct the problems of poverty and unemployment that give rise to the disturbances. And, of course, other people have opinions in between. Suppose the people who stress the use of force are at one end of this scale

> – at point number 7. And suppose the people who stress doing more about the problems of poverty and unemployment are at the other end – at point number 1. Where would you place ... on this scale?

The Vietnam War scale uses the endpoints "Immediate Withdrawal" and "Complete Military Victory":

> There is much talk about "hawks" and "doves" in connection with Vietnam, and considerable disagreement as to what action the United States should take in Vietnam. Some people think we should do everything necessary to win a complete military victory, no matter what results. Some people think we should withdraw completely from Vietnam right now, no matter what results. And, of course, other people have opinions somewhere between these two extreme positions. Suppose the people who support an immediate withdrawal are at one end of this scale at point number 1. And suppose the people who support a complete military victory are at the other end of the scale at point number 7. At what point on the scale would you place ...?

We first analyze the Urban Unrest data, setting Hubert Humphrey as the left-leaning stimulus. The data fit a one-dimensional spatial model well (with an overall R^2 value of 0.78 and an AM fit statistic of 0.09). Hubert Humphrey and President Lyndon Johnson are on the left end of the scale, stemming from the Johnson Administration's work on civil rights legislation and its "War on Poverty" programs. George Wallace, who carried five southern states in 1968, is the most rightward stimuli. Wallace ran on "law and order" platform that explicitly appealed to racial animus. Richard Nixon also ran on a "law and order" issues, but his rhetoric was more moderate (Vavreck, 2009, pp. 86-90) and thus is estimated near the middle of the scale. Given George Wallace's extreme position, very few respondents (81) have negative weights (ranked Wallace to the left of Humphrey).

```
# Loading 'urbanunrest' data
data(nes1968_urbanunrest)
# Creating object with US president left-right dimensions
urban <- as.matrix(nes1968_urbanunrest[,-1])
#Running Bayesian Aldrich-Mckelvey scaling on President positions
result.urb <- aldmck(urban, polarity=2, respondent=5,
    missing=c(8,9), verbose=FALSE)
summary(result.urb)
```

```
SUMMARY OF ALDRICH-MCKELVEY OBJECT
----------------------------------

Number of Stimuli: 4
```

```
Number of Respondents Scaled: 1191
Number of Respondents (Positive Weights): 1110
Number of Respondents (Negative Weights): 81
Reduction of normalized variance of perceptions: 0.09

         Location
Humphrey   -0.428
Johnson    -0.399
Nixon       0.015
Wallace     0.811
```

Because we have ideal points for both candidates and voters, we can show the sources of support for each presidential candidate along the recovered dimension. The code below produces the plot in Figure 2.5. This code fragment divides voters with positive `weight` values by presidential vote choice: Humphrey (`vote`=3), Nixon (`vote`=5) and Wallace (`vote`=6). We calculate each candidate's share of the vote, and plot the distribution of each group of voters in proportion to the candidate's vote share.

Figure 2.5 is satisfying from the standpoint of spatial voting theory. Each of the three presidential candidates are located near the center-peak of the distribution of their supporters. Except for some random error and valence (non-policy) effects, voters should support the candidate nearest their ideal point. Likewise, candidates should draw the most support from voters who share their ideal point and steadily garner less support moving away from the ideal point in both directions.

```
# Extracting vote.choice column
# recode so that only Humphrey, Nixon and Wallace are present
vote <- car::recode(nes1968_urbanunrest[,1],
    "3='Humphrey'; 5 = 'Nixon'; 6 = 'Wallace'; else=NA",
    as.factor=FALSE)
# Convert vote to factor with appropriate levels
vote <- factor(vote, levels=c("Humphrey", "Nixon", "Wallace"))
# Plot population distribution by vote choice
plot_resphist(result.urb, groupVar=vote, addStim=TRUE,
  xlab="Liberal-Conservative")  +
  theme(legend.position="bottom", aspect.ratio=1)  +
  guides(shape = guide_legend(override.aes =
      list(size = 4, color=c("gray25", "gray50", "gray75"))),
    colour = "none") +
  xlim(c(-2,2)) +
  labs(shape="Candidate")
```

Figure 2.6 shows the recovered locations of the candidates from the Vietnam War issue scale data. Hubert Humphrey is on the left ("dove") end of the scale, and George Wallace on the right ("hawk") end. Wallace is the outlier on the

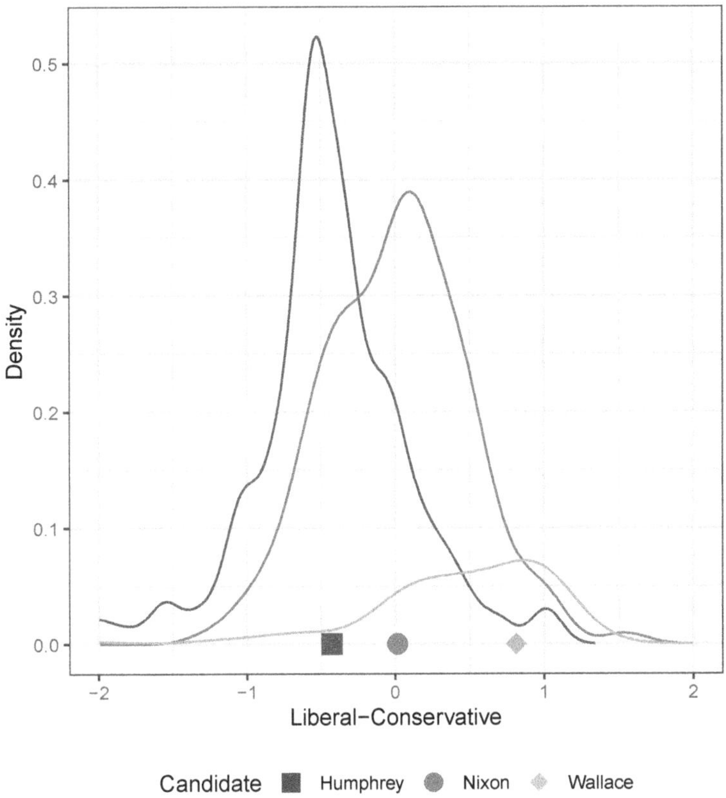

FIGURE 2.5: Aldrich-McKelvey Scaling of Urban Unrest Scale: Candidates and Voters (1968 American National Election Study)

scale, with the distance between Wallace and Nixon more than double that between Humphrey and Nixon.[†]

The recovered locations of the candidates from the Vietnam War data essentially square with our understanding of the 1968 campaign. However, the results provide a good illustration of the dangers involved in generating spatial models when the assumptions do not hold. Namely, the assumption of single-peaked utility may well be violated with this data, since many voters held a "win or get out" attitude regarding US involvement in Vietnam (Shapiro and Page, 1988). Such voters would have bi-modal utility functions, with peaks at

[†]Indeed, George Wallace's running mate—General Curtis LeMay—advocated the use of nuclear weapons in Vietnam, which led Humphrey to refer to Wallace and LeMay as the "Bombsey Twins" (Carter, 1995).

or near the two poles of the scale. Moreover, Humphrey's ties to the Johnson Administration and its handling of the Vietnam War apparently generated confusion among respondents, as evidenced by a high number of respondents with negative weights (231, compared to 81 for the Urban Unrest data), and a poorer fit of the model (an R^2 value of 0.67 and an AMfit statistic of 0.19) to the Vietnam War data than the Urban Unrest data.

```
data(nes1968_vietnam)
vietnam <- as.matrix(nes1968_vietnam[,-1])
#Aldrich-Mckelvey function for vietnam dataset
result.viet <- aldmck(vietnam, polarity=2, respondent=5,
    missing=c(8,9), verbose=FALSE)
summary(result.viet)
```

```
SUMMARY OF ALDRICH-MCKELVEY OBJECT
----------------------------------

Number of Stimuli: 4
Number of Respondents Scaled: 1031
Number of Respondents (Positive Weights): 800
Number of Respondents (Negative Weights): 231
Reduction of normalized variance of perceptions: 0.19

          Location
Humphrey    -0.436
Johnson     -0.330
Nixon       -0.068
Wallace      0.834
```

2.1.4 Estimating Bootstrapped Standard Errors for Aldrich-McKelvey Scaling

Below we show how to use the nonparametric bootstrapping method (Efron and Tibshirani, 1993) to estimate standard errors for the A-M results. The logic behind the bootstrap is that we can take repeated, random draws from a single data set, and treat these draws as random samples themselves. In this case, we can take a random sample (with replacement) of respondents, perform the A-M scaling procedure on that data, save those results, and repeat the process a specified number of times (often, 100 or 1,000).[‡]

This will give us a distribution of point estimates for the locations of the stimuli. This means that we will have 100 estimates of, for example, the French National Front Party's location. This distribution will of course have

[‡]There are no set rules for how many bootstrap replications are needed to produce good estimates of the standard error of $\hat{\theta}$. Efron and Tibshirani (1993, p. 52) suggest that 100 trials are sufficient in most cases, and that very seldom are more than 200 trials needed.

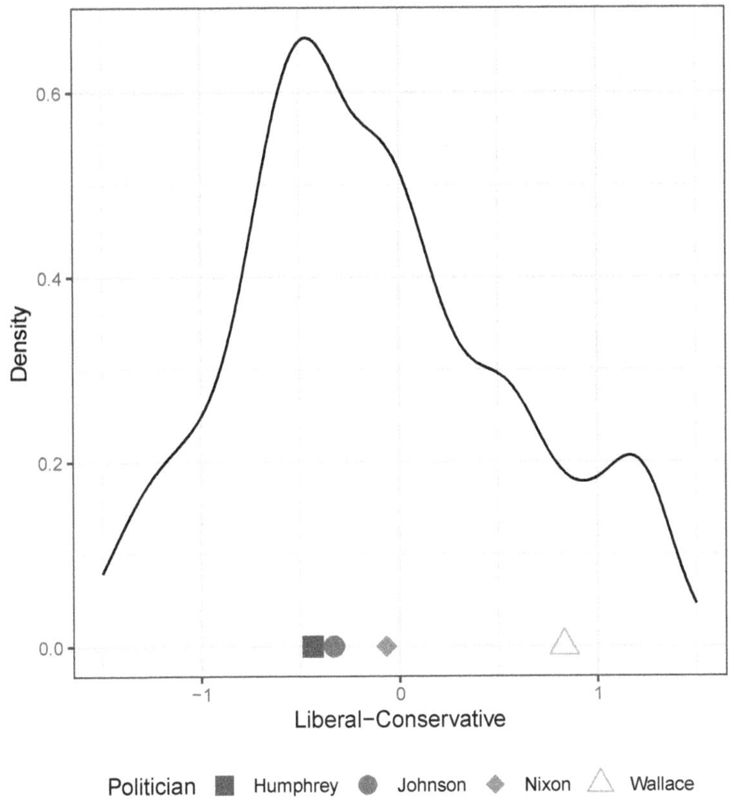

FIGURE 2.6: Aldrich-McKelvey Scaling of Vietnam War Scale: Candidates and Voters (1968 American National Election Study)

the standard moments: mean, variance, etc., which can be used to calculate standard errors of the point estimate $\hat{z}_j$.

Below we show how to implement the nonparametric bootstrap with the A-M scaling procedure using data from the 2009 EES French module (the example in Section 2.1.2). We first program the function `boot.fun()`, which allows for A-M scaling to be run a specified number of times within the `boot()` function. The command `apply(boot.aldmck$t, 2, sd)` calculates the standard error of each party's estimate from the `result` matrix. Finally, we combine the stimuli point estimates from the matrix `result`, the bootstrapped standard errors, and the upper and lower 95% confidence intervals in the matrix `boot.out`, ordering the parties from left to right.

```
boot.france  <- boot.aldmck(franceEES2009,
```

```
        polarity=2, respondent=1, missing=c(77,88,89),
        verbose=FALSE, boot.args = list(R=100))
```

The following code is used to plot the bootstrap results in Figure 2.7. The point estimates of the French parties are shown as dots, with 95% confidence intervals in bars. Note that the uncertainty bounds are quite small, which speaks to the precision of the A-M estimator.

```
library(ggplot2)
plot(boot.france$sumstats) +
  ylab(NULL) +
  xlab("Left-Right")
```

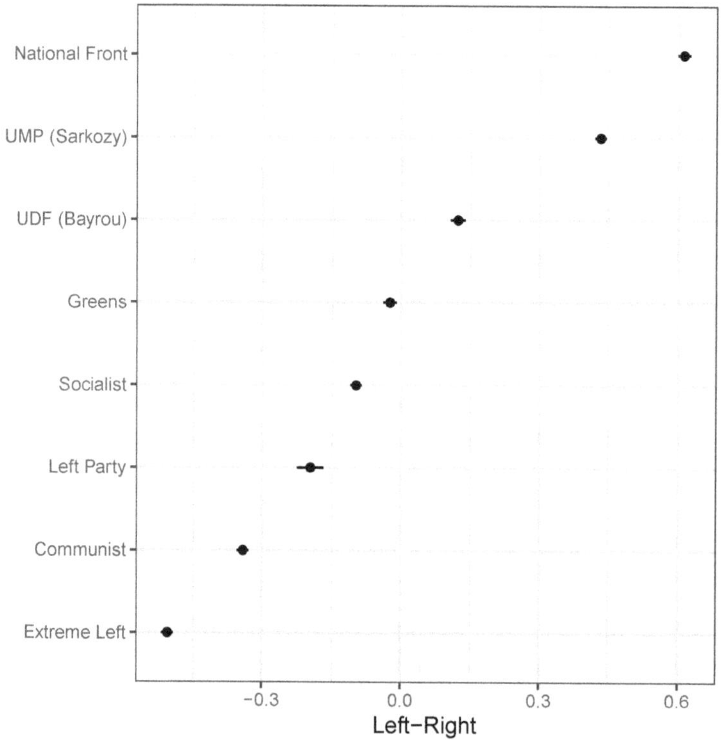

FIGURE 2.7: Aldrich-McKelvey Scaling of Left-Right Placements of French Political Parties (2009 European Election Study) with Bootstrapped Standard Errors

2.2 Basic Space Scaling: The blackbox Function

Poole (1998*a*,*b*) developed the Basic Space scaling procedure (implemented in the `blackbox()` and `blackbox_transpose()` functions in `R`) as a generalization of Aldrich-McKelvey (A-M) scaling in that it estimates a series of weight and constant terms that map issue scale data onto a latent space. Two advantages of the Basic Space method are especially worthy of emphasis. First, the procedure permits the inclusion of missing data (for example, if a respondent is unable or refuses to answer a survey item, a common occurrence when dealing with public opinion data). As with Bayesian A-M scaling, this feature is necessary in order to bridge across time or geographic units since the respondents cannot answer all of the items. For example, Bakker et al. (2013) use Basic Space scaling to estimate comparable ideological positions of European political parties by asking experts to place the parties in their country of expertise and three hypothetical parties (whose platform is described in a vignette) on a left-right scale. Missing data will be present in all of the parties because French political parties, for example, are ranked only by the French experts.

The second difference between the Basic Space and A-M scaling methods is that Basic Space scaling allows for the analysis of multiple issue scales in multiple dimensions. A-M scaling, in contrast, restricts each issue scale to a separate dimension. This feature of Basic Space scaling more closely accords with Converse's (1964) notion of "constraint," the process by which political belief systems bind together multiple issue positions.

The geometric extension of Converse's theory is the basic space or "two-space" theory (see Section 1.1.3), which posits that the issue or action space (represented by the issue scales) maps onto a low-dimensional basic space. Accordingly, the goal of the Basic Space scaling method is to estimate a vector of parameters that map the observed issue scale data onto the individuals' true coordinates in the latent (basic) space. This allows for the recovery of an s-dimensional basic space *directly from the data*, rather than from a correlation or covariance matrix as with alternative data reduction methods like factor analysis. Response-level data provides a richer picture of the political attitudes and perceptions of survey respondents, while correlation/covariance matrices, as noted by Jackman (2001, p. 230), "[discard] information about the means and the variances of the input variables. Information is necessarily lost in this way." Simply put, the use of individual choice data is more appropriate for the analysis of individual behavior within the spatial model of choice.

Generally, the `blackbox()` function is used to scale individuals from preference data. For example, survey respondents state their preferred policy outcome on multiple issue scales, and the `blackbox()` procedure recovers the ideal points of these individuals in an s-dimensional basic space. Conversely, the `blackbox_transpose()` function is used to estimate the latent coordinates of stimuli that are rated by individuals (i.e., based on perceptual data). For example, a set of party experts rank the positions of European political parties across a set of issue scales. The `blackbox_transpose()` function transposes the matrix, placing the stimuli on the rows and the individuals on the columns—since in these cases we want to scale the stimuli. Despite this distinction, both functions are applications of the same underlying method, which we detail below.

Let x_{ij} be the ith individual's ($i = 1,...,n$) reported position on the jth issue ($j = 1,...,q$) and let X_0 be the n by q matrix of observed data where the 0 subscript indicates that elements are missing from the matrix—not all individuals report their positions on all issues. Let Ψ_{ik} be the ith individual's position on the kth ($k = 1,...,s$) basic dimension. The model estimated is:

$$X_0 = [\Psi W' + J_n c']_0 + E_0 \tag{2.10}$$

where Ψ is the n by s matrix of coordinates of the individuals on the basic dimensions, W is an q by s matrix of weights, c is a vector of constants of length q, J_n is an n length vector of ones, and E_0 if a n by q matrix of error terms. W and c map the individuals from the basic space onto the issue dimensions. Put differently, the observed data is treated as a function of individuals' true coordinates in the basic space multiplied by the weights (W) plus a constant (c) and an error term (E_0).

Equation (2.10) can be written as the product of partitioned matrices

$$X_0 = [\Psi | J_n][\frac{W'}{c'}]_0 + E_0 \tag{2.11}$$

where $[\Psi|J_n]$ is a n by $s+1$ matrix and $[W|c]$ is a q by $s+1$ matrix. If $n > q$ and there is no error or missing data, then the rank of X is s and the rank of $X - J_n c'$ is less than or equal to s.

2.2.1 Example 1: 2000 Convention Delegate Study

To demonstrate how to use the `blackbox()` function to recover a basic space from issue scale data, we use data from the 2000 Convention Delegate Study (CDS), which interviewed delegates to the Republican and Democratic National Conventions. Studies that have analyzed CDS data include Stone and Abramowitz (1983), Layman (2001), and Layman et al. (2010). In 2000, the CDS received completed questionnaires from 1,907 delegates to that year's Democratic National Convention and 985 delegates to the Republican National Convention. The survey included a battery of issue scales on which delegates were asked to place their policy preference and those of major political stimuli (e.g., Al Gore and George W. Bush).

We use the `blackbox()` function to analyze issue scale questions where delegates expressed their own preferences. Specifically, we include ten issue scales stored in columns 5-14 in the matrix `CDS2000`: liberal-conservative placement, abortion, government services, defense spending, aid to blacks, government health insurance, employment protection for homosexuals, affirmative action, use of the budget surplus for tax cuts, and support for free trade. All questions use a seven-point scale except abortion (a four-point scale) and affirmative action, using the budget surplus for tax cuts, and free trade (all five-point scales). Missing responses are coded as 99.

```
library(basicspace)
data(CDS2000)
head(CDS2000[,5:8])
```

```
     Lib-Con Abortion Govt Services Defense Spending
[1,]       3        4             5                4
[2,]       2        4             5                6
[3,]       1        4             7                6
[4,]       2       99             6                5
[5,]       4        4             7                1
[6,]       4        4             7                4
```

The `blackbox()` function requires five arguments: the matrix to be analyzed (`issues`), `missing` (a vector or matrix of missing values in the matrix), `dims` (the number of dimensions to be estimated), `minscale` (the minimum number of valid responses required for a respondent to be included in the scaling), and `verbose` (a logical argument (`TRUE/FALSE`) that specifies whether verbose output is desired as the function is executed).

```
issues <- as.matrix(CDS2000[,5:14])
#Blackbox syntax of Republican-Democrat left-right scale
result.repdem <- blackbox(issues,
           missing=99, dims=3, minscale=5, verbose=TRUE)
```

```
       Beginning Blackbox Scaling...10 stimuli have been provided.

       Blackbox estimation completed successfully.
```

The `blackbox()` function stores nine objects in the dataframe `result`. Note that separate estimates are provided for each dimensional configuration. In this example, we specify `dims=3` in the `blackbox()` call, estimating three dimensions in order to test for the possibility of a high-dimensional solution. Thus, one-, two- and three-dimensional solutions are calculated. Dimensionality is specified in hard brackets for `result$stimuli` and `result$individuals` (e.g., `result$individuals[[1]]`, `result$individuals[[2]]`, etc.). Each of the nine objects can be accessed using the commands `result$stimuli`, `result$individuals`, etc:

stimuli Estimates of the stimuli:

N Number of individuals providing a valid placement of the stimulus.

c Constant term (c from Equation 2.10).

w1, ..., ws Weight term on the kth dimension (w from Equation 2.10).

R2 Percent of variance explained for stimulus.

individuals Estimates of the individuals:

Ψ1, ..., Ψs Respondent ideal point on the kth dimension (Ψ from Equation 2.10); missing values are coded as `NA`.

fits Fit statistics for each of the k-dimensional solutions:

SSE Sum of squared errors.

SSE.explained Explained sum of squared errors.

percent Percentage of total variance explained.

SE Standard error of the estimate.

singular Singular value for the dimension.

Nrow Number of rows (individuals).

Ncol Number of columns (issues).

Ndata Number of total entries.

Nmiss Number of missing entries.

SS_mean Sum of squares grand mean.

dims Number of dimensions estimated.

We first want to determine how well delegates' policy attitudes can be explained by the latent dimensions of the basic space. To examine the fit statistics, we use the `result$fits` command:

```
result.repdem$fits
```

```
                 SSE SSE.explained   percent       SE
Dimension 1 31388.39      56868.27 64.435103 1.119187
Dimension 2 24890.89      63365.78  7.362055 1.059063
Dimension 3 20213.13      68043.54  5.300175 1.022721
             singular
Dimension 1 235.21101
Dimension 2  82.83445
Dimension 3  70.64624
```

There is clearly a dominant first dimension to this data, with a single dimension explaining just over 64% of the total variance in responses to the issue scales. A second dimension contributes an additional 7.36% to the proportion of explained variance. The second dimension could represent attitudes on cross-cutting issues (for example, social or foreign policy issues), but we should be cautious about substantively interpreting this dimension because it could simply be picking up noise in the data. We can examine the substantive meaning of the dimensions by looking at how well each dimension taps into the issue attitudes with the issue-specific weight terms (W) and R^2 values on each dimension.[§] The weight terms and R^2 values for each dimensional configuration can be accessed with the **result$stimuli** command:

```
result.repdem$stimuli

[[1]]
                        N     c     w1    R2
Lib-Con              2804 3.700  4.980 0.736
Abortion             2693 3.239 -2.439 0.467
Govt Services        2805 4.267 -5.579 0.779
Defense Spending     2816 3.525 -4.224 0.507
Aid to Blacks        2789 3.633  4.944 0.587
Health Insurance     2804 3.476  7.010 0.760
Protect Homosexuals  2803 3.452  6.753 0.730
Affirmative Action   2806 3.213  3.797 0.563
Surplus for Tax Cuts 2811 3.284 -4.503 0.605
Free Trade           2805 3.168 -1.315 0.078

[[2]]
                        N     c     w1     w2    R2
Lib-Con              2804 3.700  4.981  0.552 0.739
Abortion             2693 3.239 -2.441 -0.143 0.467
Govt Services        2805 4.268 -5.581 -0.985 0.787
Defense Spending     2816 3.526 -4.219 -1.564 0.530
Aid to Blacks        2789 3.629  4.947 -4.179 0.731
Health Insurance     2804 3.477  7.004  4.275 0.857
Protect Homosexuals  2803 3.447  6.758 -4.358 0.834
Affirmative Action   2806 3.213  3.800 -1.927 0.614
Surplus for Tax Cuts 2811 3.283 -4.504 -1.509 0.628
Free Trade           2805 3.165 -1.303 -4.230 0.355

[[3]]
                        N     c     w1     w2     w3    R2
Lib-Con              2804 3.701  4.977 -0.632  0.688 0.744
Abortion             2693 3.236 -2.436  0.264 -1.284 0.508
```

[§]Note that the weight terms in Basic Space scaling are analogous to factor loadings in factor analysis or discrimination parameters in IRT models (discussed in Chapter 6) in that they are measures of how much variation in a given stimuli is captured by a latent dimension.

```
Govt Services         2805 4.268 -5.588  0.905  0.899 0.793
Defense Spending      2816 3.526 -4.221  1.553  0.679 0.530
Aid to Blacks         2789 3.623  4.949  4.682 -4.825 0.921
Health Insurance      2804 3.475  7.001 -4.287  0.077 0.856
Protect Homosexuals   2803 3.447  6.747  3.954  4.409 0.907
Affirmative Action    2806 3.211  3.806  2.239 -2.628 0.705
Surplus for Tax Cuts  2811 3.281 -4.502  1.753 -2.142 0.675
Free Trade            2805 3.169 -1.303  3.888  3.609 0.499
```

The weight terms (`w1`, `w2`, and `w3`) and R^2 values for each issue scale show that the first dimension is doing a good job at capturing attitudes for all nine issues. The highest first-dimension R^2 values are for the government services scale (0.779) and the government health insurance scale (0.760). The highest first dimension weight terms (`w1`) for each of the dimensional configurations are for the government health insurance (7.010) and homosexual employment protection (6.753) scales. The first dimension, then, appears to be tapping liberal-conservative orientations on a range of issues (economic as well as social and national defense issues). Indeed, the first-dimension R^2 value for self-placement on the liberal-conservative scale is 0.736. Since these survey respondents are highly informed national convention delegates, it makes sense that they would exhibit a high degree of ideological constraint. This is precisely what the basic space results indicate.

Interpreting the substantive meaning of the second dimension is frequently more challenging. In this case, the second dimension appears to explain attitudes on mainly one cross-cutting issue: free trade. Free trade is the worst-fitting issue scale on the first dimension (with an R^2 value of 0.078), and its second-dimension improvement in fit is the largest of the issue scales (with an R^2 value of 0.355 in two dimensions). This is not surprising, since trade agreements (e.g., NAFTA and WTO) divided the parties—particularly the Democrats—during the 1990s (Wink, Livingston and Garand, 1996).

Finally, the third dimension appears to be solely fitting noise in the data. The improvement in fit statistics produced by the third dimension is dispersed across a number of unrelated issue scales (e.g., aid to blacks and free trade). We thus limit our analysis to the results from the two-dimensional configuration (`result.repdem$stimuli[[2]]` and `result.repdem$individuals[[2]]`).

We now plot the estimated basic space coordinates of the delegates in Figure 2.8. Figure 2.8 shows that the Republican and Democratic delegates are evenly mixed on the second dimension, providing further evidence that the ideological differences dividing the parties are represented by the first dimension.

```
# Party: Democrats = 1; Republicans = 2
party <- car::recode(CDS2000[,1],
    "1='Democrat'; 2='Republican'; else=NA",
    as.factor=TRUE)
# Make the plot
```

```
plot_blackbox(result.repdem, dims=c(1,2), groupVar=party,
              xlab= "First Dimension\n(Left-Right)",
              ylab="Second Dimension") +
  theme(legend.position="bottom", aspect.ratio=1) +
  guides(shape=guide_legend(override.aes=list(size=4))) +
  labs(colour="Party")
```

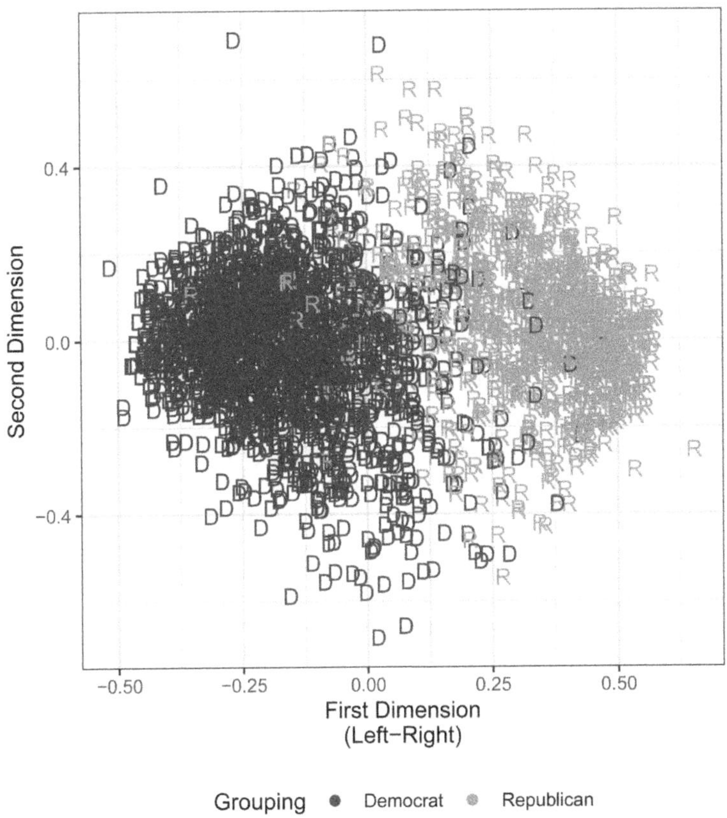

FIGURE 2.8: Basic Space (Blackbox) Scaling of US Party Convention Delegates (2000 Convention Delegate Study)

In order to examine how specific issues map onto the latent space and verify our understanding of the substantive meaning of the dimensions, we next plot the normal vectors of delegates' responses to the abortion and free trade survey items Poole (2005, pp. 152-155). The normal vector runs from the origin and indicates the direction of an observed quantity through the recovered space, and so it is useful for assessing dimensionality. In Chapter 5, we discuss the geometry of normal vectors in greater detail.

To calculate the normal vectors of the abortion and free trade issues, we first regress the survey responses onto the latent dimensions. The coefficient values (β) can then be used to calculate the normal vector for each dimension k ($k = 1, ..., s$) using Equation 2.12 (note that the intercept term, β_0, is not used in the computation — see Poole (2005, pp. 37-40) for the mathematics). As denoted below, each β_k is divided by the Euclidian norm (the square root of the sum of all squared $\beta_{1,...,s}$ values).

$$NV_k = \frac{\beta_k}{\sqrt{\beta_1^2 + \ldots + \beta_s^2}} \tag{2.12}$$

For example, below we use ordered probit (since the responses are ordinal) using the `polr()` function in the `MASS` package in `R` to regress respondents' abortion responses onto their first and second dimension scores (Venables and Ripley, 2002). Consistent with the abortion issue's high first dimension weight term, the first dimension is most important in explaining variation in abortion attitudes as $\beta_1 = -3.558$ (standard error $= 0.099$) and $\beta_2 = -0.188$ (0.141).

With these coefficients, we can use Equation 2.12 to calculate the normal vector (in two dimensions) of the abortion issue. As shown below in Equation 2.13, this produces the point coordinate of $(0.999, -0.053)$; hence, the normal vector runs between the origin and the point $(-0.999, -0.053)$. Its reflection (`-N1`, `-N2`) runs between the origin and the point (0.999,0.053).

$$\begin{bmatrix} \frac{-3.558}{\sqrt{-3.558^2 + -0.188^2}} \\ \frac{-0.188}{\sqrt{-3.558^2 + -0.188^2}} \end{bmatrix} = \begin{bmatrix} -0.999 \\ -0.053 \end{bmatrix} \tag{2.13}$$

We repeat this process for the free trade issue and obtain the coordinate of $(-0.291, -0.957)$ for its normal vector. Both normal vectors and their reflections are plotted in Figure 2.9 with the commands below. The `exp.factor` value is simply used to expand or, in this case, contract the length of the normal vector for illustration purposes. Substantively, the angles of these normal vectors further validate our understanding that the first dimension taps into issues like abortion that are tied to the contemporary liberal-conservative divide while the second dimension picks up attitudes on cross-cutting issues like free trade.

```
plot_blackbox(result.repdem, dims=c(1,2), groupVar=party,
    issueVector=c(2,10), data=issues,
    nudgeX= c(0,.125), nudgeY=c(-.05,0)) +
  theme(legend.position="bottom", aspect.ratio=1) +
  labs(colour="Party") +
  guides(shape = guide_legend(override.aes = list(size = 4)))
```

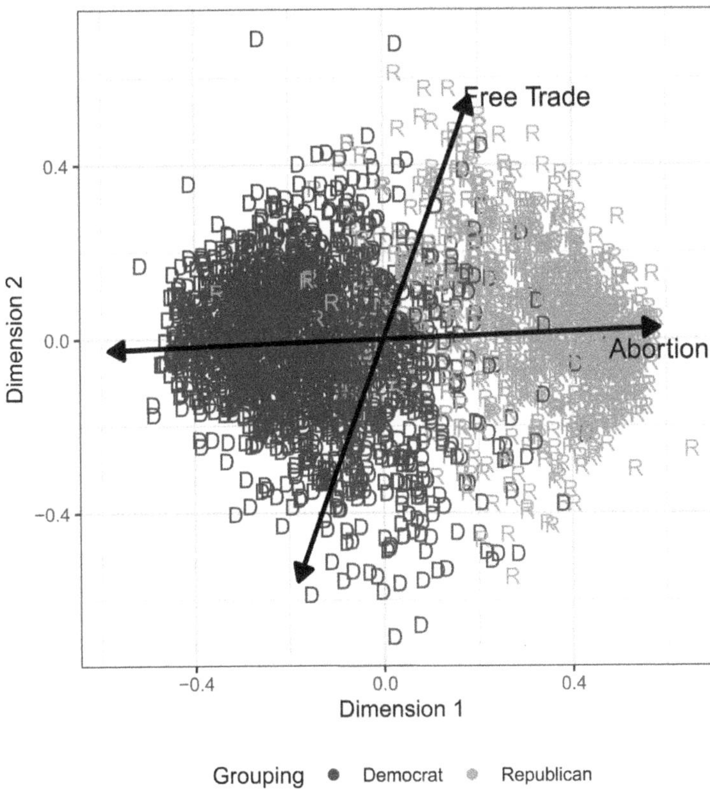

FIGURE 2.9: Basic Space (Blackbox) Scaling of US Party Convention Delegates (2000 Convention Delegate Study) with Issue Normal Vectors

In Figure 2.10 we plot a smoothed histogram of Republican and Democratic delegates' first dimension scores. There is very little ideological overlap between the two partisan groups, which is to be expected given the state of polarization among political elites in contemporary American politics (McCarty, Poole and Rosenthal, 2006).

```
#Weight Rep/Dem data by number of delegates
plot_resphist(result.repdem, groupVar=party,
  addStim=FALSE, xlab = "Liberal-Conservative", dim=1, whichRes=2) +
  theme(legend.position="bottom", aspect.ratio=1) +
  labs(colour="Party") +
  guides(colour = guide_legend(override.aes = list(size = 2)))
```

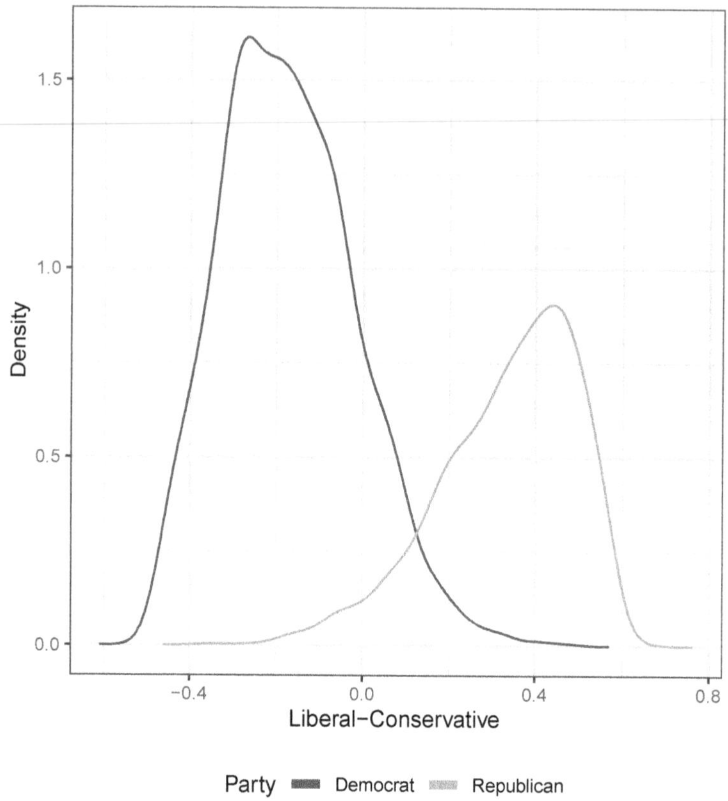

FIGURE 2.10: Density Plot of First Dimension Basic Space (Blackbox) Scores of US Party Convention Delegates (2000 Convention Delegate Study)

2.2.2 Example 2: 2010 Swedish Parliamentary Candidate Survey

In 2010, the Swedish public broadcasting network Sveriges Television (SVT) conducted a survey of that country's 5,627 parliamentary candidates, completing interviews with 2,830 candidates (including 289 of the 349 candidates who won election). Candidates were asked 50 Likert-type questions, in which candidates used a 4-point scale (from "strongly disagree" to "strongly agree") to register their opinion on a series of policy statements (e.g., "Those who are 58 years old should be eligible for early retirement"). Most of the issue scales focus on economic/social welfare issues, but questions that deal with foreign policy, social/cultural, law and order, immigration, and environmental issues are also included. This type of elite survey is valuable in measuring the ideological preferences of candidates and elected officials, particularly in parliamentary systems where high levels of party discipline limit our ability to recover legislators' ideological positions from roll call data.

The columns of `Sweden2010` are arranged in the following order: case ID, whether the candidate was elected (1) or not (0), name of the political party, code of the political party, whether the candidate is a member of one of the parties in the government coalition (1) or not (0), left-right self placement (a 1-5 scale), and the fifty issue scale questions. Missing responses are coded as 8. We include a conditional command after the `blackbox` call to reverse the signs if Candidate #13—a right-wing candidate—has a negative first dimension coordinate.

```
library(basicspace)
data(Sweden2010)
head(Sweden2010[,1:8])
```

```
     id elected          party.name party.code govt.party
1 39681       0 Conservative Party        300          1
2 17735       1 Conservative Party        300          1
3 41923       0 Conservative Party        300          1
4 43665       0 Conservative Party        300          1
5 15867       0 Conservative Party        300          1
6 39829       0 Conservative Party        300          1
  left.right.self.fivept congestion.taxes highspeed.trains
1                      5                3                2
2                      5                3                3
3                      4                4                1
4                      5                4                2
5                      5                3                3
6                      4                2                2
```

```
#Extract issues scales and convert to numeric
issues.sweden <- as.matrix(Sweden2010[,7:56])
mode(issues.sweden) <- "numeric"
#Blacbox syntax for Sweden issue scale
result.sweden <- blackbox(issues.sweden, missing=8,
                          dims=3, minscale=5, verbose=FALSE)
# change polarity of scores
if(result.sweden$individuals[[1]][13,1] < 0)
result.sweden$individuals[[1]][,1] <-
    result.sweden$individuals[[1]][,1] * -1
result.sweden$fits
```

```
                 SSE SSE.explained   percent        SE
Dimension 1 81259.51      91466.73 52.954739 0.8069252
```

```
Dimension 2 73388.39       99337.86  4.556993 0.7755642
Dimension 3 67238.03      105488.22  3.560756 0.7509879
             singular
Dimension 1 255.24828
Dimension 2  92.78285
Dimension 3  85.00461
```

The first dimension is most important in modeling candidate attitudes, with a single dimension explaining about 53% of the total variation. Additional dimensions make only minor contributions to the overall fit. A quick inspection of the stimuli fits reveals that the first dimension is picking up primarily economic left-right issues. Consider, for example, the fit statistics for ten of the issue scales below (numbers 16-25). The first six concern taxation and have high R^2 values, with only the scale concerning pension and wage taxes having an R^2 value below 0.588. Non-economic issues (e.g., repealing legal prostitution and increasing criminal sentences) generally have poor first-dimensional fits, although there are some non-economic issues (like state wiretaps) that map strongly onto the left-right divide. Generally, though, we are comfortable in labeling the first dimension as representing economic left-right issues.

```
result.sweden$stimuli[[1]][16:25,]
```

```
                              N     c     w1    R2
property.taxes.wealthy     2594 2.271  3.615 0.787
wealth.tax                 2621 1.983  3.420 0.793
tax.wealthy                2652 2.337  3.635 0.820
tax.pensions               2561 3.013  1.951 0.364
household.services.deduction 2663 2.956 -3.726 0.809
work.income.tax            2617 2.992 -2.553 0.588
sex.purchase               2544 1.454 -0.736 0.084
DUI.penalty                2537 3.384 -0.437 0.033
criminal.sentences         2497 3.103 -1.461 0.278
wiretaps                   2393 2.786  2.732 0.553
```

Figure 2.11 shows the distribution of candidates' first dimension scores by party.

```
elected <- as.numeric(Sweden2010[,2])
party.name.sweden <- as.factor(Sweden2010[,3])
plot_resphist(result.sweden, groupVar=party.name.sweden, dim=1,
  scaleDensity=FALSE) +
  facet_wrap(~stimulus, ncol=2) +
  theme(legend.position="none") +
  scale_color_manual(values=rep("black", 10))
```

Figure 2.12 shows only candidates if they were elected. Note that no candidates who were affiliated with the Feminist Initiative or Pirate Party won election.

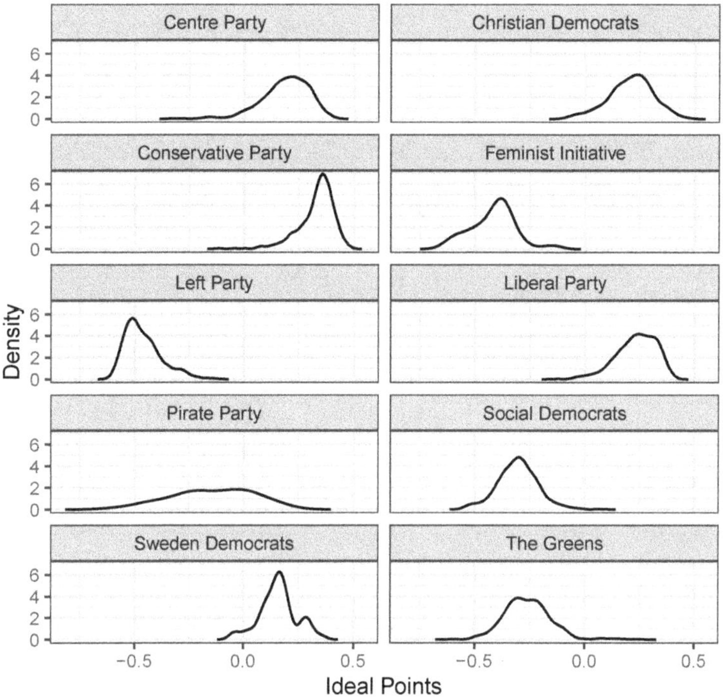

FIGURE 2.11: Basic Space (Blackbox) Scaling of 2010 Swedish Parliamentary Candidate Data (Candidates by Party)

```
#Density plot syntax and comparison of defeated/elected candidates
# Keep only the parties of elected candidates, set others to NA
party.name.sweden[which(elected == 0)] <- NA
plot_resphist(result.sweden, groupVar=party.name.sweden, dim=1,
   scaleDensity=FALSE) + facet_wrap(~stimulus, ncol=2) +
  theme(legend.position="none") +
  scale_color_manual(values=rep("black", 10))
```

2.2.3 Estimating Bootstrapped Standard Errors for Black Box Scaling

To estimate standard errors for the point estimates produced by `blackbox` scaling, we use the same nonparametric bootstrapping method as with Aldrich-McKelvey scaling (see Section 2.1.4). However, in this case, we sample (with replacement) on *columns* (the issue scales), rather than on *rows* (individuals). This is because in `blackbox` scaling we are concerned with

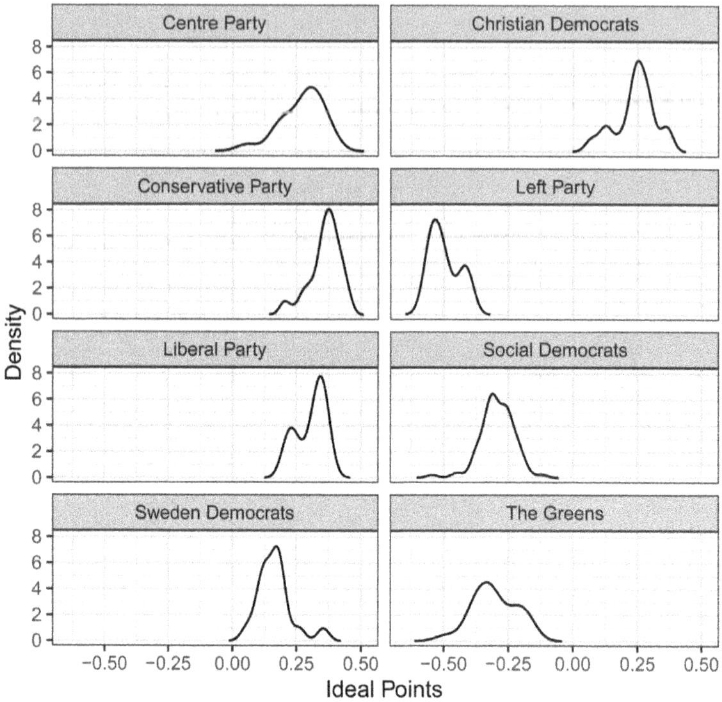

FIGURE 2.12: Basic Space (Blackbox) Scaling of 2010 Swedish Parliamentary Candidate Data (Elected and Defeated Candidates by Party)

estimating the locations of individuals rather than stimuli.

Using the 2010 Swedish parliamentary candidate data, we estimate standard errors and 95% confidence intervals via the nonparametric bootstrap for the point estimates of the candidates. We can use this measure to ask substantively interesting questions; for example, is there greater uncertainty associated with the ideological positions of losing candidates than winning candidates?

We first estimate the original, non-bootstrapped candidate point estimates in the matrix `original`. Next, we generate a specified number (`Ntrials`) of bootstrapped data sets in the list (`sample.data`). We then run `blackbox` on each of the `sample.data` matrices, storing the output in the list `result`. Note our use of for loops to create both `sample.data` and `result`, where the for loop runs from 1 to the specified number of trials (`Ntrials`). We then use a sign check to reverse signs if a right-wing candidate has a negative score.

```
#Candidate point estimates blackbox syntax
```

```
outbb <- boot.blackbox(issues.sweden, missing=8, dims=3, minscale=5,
    verbose=FALSE, posStimulus=13)
```

We then create a matrix (`final`) to store the bootstrapped estimates of the 2,830 candidates' first dimension scores. We store the standard deviation of those scores in `boot.se` with the command `apply(final, 1, sd, na.rm=TRUE)`. This command uses the helpful `apply()` function, which returns the results of a specified operation (in this case, `sd` for "standard deviation") on either the rows (1) or columns (2) of an array or matrix (`final`). The argument `na.rm=TRUE` tells `R` to ignore `NA` values in the data. Finally, we assemble the point estimates, standard errors, and the lower and upper 95% confidence intervals (calculated by subtracting or adding the standard error multiplied by 1.96 from or to the candidates' point estimates) in the matrix `out`.

```
#Matrix creation for Swedish candidates
first.dim <- data.frame(
  point = result.sweden$individuals[[3]][,1],
  se = apply(outbb[,1,], 1, sd)
)
first.dim$lower <- with(first.dim, point - 1.96*se)
first.dim$upper <- with(first.dim, point + 1.96*se)
first.dim$elected <- factor(elected, levels=c(0,1),
    labels=c("Not Elected", "Elected"))
head(first.dim)
```

```
   point         se      lower       upper     elected
1 -0.355 0.06962347 -0.4914620 -0.21853800 Not Elected
2 -0.383 0.07595791 -0.5318775 -0.23412250     Elected
3 -0.335 0.08639104 -0.5043264 -0.16567355 Not Elected
4 -0.359 0.07413758 -0.5043097 -0.21369034 Not Elected
5 -0.206 0.08284983 -0.3683857 -0.04361433 Not Elected
6 -0.072 0.06556910 -0.2005154  0.05651544 Not Elected
```

Our research question in this example is whether there is a difference between the uncertainty in ideological positions between winning and losing candidates. If voters are risk averse, then they would be less likely to elect ambiguous candidates (Alvarez, 1997). Conversely, a risk neutral or even risk acceptant electorate would be less concerned about policy ambiguity (Berinsky and Lewis, 2007). Our goal here is not to weigh in on this debate (indeed, we should not, since we lack the requisite covariates to control for the possibility of a spurious relationship). Rather, we use this example to illustrate how these measures of uncertainty can be used to address substantive questions.

We first examine the distribution of first dimension bootstrapped standard errors for elected and defeated candidates in Figure 2.13.

```
#Plot for the distribution of first dimensions bootstrapped SE
ggplot(first.dim, aes(x=se, group=elected)) +
```

```
stat_density(geom="line", bw=.005) +
facet_wrap(~elected) +
theme(aspect.ratio=1) +
xlab("Standard Error") +
xlim(c(0,.2)) +
theme_bw()
```

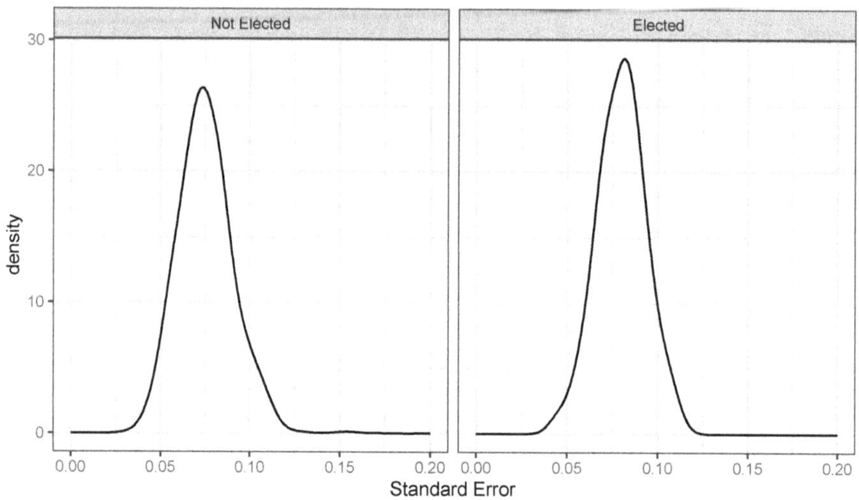

FIGURE 2.13: Basic Space (Blackbox) Scaling of 2010 Swedish Parliamentary Candidate Data with Boostrapped Standard Errors (Elected and Defeated Candidates)

The two groups of candidates appear to be fairly similar, but there is a definite right skew to the distribution of standard errors for defeated candidates, with an especially long tail stretching out to high standard error values. That is, defeated candidates as a group appear to have greater uncertainty associated with their ideological positions. To determine whether the difference between the two distributions is statistically significant, we compute a permutation test for the difference in standard errors variances. Since $p > .05$, we cannot reject the null that, on average, the two groups of candidates have similarly sized standard errors.

```
#Variance test syntax
library(perm)
levels(first.dim$elected) <- c("No", "Yes")
permTS(se ~ elected, data=first.dim,
       alternative="greater", method="exact.mc",
       control=permControl(nmc=10^4-1))
```

```
Exact Permutation Test Estimated by Monte Carlo

data:  se by elected
p-value = 1
alternative hypothesis: true mean elected=No -
mean elected=Yes is greater than 0

sample estimates:
mean elected=No -mean elected=Yes
-0.004336414

p-value estimated from 9999 Monte Carlo replications
99 percent confidence interval on p-value:
0.9994703 1.0000000
```

2.3 Basic Space Scaling: The blackbox_transpose Function

The `blackbox_transpose()` function is used to estimate the locations of political stimuli in the latent space from respondent evaluations of those stimuli. The `blackbox_transpose()` function implements the same procedure as the `blackbox()` function, but on a transposed matrix arranged such that the stimuli are on the rows and the respondents are on the columns. For example, for a survey in which 1000 respondents place 8 presidential candidates on a liberal-conservative scale, the dimensions of the matrix analyzed by `blackbox_transpose()` would be 8×1000. Because the processing time increases as the square of the number of columns, doubling the number of respondents increases CPU time by a factor of four. Hence, the analysis of matrices with more than 1500 respondents is quite computationally intensive for the `blackbox_transpose()` function. We discuss methods for dealing with this constraint in Section 2.3.3.

2.3.1 Example 1: 2000 and 2006 Comparative Study of Electoral Systems (Mexican Modules)

The results from the `blackbox_transpose()` function are in the same format as those for the `blackbox` function, with the important exception that the estimates of interest (the stimuli coordinates) are stored in the **`$stimuli`** object. This is because we are transposing the matrix so that we are estimating the locations of the stimuli, rather than the respondents. The call to the `blackbox_transpose()` function itself is identical to that for the `blackbox()` function. This also includes a **`minscale`** argument that specifies the minimum number of valid responses for a respondent to be included in the scaling, *not* the number of valid placements required for a stimulus to be included.

To demonstrate the `blackbox_transpose()` function, we use data from the 2000 and 2006 Mexican modules of the Comparative Study of Electoral Systems (CSES). In these surveys, Mexican citizens were asked to place the major political parties on a eleven-point left-right scale. It is important to note that the transposing takes place within the `blackbox_transpose()` function, so that the matrix to be processed is in standard form (with respondents on rows and stimuli on columns). This can be seen in the `mexicoCSES2000` data matrix below.

```
library(basicspace)
data(mexicoCSES2000)
data(mexicoCSES2006)
head(mexicoCSES2000)
```

```
     PAN PRI PRD PT Greens PARM
[1,]  11   6   1 99      6   99
[2,]  11   6   5  5      5    5
[3,]  10   4   3  3      8    3
[4,]   8   9   7 99     99   99
[5,]   9   5   3  4     99   99
[6,]   9   6   1  1      9   99
```

```
#Blackbox syntax for two datasets, with data cleaning arguments
result_2000 <- blackbox_transpose(mexicoCSES2000, missing=99,
    dims=3, minscale=5, verbose=TRUE)
result_2006 <- blackbox_transpose(mexicoCSES2006, missing=99,
    dims=3, minscale=5, verbose=TRUE)
```

The `result$stimuli` object is like the `result$individuals` object produced by the `blackbox()` function in that both include the estimated scores of the objects of interest (i.e., individuals in `blackbox` and stimuli in `blackbox_transpose`). However, `result$stimuli` also includes two additional columns: `N` (the number of valid responses) and `R2` (the explained variance in respondents' placements of the stimuli). Below we show the estimated two-dimensional configuration of the parties in 2000.

```
print(result_2000$stimuli[[2]])
```

```
          N coord1D coord2D    R2
PAN    1060  -0.810  -0.295 0.959
PRI    1059  -0.123   0.903 0.999
PRD    1062   0.269  -0.099 0.542
PT     1052   0.338  -0.182 0.711
Greens 1059  -0.048  -0.194 0.583
PARM    956   0.374  -0.131 0.595
```

In order to set the polarity of the space (placing left-wing parties on the left by assigning them negative scores and the reverse for right-wing parties), we multiply the parties' first dimension by -1 because the right-wing National Action Party's (PAN) first dimension scores in 2000 and 2006 are negative. If PAN's first dimension scores were positive, there would be no need to flip the space. We also flip the second dimension of the 2006 scores so that the ordering is most consistent. We retain the parties' original second dimension coordinates for 2000.

```
#Multiplying here to avoid negative scores
first.dim.2000 <- -1 * result_2000$stimuli[[2]][,2]
second.dim.2000 <-  result_2000$stimuli[[2]][,3]
first.dim.2006 <- -1 * result_2006$stimuli[[2]][,2]
second.dim.2006 <- -1*result_2006$stimuli[[2]][,3]
plot.df <- data.frame(
  dim1 = c(first.dim.2000, first.dim.2006),
  dim2 = c(second.dim.2000, second.dim.2006),
  year = rep(c(2000, 2006), c(length(first.dim.2000),
                              length(first.dim.2006))),
  party = factor(c(rownames(result_2000$stimuli[[2]]),
                  rownames(result_2006$stimuli[[2]]))))
plot.df$nudge_x <- c(0, 0,  0, 0, 0, -.125,
                     0, 0, 0, 0.13,   0, 0, -.225, 0)
plot.df$nudge_y <- c(-.05, -.05,.05, -.05, -.05,    0,
                     .05, .05, .05, -.03, -.05, .05,  .025, -.05)
head(plot.df)
```

```
   dim1   dim2 year  party nudge_x nudge_y
1 0.810 -0.295 2000    PAN   0.000   -0.05
2 0.123  0.903 2000    PRI   0.000   -0.05
```

```
3 -0.269 -0.099 2000    PRD   0.000    0.05
4 -0.338 -0.182 2000     PT   0.000   -0.05
5  0.048 -0.194 2000 Greens   0.000   -0.05
6 -0.374 -0.131 2000   PARM  -0.125    0.00
```

Next, we plot the first and second dimension scores of the parties in 2000 and 2006. Scholars of Mexican politics argue that the 2006 Mexican presidential elections marked a sea change in the relevance of left-right conflict (Moreno, 2007; McCann, 2012). Stemming from the Institutional Revolutionary Party's (PRI) seven-decade majority status (which ended in 2000), prior presidential elections were primarily referenda on the governing performance of the PRI, which were heavily weighted in the PRI's favor due to its "hyper-incumbency advantages" (Greene, 2007). Following PAN's victory over the PRI in 2000, ideological conflict (both on economic and social issues) emerged as the main line of cleavage in the 2006 election between the rightist PAN and the leftist Party of the Democratic Revolution (PRD), with the debate over NAFTA and NAFTA-style trade agreements occupying a central place in the campaign (Moreno, 2007). The PRI continued its tradition of touting centrist and ambiguous policy positions (McCann, 2012). The 2006 results had Calderón and the PAN edging out López Obrador by about one-half of one percent of the vote.

```
#Year 2000 plot syntax
ggplot(plot.df, aes(x=dim1, y=dim2, group=year)) +
  geom_point() +
  geom_text(aes(label=party), nudge_y=plot.df$nudge_y, size=3,
            nudge_x=plot.df$nudge_x, group=plot.df$year) +
  facet_wrap(~year) +
  xlim(-.55,1) +
  ylim(-.55,1) +
  theme_bw() +
  labs(x="First Dimension", y="Second Dimension")
```

The plots in Figure 2.14 are suggestive that the ideological divide between PAN and PRD was more pronounced in 2006 than in 2000, but we cannot establish this assertion definitively because the scalings are independent. If there were a panel study in which the same respondents were asked to rate the parties in both years, we could bridge across the surveys and place the parties in a common space spanning this period. This would allow us to draw more definitive conclusions. Looking at the plots separately, then, it appears that the first dimension represents left-right ideology and explains most of the variance in the data in both years (we show the R^2 values in the axis titles). The meaning of the second dimension is less clear. It is possible that it represents a "change versus establishment" cleavage, but it is unlikely that this divide would emerge from left-right party placements. Further, several

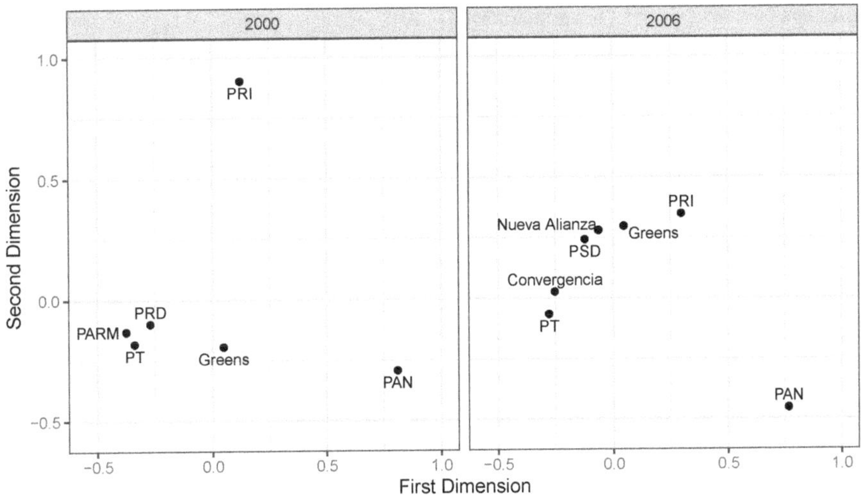

FIGURE 2.14: Basic Space (Blackbox Transpose) Scaling of Left-Right Placements of Mexican Political Parties (2000 and 2006 Comparative Study of Electoral Systems)

minor parties are clustered near the PRI on the second dimension in 2006. In this case, then, we do not reject the null that the second dimension is fitting noise in the data. Also note that even if there was greater ideological disparity between the PAN and PRD in 2006 (when they are the two parties furthest towards the edges of the first dimension) than 2000, ideological structure is not absent in 2000. Mexican voters still perceived the correct left-right ordering of the three major parties in 2000, even though ideological divides may have been less salient during this period.

2.3.2 Estimating Bootstrapped Standard Errors for Black Box Transpose Scaling

We can estimate standard errors for the locations of political stimuli from `blackbox_transpose` scaling with the same nonparametric bootstrapping method we have used previously. As with Aldrich-McKelvey scaling, we sample (with replacement) on individuals.

In this example, we return to data from the French module of the 2009 European Election Study, where respondents were asked to place seven French political parties on an eleven-point left-right scale. A quick inspection of the fit statistics for the result indicates that a one-dimensional model is appropriate (explaining nearly 79% of variation in the data). We include a sign check to reverse the scale if the first stimuli (the Extreme Left party) has a positive score.

```
rankings <- as.matrix(franceEES2009[,2:9])
mode(rankings) <- "numeric"
original <- blackbox_transpose(rankings,
                    missing=c(77,88,89),
    dims=3, minscale=5, verbose=FALSE)
#Reverse check
if(original$stimuli[[1]][1,2] > 0)
    original$stimuli[[1]][,2] <- -1 *
  original$stimuli[[1]][,2]
original$fits

outbbt <- boot.blackbox_transpose(rankings, missing=c(77,88,89),
                dims=3, minscale=5, verbose=FALSE, R=5)
```

As can be seen from the small size of the standard errors, the stimuli locations are estimates very precisely (as was also the case with A-M scaling of the same data).

```
france.boot.bbt <- data.frame(
  party = colnames(rankings),
  point = original$stimuli[[1]][,2],
  se = apply(outbbt[,1,], 1, sd, na.rm=TRUE)
)
france.boot.bbt$lower <- with (france.boot.bbt, point-1.96*se)
france.boot.bbt$upper <- with (france.boot.bbt, point+1.96*se)
france.boot.bbt
```

```
           party  point         se       lower       upper
1   Extreme Left -0.453 0.04401477 -0.53926895 -0.36673105
2      Communist -0.327 0.01286468 -0.35221477 -0.30178523
3      Socialist -0.089 0.03544432 -0.15847087 -0.01952913
4         Greens -0.025 0.05044502 -0.12387224  0.07387224
5   UDF (Bayrou)  0.115 0.01181948  0.09183383  0.13816617
6  UMP (Sarkozy)  0.456 0.03494710  0.38750368  0.52449632
7 National Front  0.612 0.04149458  0.53067063  0.69332937
8     Left Party -0.290 0.01423728 -0.31790506 -0.26209494
```

2.3.3 Using the blackbox_transpose Function on Data sets with Large Numbers of Respondents

The `blackbox_transpose()` function becomes computationally intensive when the number of respondents becomes very large. The maximum for the function is set to 1,500 respondents. As a technical matter, this restriction can be overridden (see the `big_blackbox_transpose.r` code on the book website for details), but the processing time tends to increase with the square of the number of respondents. Since many survey data sets exceed the 1,500 limit, we discuss two alternative approaches for using `blackbox_transpose` on matrices with very large numbers of respondents.

To demonstrate and test both methods, we use data from the California module of the 2010 Cooperative Congressional Election Study (CCES). The CCES surveyed 5,433 California residents and asked them to place several national and state-level political stimuli on a 7-point ideological scale.[¶] The California example is useful because we can compute the "true" stimuli locations and compare them with those estimated from alternative methods that stay within the 1,500-respondent limit. This allows us to compare how well these approaches approximate the "true" stimuli positions when scaling the full matrix is infeasible (for example, the complete CCES data set includes 55,400 respondents).

We ran `blackbox_transpose` on the complete, $5{,}433 \times 8$ matrix in order to compare the "true" stimuli positions with those estimated from these alternative approaches.

The first method is a partial scaling approach of running `blackbox_transpose` on random samples (without replacement) of 100, 250, 500, 1,000 and 1,500 respondents from the universe of 5,433 total respondents. We then save the stimuli estimates and compare them to the results obtained from the complete matrix. We sum all errors (the difference between the estimated and "true" stimuli positions) in the bottom row for each sample.

Table 2.1 summarizes the point estimates obtained with each sample. The results suggest that the partial scaling approach yields reasonable estimates of $\hat{z}$, the estimates of stimuli positions based on the complete data matrix. This is especially true once 500 respondents, or about 10% of the population, are sampled. However, a sample of 10% of the respondents may still exceed the 1,500 limit (for instance, the full 2010 CCES includes more than 50,000 respondents).

We next assess the feasibility of a bootstrapping solution. We have thus far used the bootstrap to estimate uncertainty bounds, but we can also use it to generate point estimates. We use the standard error of the distribution of results from the bootstrap replications to estimate confidence intervals around the point estimates. Likewise, we simply use the mean of this distribution as the point estimate. We perform 100 bootstrap replications of 100 random samples (with replacement) of the specified number of respondents.

The results presented in Table 2.2 indicate that the bootstrapping approach produces very precise estimates of $\hat{z}$, even when the sample size is small (e.g., 100, or about 2% of the respondents). This is because 100 samples of 100 respondents yields 10,000 draws of the population of 5,433 respondents. This means that any single individual is very likely (an 84.4% chance) to be included in the analysis. We conclude that the bootstrapping approach offers a superior means of bypassing the 1,500 respondent constraint in `blackbox_transpose`. However, one should be cautious when there are high rates of missing data,

[¶]The CCES did not ask respondents to place the gubernatorial candidates on an ideological scale, so Democratic nominee Jerry Brown and Republican nominee Meg Whitman are not included.

Table 2.1: Blackbox Transpose Results for Stimuli Locations: Partial Scaling Approach

	RESPONDENTS					$\hat{z}$
	100	250	500	1,000	1,500	(5,433)
Democratic Party	-0.331	-0.346	-0.339	-0.338	-0.344	-0.341
Republican Party	0.346	0.399	0.382	0.376	0.378	0.380
Tea Party	0.544	0.485	0.501	0.497	0.505	0.506
Barack Obama	-0.342	-0.350	-0.340	-0.342	-0.342	-0.341
A. Schwarzenegger	0.060	0.078	0.075	0.080	0.076	0.074
Dianne Feinstein	-0.256	-0.267	-0.271	-0.276	-0.272	-0.272
Barbara Boxer	-0.390	-0.378	-0.387	-0.384	-0.379	-0.382
Carly Fiorina	0.369	0.379	0.379	0.387	0.377	0.376
Σ ERRORS	0.128	0.070	0.020	0.040	0.013	

as is the case when using anchors to bridge across units or time (e.g., surveys including multiple states, where respondents are asked about national figures [the anchors] and stimuli from their state only). In such cases, partial draws from the national sample will include only a small number of placements of figures from small states (e.g., a Senate candidate from Wyoming). Here, the use of Bayesian A-M scaling may be more appropriate since it can include all respondents.

Table 2.2: Blackbox Transpose Results for Stimuli Locations: Bootstrap Approach

	RESPONDENTS					$\hat{z}$
	100	250	500	1,000	1,500	(5,433)
Democratic Party	-0.340	-0.341	-0.341	-0.341	-0.341	-0.341
Republican Party	0.380	0.379	0.380	0.380	0.380	0.380
Tea Party	0.507	0.507	0.506	0.506	0.507	0.506
Barack Obama	-0.341	-0.341	-0.341	-0.340	-0.341	-0.341
A. Schwarzenegger	0.072	0.075	0.074	0.073	0.074	0.074
Dianne Feinstein	-0.271	-0.272	-0.272	-0.272	-0.271	-0.272
Barbara Boxer	-0.382	-0.382	-0.382	-0.382	-0.383	-0.382
Carly Fiorina	0.375	0.375	0.376	0.376	0.375	0.376
Σ ERRORS	0.006	0.004	0.000	0.002	0.004	

Note: Cell entries are μ of recovered stimuli locations from 100 bootstrap replications with specified number of randomly sampled respondents (with replacement).

2.4 Ordered Optimal Classification

A nonparametric alternative to Basic Space scaling can be found in the Ordered Optimal Classification (OOC) method. The original (or standard) Optimal Classification method—discussed at length in Chapter 5—is a novel scaling technique for binary (usually roll call) choice data that estimates a latent ideological configuration of legislators without imposing strict parametric assumptions about individual utility functions or the error term (Poole 2000, 2005). OOC uses the same underlying machinery of the Optimal Classification method, but generalizes the method to allow for ordinal or combined ordinal/binary issue scale data (Hare, Liu and Lupton, 2018).

Below we demonstrate how to perform OOC in `R` with use of the `ooc()` function in the `ooc` library. To do so, we use issue scale data from the 2004 American National Election Study (ANES). The 2004 ANES asked respondents about their policy preferences on issues ranging from diplomacy and defense spending to government spending and abortion.

We load the `ooc` library and extract fourteen of these issue scales below:

```
#install.packages("devtools")
#devtools::install_github('tzuliu/ooc')
library(ooc)
data("ANES2004_OOC")
issuescales <- ANES2004[,1:14]
head(issuescales)
```

Note that these scales have different numbers of response categories. For instance, `libcon` (liberal-conservative self-identification) is a seven-point scale, while `govtfundsabortion` (government funding for abortion) is a four-point scale. This does not present a problem for the OOC method. However, the data matrix that is fed into the `ooc()` function should always include non-zero integer values in which categories are ordered sequentially from 1 to C_k—the total number of categories for issue scale k. For instance, binary data should be coded using the values 1 and 2. Missing data should be coded as `NA`.

The other arguments in the `ooc()` function include `dims` (the number of latent dimensions to be estimated), `minvotes` (the minimum number of non-missing responses required for an individual to be included in the scaling), `lop` (the cutoff value used to exclude near-unanimous lopsided issue scales with a proportion of non-modal responses less than the specified amount), and `polarity` (used to orient the space by specifying the row number of the respondent whose score is constrained to be positive on the corresponding dimension). The three remaining arguments (`iter`, `nv.method`, and `cost`) concern the estimation procedure itself. OOC, like Optimal Classification, iterates between separate heuristic searches for the optimal configuration of individuals, the issue normal vectors, and the cutpoints along the normal vector that define predicted choices and collectively form the Coombs mesh.

In the `ooc()` function, `iter` specifies the number of iterations to loop through all three procedures, while `nv.method` specifies which method should be used to conduct the normal vector search procedure. By default, `svm.reg` is used to conduct support vector machine (SVM) regression by projecting issue scale responses onto the (most recent) configuration of respondent coordinates. SVM regression is the default choice because it is both nonparametric and computationally efficient, but other choices for `nv.method` include `oprobit` (ordered probit regression) and `krls` (kernel regularized least squares).[‖] The argument `cost` represents the cost parameter C in SVM regression, which is set to 1 by default.

The command below performs OOC on the 2004 ANES issue scale data in two dimensions:

```
ooc.result <-ooc(issuescales,dims=2,minvotes=10,lop=0.001,
polarity=c(1,1),iter=25,nv.method="svm.reg",cost=1)
```

We begin by examining the fit and projection of each of the issue scales in two-dimensional ideological space with the command:

```
issue.result <- ooc.result$issues.unique
rownames(issue.result) <- colnames(issuescales)
print(issue.result[,c("normVectorAngle2D","wrongScale",
                     "correctScale", "errorsNull","PREScale")])
```

```
                     normVectorAngle2D wrongScale
libcon                        6.016538        370
diplomacy                   -12.530408        509
iraqwar                      -3.584880        347
govtspend                    35.806667        490
defense                       1.793264        525
bushtaxcuts                 -23.129661        229
healthinsurance              49.406286        382
govtjobs                     45.383730        336
aidblacks                    37.097474        457
govtfundsabortion           -22.946871        475
partialbirthabortion        -34.539262        296
environmentjobs             -20.761694        503
deathpenalty                -23.400710        554
gunregulations               -9.833019        411
                     correctScale errorsNull   PREScale
libcon                        550        623 0.40609952
diplomacy                     532        774 0.34237726
iraqwar                       842        596 0.41778523
```

[‖]See Hainmueller and Hazlett (2014) for more details on the KRLS method.

```
govtspend                    570     776 0.36855670
defense                      536     761 0.31011827
bushtaxcuts                  550     456 0.49780702
healthinsurance              730     892 0.57174888
govtjobs                     767     878 0.61731207
aidblacks                    616     798 0.42731830
govtfundsabortion            666     603 0.21227197
partialbirthabortion         829     503 0.41153082
environmentjobs              516     744 0.32392473
deathpenalty                 609     599 0.07512521
gunregulations               790     669 0.38565022
```

The normal vector angle for each issue scale (in two dimensions) is provided by `normVectorAngle2D`. Overlapping issues with a common ideological orientation will have similar normal vector angles; for instance, the issue scales `healthinsurance` and `govtjobs` have normal vector angles within a few degrees of each other. In two dimensions, ideologically orthogonal issues will have normal vector angles separated by 90 degrees. This includes, for example, the issue scales `govtjobs` and `partialbirthabortion`, which are separated by 89.8 degrees. This result seems sensible, given that these issues are drawn from separate economic and social/cultural contexts.

The other columns above provide information about the fit of each issue scale in two-dimensional ideological space. `wrongScale` provides the number of incorrectly classified responses and `correctScale` provides the number of correctly classified responses. `errorsNull` lists the number of incorrectly classified responses from a null model that classifies every response the same as the modal (most frequent) response to the issue scale. The proportional reduction in error (PRE) statistic measures OOC's proportional improvement in correct classification over such a null model, with higher values indicating greater improvement in fit. PRE is calculated as:

$$\frac{E_{null} - E_{model}}{E_{null}} \tag{2.14}$$

where E_{null} and E_{model} are the number of classification errors by the null and alternative models, respectively.

The results above indicate that the issue scales `gaymarriage` and `libcon` have the best fit in two-dimensional space while `govtfundsabortion` and `womensrole` have the poorest spatial fit.

Finally, we can extract and plot the estimated respondent ideal points with the code below. In Figure 2.15, we use presidential vote intention (Bush or Kerry) to examine the ideological distinctiveness of the candidates' electoral coalitions. The results suggest that the first recovered ideological dimension does an effective job of separating Bush from Kerry voters, with the second dimension playing a more muted role in dividing respondents by vote choice.

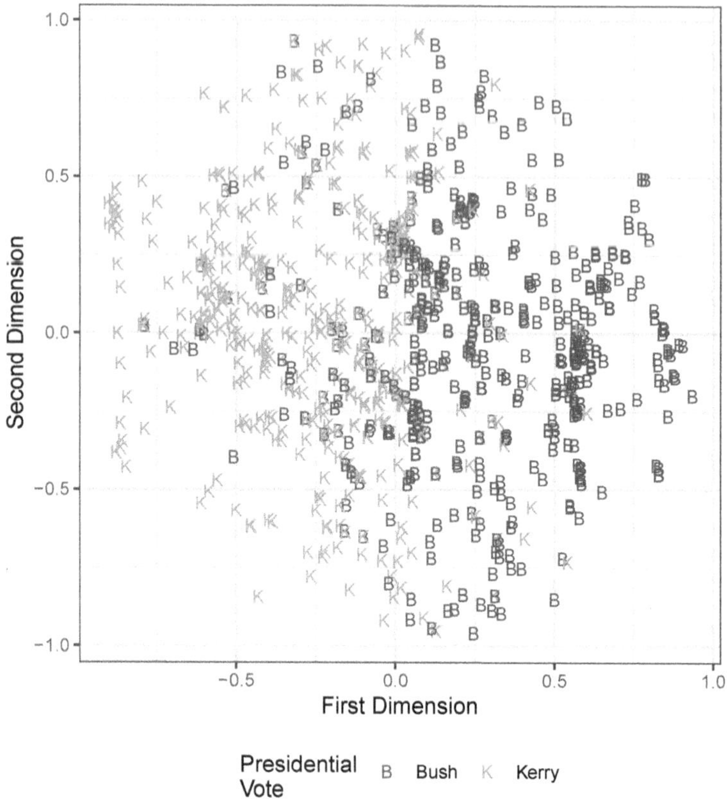

FIGURE 2.15: Ordered Optimal Classification Scaling of 2004 ANES Issue Data, By Vote Choice

2.5 Using Anchoring Vignettes

In the previous examples, respondents placed themselves and stimuli within the same context (i.e., public opinion and major candidates in the US, political parties in Mexico). It is often the case, however, that we will have data from across a variety of contexts. The traditional approaches to scaling assume that respondents across different contexts have comparable perceptions of the underlying scale. When respondents from different countries, for example, interpret the underlying scales from a common survey differently, it is difficult, at best, to assume comparability. While the Basic Space scaling procedures described in this chapter are ideally suited for addressing such differences in scale interpretation, they require that all respondents place at least some common stimuli in order to yield a cross-contextually comparable scale. When surveys are administered across a variety of countries, for example, we often lack the 'bridging' information that allows us to overcome cross-contextual differences in scale interpretation, commonly referred to as differential item functioning (DIF).

When DIF is present, what the different scale points mean on a Likert-type scale vary across respondents. King and Wand (2007) illustrate this problem by analyzing data on political efficacy from a survey administered in Mexico and China. They demonstrate that, according to the raw survey data, Chinese citizens display higher levels of political efficacy than do Mexican citizens and attribute this counterintuitive result to DIF between the two sets of respondents. In order to overcome this problem, they introduce a series of anchoring vignettes to the survey. The vignettes describe the views of hypothetical people through the use of concrete examples about their views of their own efficacy. King and Wand (2007) develop software that corrects for any DIF present in the data and produces a cross-nationally comparable scale of political efficacy. After correcting for the DIF in the data, the Mexican respondents display higher levels of political efficacy than do their Chinese counterparts. Although the techniques implemented by King and Wand are very much in the spirit of Aldrich-McKelvey scaling and produce DIF-corrected scales, they do not produce an interval-level, exportable variable that can be used in subsequent analyses. However, surveys that include anchoring vignettes are well-suited for analysis via the `blackbox_transpose()` function described above.

For the following example, we will use data from the Chapel Hill Expert Survey (CHES). In this survey, country-specific experts place political parties in their own countries on a variety of issues and dimensions, including the general left-right ideological dimension; that is, British experts place British parties, Polish experts place Polish parties, etc. When all countries are combined, the resulting data set is a block diagonal matrix with large amounts of missingness on the off-diagonal (as British experts do not place any parties outside of the UK). Without testing for DIF, we have no idea if a British expert placing a party as a 4 on an 11-point scale has the same meaning as a Polish expert placing a party as a 4. Although experts only place actual parties from one country, all experts place hypothetical parties, described by a series of anchoring vignettes. As all experts place the 3 hypothetical parties, regardless of their countries of expertise, we can use this bridging information and the `blackbox_transpose()` function to estimate a cross-nationally comparable measure of the left-right ideological positions of political parties from the CHES data.

The 2010 wave of the CHES consists of 118 parties and 224 experts in 14 member countries of the European Union. Over 160 experts placed all 3 vignette parties and between 8 and 17 experts placed each of the actual parties. As described above, the `blackbox_transpose()` routine allows us to estimate several dimensions from the data as well as allowing for missing data in the inputs. Most of the steps in this example have already been explained in Section 2.3.1, but here we will graphically compare the results of the DIF-corrected scale to the raw data and assess the degree to which the data are cross-nationally comparable. To begin, we read in the file `eu_common_space_lr_transposed_min_8.dta` and perform several steps of data management/cleaning:

```
data("ches_eu")
# calculate party means and standard deviations across all CHES experts
means <- colMeans(sub.europe, na.rm=TRUE)
sds2 <- apply(sub.europe, 2, sd, na.rm=TRUE)
```

Next we run the `blackbox_transpose()` function.

```
#Convert to matrix to numeric
sub.europe <- as.matrix(sub.europe)
mode(sub.europe) <- "numeric"
#Call the blackbox_transpose routine
result <- blackbox_transpose(sub.europe,dims=3,
                             minscale=5,verbose=TRUE)
```

```
Beginning Blackbox Transpose Scaling...118 stimuli have been provided.

Blackbox-Transpose estimation completed successfully.
```

After computing the result, we can organize the results and the original mean placements into a data set that we can use later for plotting.

```
europe.dat <- data.frame(
  x = -result$stimuli[[2]][,2],
  y = result$stimuli[[2]][,3],
  means = means,
  party = colnames(sub.europe),
  type = car::recode(means,
         "lo:3 = 'Left'; 3:7='Moderate'; 7:hi = 'Right'",
         as.factor=TRUE)
   )
parties.dat <- europe.dat[-(1:3), ]
vignette.dat <- europe.dat[(1:3), ]
onedim <- result$fits[1,3]
twodim <- result$fits[2,3]
```

The result is Figure 2.16, with leftist parties shown as circles, rightist parties shown as squares, and moderate parties shown as triangles. The locations of the vignette parties are indicated as A, B and C.

```
ggplot(parties.dat, aes(x=x, y=y)) +
  geom_point(aes(shape=type, color=type), size=3) +
```

```
scale_color_manual(values=gray.palette(3)) +
theme_bw() +
geom_text(data=vignette.dat, label=c("A", "B", "C"),
  show.legend=FALSE, size=10, color="black") +
xlab(paste0("First Dimension (fit = ", round(onedim,1), "%)")) +
ylab(paste0("Second Dimension (fit = ", round(twodim,1), "%)")) +
theme( legend.position="bottom", aspect.ratio=1) +
labs(colour="Party Group", shape="Party Group")
```

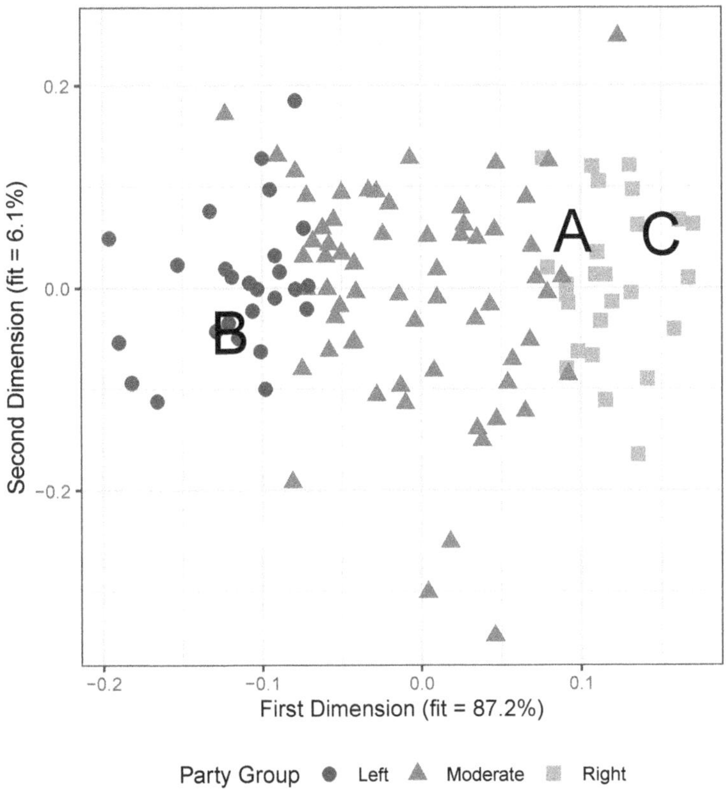

FIGURE 2.16: Result of Blackbox Transpose with Anchoring Vignettes

In Figure 2.17, we plot both the first and second dimension coordinates against the mean expert left-right placements for comparison.

```
ggplot(parties.dat, aes(x=x, y=means)) +
  geom_smooth(method="loess", color="black", lwd=.5, se=FALSE) +
  geom_point(aes(shape=type, color=type), size=3) +
  scale_color_manual(values=gray.palette(3)) +
  theme_bw() +
  xlab("First Dimension Coordinates") +
  ylab("Mean Party Placement") +
  theme(legend.position="bottom", aspect.ratio=1) +
  labs(shape="Party Group", colour="Party Group")
```

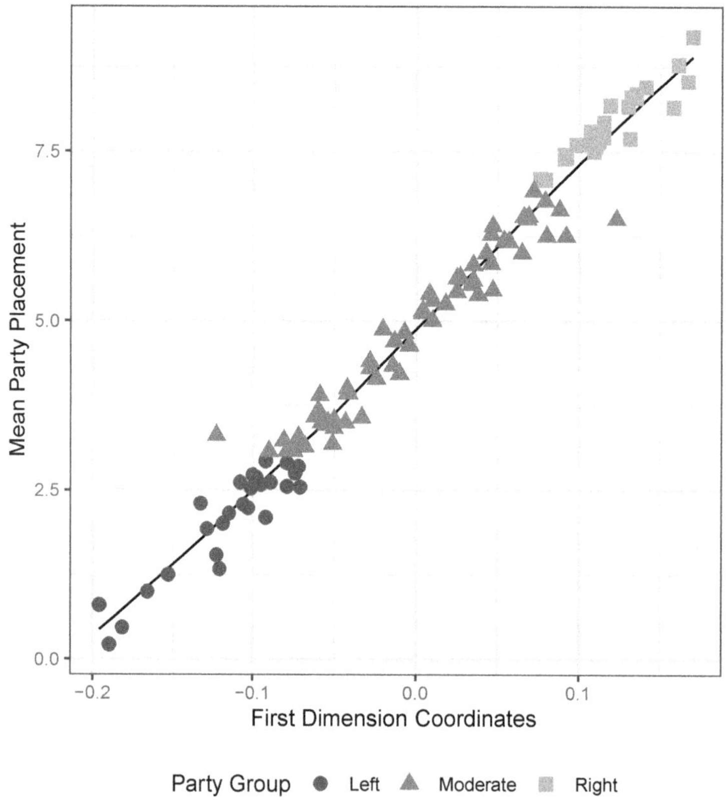

FIGURE 2.17: Result of Blackbox Transpose versus Mean Placements

2.6 Conclusion

The utility of issue scales lies in their spatial organization. Individuals are asked to picture an abstract policy continuum and place themselves and/or stimuli on it. This is consistent with the spatial (geometric) model of choice: policy alternatives are ordered spatially, and individuals have single-peaked preferences around their ideal points over the latent dimensions.

The methods discussed in this chapter—Aldrich-McKelvey (A-M) scaling and Basic Space scaling—are effective remedies for the problem of interpersonal incomparability or differential item functioning (DIF) that is often found in issue scale data (Brady, 1985; King et al., 2004).

A-M scaling is appropriate when working with unidimensional concepts such as a single issue or the left-right (or liberal-conservative) ideological continuum. Classical maximum likelihood (ML) A-M scaling is available via the `aldmck()` function in the `basicspace` package in `R`. As we will show later, Bayesian A-M scaling adds the attractive features of accepting missing data (especially essential when "bridging" across units), the simultaneous estimation of stimuli locations and the individual distortion parameters, and reliable estimates of uncertainty for the stimuli estimates.

The Basic Space scaling methods (the `blackbox()` and `blackbox_transpose()` functions in the `basicspace` package in `R`) can be used to analyze multiple issue scales on multiple dimensions. This procedure is consistent with the theoretical notion of ideological "constraint," where many different policy attitudes reduce to positions in low-dimensional space (Converse, 1964). Basic Space scaling does not explicitly estimate respondent and stimuli locations simultaneously in the manner that Aldrich-McKelvey does, but the locations recovered by scaling one group can be overlaid on the estimated locations of the other group. Ordered Optimal Classification, meanwhile, offers a non-parametric approach to multiple issue scales.

Finally, Basic Space scaling can also be used with "anchoring vignettes" to adjust for DIF in issue scale data — variation in respondents' interpretations of the issue scales (e.g., cross-cultural differences, demographic variables, education or political sophistication, etc.).

2.7 Exercises

1. Use the aldmck() function to perform Aldrich-McKelvey scaling on the Pennsylvania module of the 2010 Cooperative Congressional Election Study (the `CCES2010.PA.Rda` data set). `CCES2010.PA` is a list that

stores respondents' 7-point ideological placements of themselves, the Democratic and Republican Parties, President Obama, the Tea Party movement, and Democratic Senators Bob Casey and Arlen Specter (`CCES2010.PA$libcon.placements`) and a validated measure of whether or not they voted in the 2010 general election (`CCES2010.PA$voted`). Missing values are coded as 9.

(a) Neatly format and report the `summary(result)` object.

i. What is the proportion of respondents with negative weights?

(b) Plot a smoothed histogram of the respondent ideal points (both positive and negative weights) and draw arrows corresponding to the locations of the stimuli.

(c) Modify the plot from Exercise 1(b) to draw separate densities for respondents who voted (`voted==1` in `CCES2010.PA`) and those who did not vote (`voted==0` in `CCES2010.PA`) in 2010. Is there an ideological difference between voters and non-voters? Explain.

2. Data from the 2012 American National Election Study (ANES) is stored in the data set `ANES2012.Rda`. `ANES2012` is a list that stores respondents' placements of themselves and four stimuli (Barack Obama, Mitt Romney, and the Democratic and Republican Parties) on seven-point liberal-conservative and government/private health insurance scales (`ANES2012$libcon.placements` and `ANES2012$healthins.placements`), 0-100 feeling thermometer ratings of eight stimuli (Barack Obama, Mitt Romney, Joe Biden, Paul Ryan, Hillary Clinton, George W. Bush, and the Democratic and Republican Parties) (`ANES2012$thermometers`) and presidential vote choice (`ANES2012$presvote`). Missing values are coded as 999.

(a) Run `aldmck()` on the government/private health insurance scale data stored in `ANES2012$healthins.placements`.

(b) Do a side-by-side plot with separate histograms showing the distribution of ideal points for respondents with positive and negative weights. Label the estimated positions of Barack Obama, Mitt Romney, and the Democratic and Republican Parties in both.

i. How do the two distributions differ? In this example, what does it mean for respondents to "flip" the space?

3. Use the `blackbox()` function to analyze the 2011 Canadian Election Study data set (`CES2011.Rda`) using the Basic Space scaling procedure. `CES2011` is a list that stores respondents' preferred party (`CES2011$party`), attitudes on 11 policy issues (`CES2011$issues`), placements of five national parties on an 11-point left-right scale (`CES2011$lrplacements`), and propensity to vote for each of the five

national parties on an 11-point scale (`CES2011$propensity`).** Missing values are coded as 999.

(a) Run blackbox() on `CES2011$issues` and report the fit statistics for the one- and two-dimensional results.

(b) In the two-dimensional estimated configuration, which issues are most strongly associated with the first and second dimensions? Based on this, what is your interpretation of the substantive meaning of each dimension?

(c) Plot smoothed histograms of respondents' first and second dimension coordinates for those who feel closest to the Conservative Party, the Liberal Party, and the NDP. What is the left-right ordering of the parties on each dimension? Is there greater overlap between the party coalitions on the first or second dimension?

4. Continuing with the `CES2011` data, use the `blackbox_transpose()` function to analyze Canadian citizens' placements of the Conservative Party, the Liberal Party, the NDP, the Bloc Quebecois and the Green Party on the left-right ideological scale in two dimensions. Recall that missing values are coded as `999`.

(a) If necessary, reverse the polarity of the party coordinates so that the Conservative Party has a positive score on the first dimension. Does the left-right order of the three largest parties correspond to the ordering estimated in Exercise 3?

(b) Report each party's R^2 value for each dimension. Which parties have the best and worst fits in one dimension?

(c) Use the bootstrapping approach to estimate standard errors for the party coordinates in two dimensions and plot the parties using cross-hairs for the 95% confidence intervals (1.96 × standard error). Which parties have the highest standard errors?

5. Load the data set `ANES2016.Rda`, which includes 4,271 responses from the 2016 American National Election Study (ANES). `ANES2016.Rda` is a list with five objects: `issues`, `thermometers`, `votechoice`, `party`, and `demographics`.

(a) Run Ordered Optimal Classification (OOC) on the 14 issue scales stored in `issues`. Estimate two dimensions and use the other default arguments.

(b) Which two issue scales have the best fit (using the `PREScale` statistic) in the configuration? Which two have the worst fit?

**Question wordings for the issue questions are available on the book website.

(c) Which two issues have the greatest difference between the normal vector angles?

(d) Produce a scatterplot of the estimated respondent ideal points, coloring or otherwise labeling the respondents by 2016 presidential vote choice.

6. Use the `anchors.order()` function to assess how well respondents in the French module of the 2009 European Election Study (the data set `franceEES2009.Rda`) perceived the left-right ordering of the Extreme Left and National Front parties. You will need to change all missing values to 0, which will first require adding 1 to all valid placements (so that the lowest category is 1 instead of 0).

   (a) What proportion of respondents perceived the Extreme Left to be to the left of the National Front, and how many reported the reverse placement? What proportion rated the two parties equally?

      i. How does the number of reverse placements compare to the number of negative weights estimated in Aldrich-McKelvey scaling (in Section 2.1)?

   (b) Test the ordering for the Socialist and Green parties. What proportion of respondents perceived the Socialist Party to the left of the Green Party? Why might this figure be different than that in Exercise 6(a)?

3

Analyzing Similarities and Dissimilarities Data

The methods discussed in this chapter deal with similarities/dissimilarities (proximities) data. Similarities data is *relational* and organized as a square matrix, where the stimuli comprise both the rows and the columns. For example, suppose a group of respondents are asked to judge how similar 10 countries are to one another. That is, each respondent is asked to compare every unique pair of countries and assign to each unique pair an integer score from zero (meaning the countries are essentially identical) to ten (the countries are completely different). This is an example of dissimilarities data, because the higher the assigned score the more dissimilar the respondent judges the stimuli to be. Consequently, this data can be treated as distances between the 10 countries.

Similarities data differs from choice data in that it measures how alike/not-alike objects are to each other. Similarities data is frequently *generated* from choice data—for example, an agreement score matrix based on how often legislators vote together is a popular type of similarities data—but does not represent the choices themselves. But this need not be the case if direct measures of similarity or dissimilarity between the stimuli are available.

Similarities and dissimilarities data are simply reflections of each other. One can always move from one to the other by adding or subtracting a constant to or from the values. For instance, a dissimilarities value of 0 in the aforementioned example would be recoded as a 10 in a similarities matrix, while a value of 10 would become a 0. Values of 5 (equidistant between the two ends) would remain the same. It is usually more convenient to work with dissimilarities data since we are modeling the *distances* (i.e., the level of dissimilarity) between stimuli.

In psychology, various methods of multidimensional scaling (MDS) have been developed during the past 60 years to analyze similarities data. MDS methods model these similarities as distances between points in a geometric space (usually simple Euclidean). In this chapter, we denote the actual or observed distances between each pair of stimuli j and m as δ_{jm} and the reproduced or approximated distances between the stimuli as d_{jm}. Each of the methods discussed in this chapter differs most fundamentally in the way it generates the d_{jm} based on the information in the δ_{jm}.

MDS programs are designed to uncover the dimensionality of a given set of data and to visually display the positions of the objects along the latent dimensions. Jacoby (1991, p. 27) provides a useful definition of dimensionality as "a set of objects is simply defined as the number of separate and interesting sources of variation among the objects." As Jacoby (1991) points out, there is no reason that the number of latent dimensions (understood as the sources of variation between the stimuli) need to be limited to the three familiar spatial dimensions of human experiences. However, MDS analysis is almost always done in three dimensions or less because the *raison d'être* of MDS methods is to produce a visual summary. Moreover, no more than three dimensions are usually required to represent the underlying structure of the data. MDS serves manifold purposes (see Borg and Groenen (2010, chap. 1) for a thorough exposition), although our focus is on the mechanics of MDS methods in `R` along with some basic best practices in the process of interpreting the results.

In the next section we show the solution for ratio scale similarities. A ratio scale is an interval scale with a true zero point. Here it simply means that the distance between a point and itself is zero. This problem was solved by Torgerson (1952, 1958) which in turn built upon work done by psychometricians in the 1930s (Eckart and Young, 1936; Young and Householder, 1938).

3.1 Classical Metric Multidimensional Scaling

Torgerson's solution for the simple metric MDS problem is (1) convert the similarities/dissimilarities to a symmetric q by q matrix of squared distances; (2) double-center the matrix of squared distances to remove the squared terms; and (3) perform an eigenvalue/eigenvector decomposition of the double-centered matrix to recover the coordinates.

Recall that q $(j = 1, ..., q)$ is the number of stimuli and where s $(k = 1, ..., s)$ is the number of dimensions). Let Z be an q by s matrix of stimuli coordinates, technically:

$$Z = \begin{bmatrix} z_{11} & z_{12} & \cdots & z_{1s} \\ z_{21} & z_{22} & \cdots & z_{2s} \\ \cdot & \cdot & \cdot & \cdot \\ \cdot & \cdot & \cdot & \cdot \\ \cdot & \cdot & \cdot & \cdot \\ z_{q1} & z_{q2} & \cdots & z_{qs} \end{bmatrix} \tag{3.1}$$

The symmetric matrix of squared distances between the stimuli is:

$$D_z = \begin{bmatrix} \sum_{k=1}^{s} (z_{1k}-z_{1k})^2 & \sum_{k=1}^{s} (z_{1k}-z_{2k})^2 & \dots & \sum_{k=1}^{s} (z_{1k}-z_{qk})^2 \\ \sum_{k=1}^{s} (z_{2k}-z_{1k})^2 & \sum_{k=1}^{s} (z_{2k}-z_{2k})^2 & \dots & \sum_{k=1}^{s} (z_{2k}-z_{qk})^2 \\ \vdots & \vdots & \vdots & \vdots \\ \sum_{k=1}^{s} (z_{qk}-z_{1k})^2 & \sum_{k=1}^{s} (z_{qk}-z_{2k})^2 & \dots & \sum_{k=1}^{s} (z_{qk}-z_{qk})^2 \end{bmatrix} \tag{3.2}$$

Or, expressed in terms of matrices:

$$D_z = diag(ZZ')J_q' - 2ZZ' + J_q diag(ZZ')' \tag{3.3}$$

where $diag(ZZ')$ is a q length vector containing the diagonal of ZZ', and J_q is a q length vector of 1's:

$$diag(ZZ') = \begin{bmatrix} \sum_{k=1}^{s} z_{1k}^2 \\ \sum_{k=1}^{s} z_{2k}^2 \\ \sum_{k=1}^{s} z_{3k}^2 \\ \vdots \\ \sum_{k=1}^{s} z_{qk}^2 \end{bmatrix} \tag{3.4}$$

$$J_q = \begin{bmatrix} 1 \\ 1 \\ 1 \\ \vdots \\ 1 \end{bmatrix} \tag{3.5}$$

D_z can also be written as the product of the q by $s+2$ partitioned matrix

$$[diag(ZZ')| -2Z|J_q] \tag{3.6}$$

multiplied by its transpose (without the -2 term):

$$D_z = [diag(ZZ')| -2Z|J_q] \left[\frac{\frac{J'_q}{Z'}}{diag(ZZ')'} \right] \tag{3.7}$$

This shows that the rank of D_z must be less than or equal to $s+2$.

The matrix D_z is double-centered as follows: from each element subtract the column mean, subtract the row mean, add the matrix mean, and divide by -2. This eliminates the squared terms and isolates the cross-product matrix, ZZ'.

That is, let the mean of the jth column of D_z be

$$d_{.j}^2 = \frac{\sum_{m=1}^{q} d_{mj}^2}{q} \tag{3.8}$$

let the mean of the mth row of D_z be

$$d_{m.}^2 = \frac{\sum_{j=1}^{q} d_{mj}^2}{q} \tag{3.9}$$

and let the mean of the matrix D_z be

$$d_{..}^2 = \frac{\sum_{m=1}^{q} \sum_{j=1}^{q} d_{mj}^2}{q^2} \tag{3.10}$$

Therefore the double centered matrix is:

$$y_{mj} = \frac{(d_{mj}^2 - d_{.j}^2 - d_{m.}^2 + d_{..}^2)}{-2} = \sum_{k=1}^{s} (z_{mk} - \bar{z}_k)(z_{jk} - \bar{z}_k) \tag{3.11}$$

In matrix notation, this produces the q by q matrix Y which is equal to the cross-product matrix of the q by s matrix Z^* multiplied by itself; namely

$$Y = Z^* Z^{*\prime} = \begin{bmatrix} z_{11} - \bar{z}_1 & z_{12} - \bar{z}_2 & \dots & z_{1s} - \bar{z}_s \\ z_{21} - \bar{z}_1 & z_{22} - \bar{z}_2 & \dots & z_{2s} - \bar{z}_s \\ . & . & . & . \\ . & . & . & . \\ . & . & . & . \\ z_{q1} - \bar{z}_1 & z_{q2} - \bar{z}_2 & \dots & z_{qs} - \bar{z}_s \end{bmatrix} \begin{bmatrix} z_{11} - \bar{z}_1 & z_{12} - \bar{z}_2 & \dots & z_{1s} - \bar{z}_s \\ z_{21} - \bar{z}_1 & z_{22} - \bar{z}_2 & \dots & z_{2s} - \bar{z}_s \\ . & . & . & . \\ . & . & . & . \\ . & . & . & . \\ z_{q1} - \bar{z}_1 & z_{q2} - \bar{z}_2 & \dots & z_{qs} - \bar{z}_s \end{bmatrix}' \tag{3.12}$$

Solving for Z is easily accomplished because, without loss of generality, we can assume that the coordinates have zero means; that is,

$$\bar{z} = \begin{bmatrix} \bar{z}_1 \\ \bar{z}_2 \\ . \\ . \\ . \\ \bar{z}_s \end{bmatrix} = \begin{bmatrix} 0 \\ 0 \\ . \\ . \\ . \\ 0 \end{bmatrix} \tag{3.13}$$

and we can use simple eigenvector/eigenvalue decomposition to solve for Z,

$$Y = U \Lambda U' \tag{3.14}$$

and set

$$Z = U \Lambda^{\frac{1}{2}} \tag{3.15}$$

3.1.1 Example 1: Nations Similarities Data

We demonstrate the double-centering procedure using a well-known data set collected by Wish (1971). In 1968, Wish (1971) asked 18 students in his psychological measurement class to rate the perceived similarity between each pair of twelve nations using a 9-point scale ranging from '1=very different' to '9=very similar'. He then constructed a matrix of average similarity ratings between the twelve nations, shown below.

```
library(asmcjr)
data(nations)
print(nations)
```

```
           Brazil Congo Cuba Egypt France India Israel
Brazil       9.00  4.83 5.28  3.44   4.72  4.50   3.83
Congo        4.83  9.00 4.56  5.00   4.00  4.83   3.33
Cuba         5.28  4.56 9.00  5.17   4.11  4.00   3.61
Egypt        3.44  5.00 5.17  9.00   4.78  5.83   4.67
France       4.72  4.00 4.11  4.78   9.00  3.44   4.00
India        4.50  4.83 4.00  5.83   3.44  9.00   4.11
Israel       3.83  3.33 3.61  4.67   4.00  4.11   9.00
Japan        3.50  3.39 2.94  3.83   4.22  4.50   4.83
China        2.39  4.00 5.50  4.39   3.67  4.11   3.00
USSR         3.06  3.39 5.44  4.39   5.06  4.50   4.17
USA          5.39  2.39 3.17  3.33   5.94  4.28   5.94
Yugoslavia   3.17  3.50 5.11  4.28   4.72  4.00   4.44
           Japan China USSR  USA Yugoslavia
Brazil      3.50  2.39 3.06 5.39       3.17
Congo       3.39  4.00 3.39 2.39       3.50
Cuba        2.94  5.50 5.44 3.17       5.11
Egypt       3.83  4.39 4.39 3.33       4.28
France      4.22  3.67 5.06 5.94       4.72
India       4.50  4.11 4.50 4.28       4.00
Israel      4.83  3.00 4.17 5.94       4.44
Japan       9.00  4.17 4.61 6.06       4.28
China       4.17  9.00 5.72 2.56       5.06
USSR        4.61  5.72 9.00 5.00       6.67
USA         6.06  2.56 5.00 9.00       3.56
Yugoslavia  4.28  5.06 6.67 3.56       9.00
```

To analyze the `nations` data set using double-centering, we must first transform the matrix. Since we will be using the data set to calculate distances between the stimuli (d_{jm}), we reverse the values in the matrix (subtracting 9 from all cells) so that higher values indicate greater dissimilarity (and consequently, greater inter-point distances). We then square each value in the distances matrix.

```
d <- (9-nations)^2
```

We next use the **doubleCenter()** function shown below. This function can be used to quickly calculate a double-centered matrix (D) from a dissimilarities matrix (d). The **eigen()** function is a base function in R (that is, it is automatically available without loading any external packages), and can be used to extract the eigenvalues (**ev$values**) and eigenvectors (**ev$vectors**) of the double-centered matrix (D):

```
D <- doubleCenter(d)
ev <- eigen(D)
```

Next, we find the point furthest from the center of space and use it to weight the eigenvectors in order to scale the space to the unit circle. The recovered coordinates for the stimuli are stored in the objects **torgerson1** and **torgerson2**.

```
torgerson.soln <- sqrt(max((abs(ev$vec[,1]))^2 +
                          (abs(ev$vec[,2]))^2))
torgerson1 <- ev$vec[,1]*(1/torgerson.soln)*sqrt(ev$val[1])
torgerson2 <- ev$vec[,2]*(1/torgerson.soln)*sqrt(ev$val[2])
```

Figure 3.1 displays the point configuration recovered by double-centering the **nations** matrix of squared distances. Note that we flip the second dimension coordinates (**torgerson2**) for display purposes by multiplying these values by -1. The standard interpretation of this result is that the substantive dimensions are aligned diagonally (shown as the dashed lines) in the recovered space. According to this view, the dimension running from the bottom-left to the top-right corner represents pro-Communist/pro-West affiliation, and the dimension running from the top-left to the bottom-right represents level of economic development (in 1968).

```
plot.df <- data.frame(
  dim1 = torgerson1,
  dim2 = torgerson2,
  country = rownames(nations)
)
ggplot(plot.df, aes(x=dim1, y=dim2)) +
  geom_point() +
  geom_text(aes(label=country), nudge_y=-.25) +
  xlab("") +
  ylab("") +
  geom_abline(slope=c(-1,1), intercept=c(0,0), lty=2) +
  xlim(-6.5,6.5) +
  ylim(-6.5,6.5) +
  theme_bw() +
  theme(aspect.ratio=1)
```

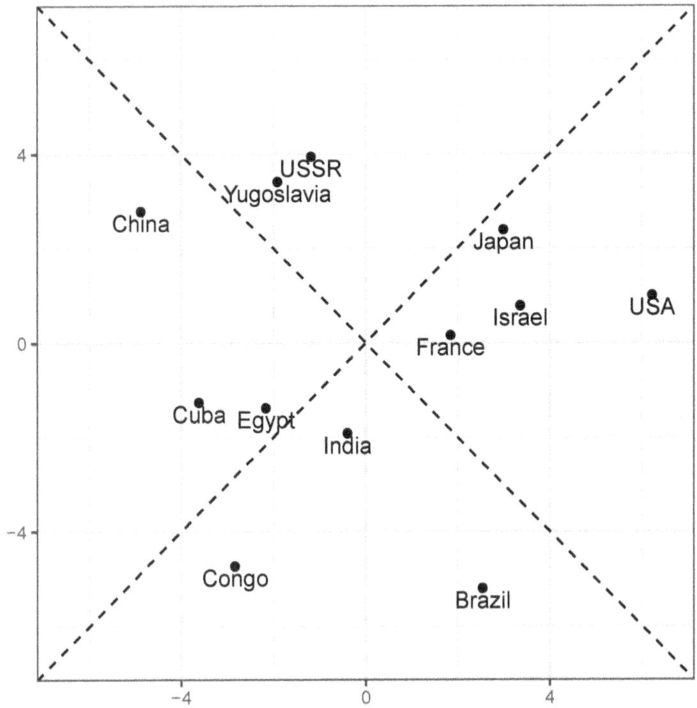

FIGURE 3.1: Double-Centered Nations Similarities Matrix - Torgerson Solution

3.1.2 Metric MDS Using Numerical Optimization

The goal of metric MDS procedures is to find a point configuration of interpoint distances (d_{jm}) that reproduces the observed distances (δ_{jm}) as closely as possible. We can do so by minimizing a loss function using numerical optimization methods (which search for values of x that minimize or maximize some function $f(x)$). If the loss function is the sum of squared errors (SSE) between the d_{jm} and the δ_{jm}, then the results will be very close to those from the double-centering procedure. However, double-centering is a linear algebra operation on a matrix of squared distances that removes the squared terms, not a statistical model that specifies a loss function.

In this example, we calculate the SSE between the δ_{jm} and the d_{jm} with the `fr()` function below. The SSE value is stored in the object `sse`.

```
fr <- function(par, obs, ndim){
        d.x <- dist(matrix(par, ncol=ndim, byrow=2))
        diff <- d.x - sqrt(obs[lower.tri(obs)])
        sse <- c(diff %*% diff)
```

```
        sse
}
```

In this example, we wish to find the point configuration that minimizes the SSE of metric MDS of the `nations` data. The Torgerson solution may reach a *local* minimum of the loss function `fr()`, but through use of a suite of numerical optimization procedures, we can conduct a more exhaustive search for the *global* minimum.

Accordingly, we use three numerical optimization methods available in the `optim()` function included in the base `stats` package in `R`. We feed the coordinates produced by one method as the starting values for the next method. This way, we combine the strengths of global and local-based optimizers (e.g., Martin and Otto, 1996). For the initial starting values of the nation coordinates, we use the eigenvectors (`ev$vector`) from the double-centered matrix. The starts are stored in the matrix `zz`.

```
ndim <- 2
Z <- c(t(ev$vector[,1:ndim]))
```

We set five arguments in the `optim()` function: `par` (the initial values); `fn` (the function to be minimized); `method` (the optimization method to be used); `control` (additional control parameters for the optimization procedures); and `hessian` (whether the Hessian matrix of second derivatives should be returned). The `control` parameter provides additional options on the optimization methods. For example, we set the maximum number of iterations at 50,000 for each of the methods.

The first optimization method we use is the simulated annealing (SANN) method. Simulated annealing is analogous to the thermodynamic process of *annealing*, in which systems (e.g., metals) at high temperatures and entropy are cooled slowly, such that they are allowed to achieve a rigid state of minimum energy (Brooks and Morgan, 1995). In its application to optimization, the simulated annealing algorithm is set at a "temperature" which decreases over the completion of a set number of iterations. At high temperatures, simulated annealing moves around the likelihood function more freely. This includes moving downhill, which means it is more likely to escape local optima and find global optima (Goffe, Ferrier and Rogers, 1994). As the temperature "cools," the procedure settles in on the region with the lowest (in the case of minimization) values.

The second optimization method we use is the BFGS (Broyden, Fletcher, Goldfarb and Shanno) procedure. BFGS is a quasi-Newton, "hill-climbing" method that updates its estimation of the shape of the function with repeated gradient calculations. Unlike Newton's method, BFGS can analyze non-differentiable functions (Fletcher, 1987, p. 49-50).

The final optimization procedure we employ is the Nelder-Mead method (Nelder and Mead, 1965), which is the default routine used by `optim`. The

Nelder-Mead method uses a simplex (a polytope with $k+1$ vertices in k-dimensional space, for example a triangle on a plane) to iteratively explore the likelihood function. Function values are calculated at each of the vertices, and when minimizing a function (as in this case), the vertex with the largest value is rejected. This point is reflected upon the existing simplex, and a new simplex is formed. This process is repeated over a set number of iterations, over which the simplex becomes smaller and converges to a local optimum.

We implement the three optimization procedures with the code below, with the final results stored in the object `model_nm`.

```
model_sann <- optim(par=Z, fn=fr, method="SANN",
        control=list(maxit=50000), hessian=TRUE, obs=d, ndim=2)
model_bfgs <- optim(par=model_sann$par, fn=fr, method="BFGS",
        control=list(maxit=50000), hessian=TRUE, obs=d, ndim=2)
model_nm <- optim(par=model_bfgs$par, fn=fr, method="Nelder-Mead",
        control=list(maxit=50000), hessian=TRUE, obs=d, ndim=2)
```

Next, we examine the SSE (sum of squared errors) produced at each stage of the optimization process. The SANN procedure produces a configuration with a SSE value very close to the final value. The Nelder-Mead and BFGS methods provide minor improvements, which makes sense given that they are most effective at finding local optima. However, the sequence of the three optimizers did not seem to matter: each ordering produced a SSE of 87.7.

```
print(SSE)
```

```
                              Sum of Squared Errors (SSE)
Stage 1 (Simulated Annealing)                    90.02549
Stage 2 (BFGS)                                   87.71188
Stage 3 (Nelder-Mead)                            87.71188
```

Even by chaining the optimizers together, though, it is entirely possible that we have not arrived at a global optimum.* To test if this is the case in this example, we re-run the three optimizers using the final configuration from `model_nm` as the starting values.

```
model_sann2 <- optim(par=model_nm$par, fn=fr, method="SANN",
        control=list(maxit=50000), hessian=TRUE, obs=d, ndim=2)
model_bfgs2 <- optim(par=model_sann2$par, fn=fr, method="BFGS",
        control=list(maxit=50000), hessian=TRUE, obs=d, ndim=2)
model_nm2 <- optim(par=model_bfgs2$par, fn=fr, method="Nelder-Mead",
        control=list(maxit=50000), hessian=TRUE, obs=d, ndim=2)
```

*Following Borg and Groenen (2010, p. 276), we refer to *a* global optimum rather than *the* global optimum because multiple configurations may produce equal values of the loss function.

Re-running the optimization procedures produced no further improvement in the SSE value. While we cannot confirm that we have reached a global optimum, we can be reasonably confident in our solution. Indeed, because computing power is so cheap in this instance, there is little disincentive in chaining and repeating optimization methods in pursuit of a global optimum.

```
print(SSE)
```

```
                               Sum of Squared Errors (SSE)
Stage 4 (Simulated Annealing)                     87.71188
Stage 5 (BFGS)                                    87.71188
Stage 6 (Nelder-Mead)                             87.71188
```

As a final check, we run an innovative optimization procedure developed by political scientists Walter R. Mebane, Jr. and Jasjeet S. Sekhon known as Genetic Optimization using Derivatives (GENOUD). The theory underlying GENOUD is detailed in Sekhon and Mebane (1998), but the basic idea is that GENOUD executes a genetic algorithm that undergoes a series of random reproductions and mutations to conduct an extremely thorough search of the objective function. GENOUD is especially effective at optimizing non-linear functions with several local optima. Mebane and Sekhon (2011) have implemented the GENOUD procedure in the `R` package `rgenoud`.

`rgenoud` includes only the function `genoud()`, the required arguments to which are essentially the same as those in the `optim()` function detailed above. The important difference, though, is the population size (the number of individuals used in the genetic algorithm) that is set with the `pop.size` argument. A larger `pop.size` value improves the performance of the algorithm but takes more time. The default value of `pop.size` is 1000; we use 3000 (which still takes less than 20 seconds to run). Also, initial values for the parameters are set with the argument `starting.values` rather than `par`.

```
library(rgenoud)
set.seed(18134)
model_genoud <- genoud(fn=fr, nvars=length(Z), pop.size=3000,
    starting.values=model_nm2$par, obs=d, ndim=2)
```

The SSE of the optimized solution is stored in the object `model_genoud$value`. The fit has not improved after the GENOUD optimization procedure, increasing our confidence in this solution.

```
print(model_genoud$value)
```

```
[1] 87.71188
```

The following code is used to plot the optimized metric MDS result (`model_nm2`) in Figure 3.2. We again find two substantive dimensions. The undeveloped-developed continuum (with Congo and Japan on the ends) and the pro-Communist to pro-West dimension (with China and Israel on the ends). The configuration in Figure 3.2 includes some desirable substantive qualities. For example, the United States is closer to one of the ends of the Communist-West dimension in the optimized configuration than in the Torgerson solution configuration shown in Figure 3.1.

```
xmetric <- data.frame(
  dim1 = scale(model_genoud$par[seq(1,23,by=2)], scale=FALSE),
  dim2 = scale(model_genoud$par[seq(2,24,by=2)], scale=FALSE),
  country = rownames(nations))
if (xmetric$dim1[8]<0) {xmetric$dim1 <- -xmetric$dim1}
if (xmetric$dim2[9]<0) {xmetric$dim2 <- -xmetric$dim2}
ggplot(xmetric, aes(x=dim1, y=dim2)) +
  geom_point() +
  geom_text(aes(label=country), nudge_y=-.25) +
  xlab("") +
  ylab("") +
  geom_abline(slope=c(-3/5,5/3), intercept=c(0,0), lty=2) +
  xlim(-4,4) +
  ylim(-4,4) +
  theme_bw() +
  theme(aspect.ratio=1)
```

3.1.3 Metric MDS Using Majorization (SMACOF)

To close our discussion of metric MDS methods, we introduce the technique of iterative majorization or SMACOF (Scaling by Majorizing a Complicated Function). SMACOF was developed by de Leeuw (1977; 1988) and de Leeuw and Heiser (1977) as an alternative method of optimization. However, in this case the function to be minimized is the *Stress* function. The Stress measure incorporates the sum of squared errors between the observed and reproduced distances for each pair of stimuli. For some point configuration X, Stress is defined as:

$$\sigma(X) = \sum_{i<j} w_{jm}(d_{jm}(X) - \delta_{jm})^2 \tag{3.16}$$

where w_{jm} is an $q \times q$ matrix of weights. Standard practice is to set w_{jm} equal to 1 if δ_{jm} is observed and 0 if δ_{jm} is missing, but w_{jm} can be assigned any non-negative value (Borg and Groenen, 2010, p. 171). Equation 3.16 can be rewritten as

$$\begin{aligned}\sigma(X) &= \sum_{i<j} w_{jm}d_{jm}^2(X) + \sum_{i<j} w_{jm}\delta_{jm}^2 - 2\sum_{i<j} w_{jm}d_{jm}(X)\delta_{jm} \\ &= \eta_\delta^2 + \eta^2(X) - 2\rho(X)\end{aligned} \tag{3.17}$$

SMACOF conducts an iterative search for the minimum of the Stress function with a majorization algorithm. Majorization approximates unwieldy and

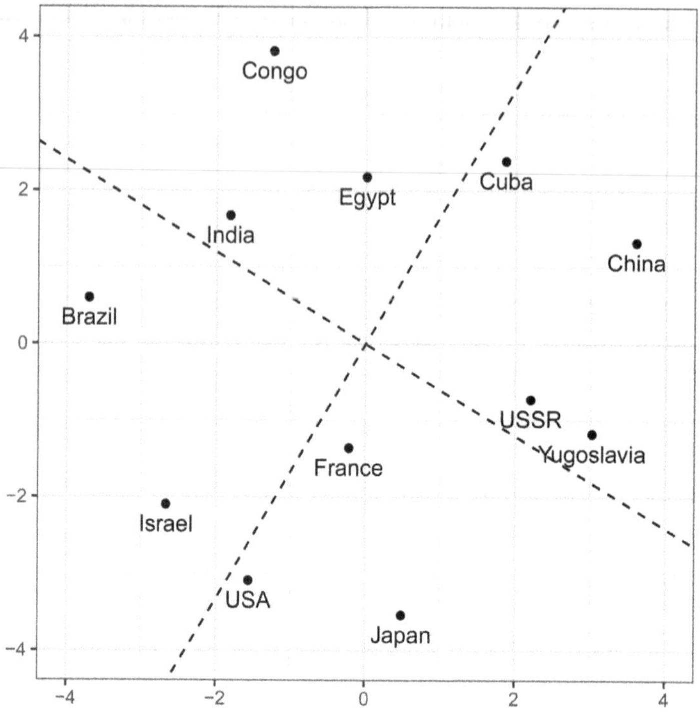

FIGURE 3.2: Numerically Optimized Metric Multidimensional Scaling of the Nations Similarities Matrix

complex functions (i.e., the Stress function) using a series of simpler, auxiliary functions (usually the quadratic function) (Borg and Groenen, 2010, pp. 178-182). The auxiliary function can be minimized more easily and is iteratively updated such that it moves toward a minimum point of the more complex function. One of the most desirable features of the majorization algorithm is that it is guaranteed to converge to a minimum, though the number of iterations required may be large. The majorization routine stops when the difference in successive function values become very small (e.g., 0.000001).

3.1.4 The smacof Package in R

The `smacof` package in `R` (de Leeuw and Mair, 2009) implements the iterative majorization algorithm in a number of scaling functions designed for use on different types of data. Namely, the `smacofSym()` function is used to analyze square dissimilarity matrices, the `smacofIndDiff()` function is used for three-way MDS of a list of dissimilarity matrices (discussed in Section 3.3), and the `smacofRect()` function performs metric unfolding of rectangular matrices (discussed in Chapter 4). Here, we demonstrate the use of the `smacofSym` on the `nations` dissimilarities matrix.[†] We set six arguments in the `smacofSym()` function: `delta` (the symmetric dissimilarities matrix), `ndim` (the number of dimensions to be estimated), `metric` (whether metric MDS should be performed; if `FALSE`, nonmetric MDS is performed), `itmax` (the maximum number of iterations to be executed), and `eps` (the convergence criterion; the SMACOF procedure stops when this value exceeds the difference in successive Stress values).

```
#install.packages('smacof')
library(smacof)
smacof_metric_result <- smacofSym(delta=d, ndim=2, itmax = 1000,
                         type = "interval", eps=0.000001)
```

The `smacofSym()` function returns fifteen objects stored in the dataframe `smacof_metric_result`:

- **delta** Observed dissimilarities, not normalized
- **dhat** Disparities (transformed proximities, approximated distances, d-hats)
- **confdist** Configuration distances
- **conf** Matrix of fitted configurations
- **stress** Stress-1 value
- **spp** Stress per point (stress contribution in percentages)
- **resmat** Matrix with squared residuals
- **rss** Residual sum-of-squares
- **weightmat** Weight matrix
- **ndim** Number of dimensions
- **init** Starting configuration
- **model** Name of `smacof` model
- **niter** Number of iterations
- **nobj** Number of objects

[†]Other functions for metric (`cmdscale()`) and nonmetric (`isoMDS()` and `sammon()`) multidimensional scaling are available in the base `MASS` and `stats` packages in `R`. We focus our attention on the MDS functions that implement the SMACOF algorithm (`smacofSym()` and `SmacofIndDiff()`). Despite the differences in estimation procedure (e.g., the `sammon()` function uses the method described in Sammon (1969)), we have found that they produce sets of interpoint distances that are highly correlated ($r > 0.9$). However, we strongly prefer the `smacof` MDS functions because they offer a superior means of convergence on the solution and can be easily adjusted to handle missing data.

type Type of MDS model

```
conf <- smacof_metric_result$conf
print(smacof_metric_result$niter)
```

```
[1] 23
```

```
print(smacof_metric_result$stress)
```

```
[1] 0.2174646
```

We see that the SMACOF procedure was completed in 23 iterations. Hence, SMACOF was able to reach convergence within the specified number of iterations. In addition, 23 is not a high iteration count, suggesting that this Stress function is not especially unwieldy. Note that the SMACOF package reports the Stress-1 statistic, as formulated by Kruskal (1964*a*), rather than the raw Stress. The two-dimensional result has a Stress-1 value of 0.21, a considerable improvement over the Stress-1 value of 43 for the one-dimensional result. Figure 3.3 shows the point configuration produced by metric SMACOF scaling of the `nations` data. The SMACOF result and the optimized result shown in Figure 3.2 are very similar.

```
smacof.dat <- data.frame(
  dim1 = conf[,1],
  dim2 = conf[,2],
  country=rownames(nations)
)
ggplot(smacof.dat, aes(x=dim1, y=dim2)) +
  geom_point() +
  geom_text(aes(label=country), nudge_y=-.05) +
  xlab("") +
  ylab("") +
  geom_abline(slope=c(-3/5,5/3), intercept=c(0,0), lty=2) +
  xlim(-1,1) +
  ylim(-1,1) +
  theme_bw() +
  theme(aspect.ratio=1)
```

Figure 3.4 displays a *scree plot* of the Stress values for each of the configurations estimated in 1-5 dimensions. A scree plot shows the proportion of the total variance in the data explained by each factor or dimension. Adding more dimensions will never produce higher Stress values, but typically there is an *elbow* after which additional dimensions make minimal contributions to decreasing Stress. The elbow in Figure 3.4 is at two dimensions. A three-dimensional configuration does reduce Stress by nearly half, but the ordering of the nations along the third dimension is unintelligible. It may be (and often is) the case that an unexpected but interpretable dimension arises from MDS procedures. We do not want to discard a dimension simply because its

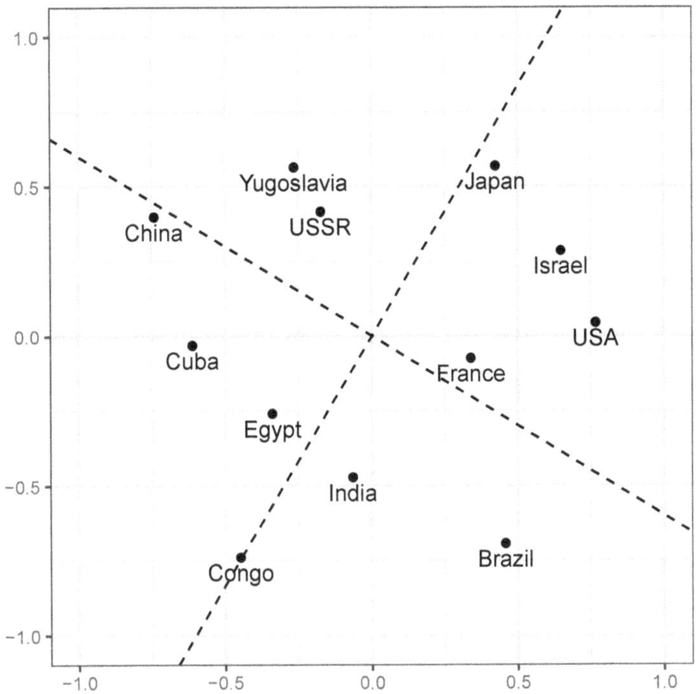

FIGURE 3.3: SMACOF (Majorization) Metric Scaling of the Nations Similarities Matrix

existence was not known in advance. However, we also want to be vigilant in separating signal from noise. Because we cannot make sense of the substantive meaning of the third dimension in this example, it seems more likely that this dimension is simply improving fit by modeling noise in the data. Hence, we remain with the two-dimensional configuration.

```
library(smacof)
ndim <- 5
result <- vector("list", ndim)
for (i in 1:ndim){
result[[i]] <- smacofSym(delta=d, ndim=i, type = "interval")
}
stress <- sapply(result, function(x)x$stress)

stress.df <- data.frame(
  stress=stress,
  dimension = 1:length(stress))
ggplot(stress.df, aes(x=dimension, y=stress)) +
```

```
geom_point() +
geom_line() +
theme_bw() +
labs(x="Dimension", y="Stress")
```

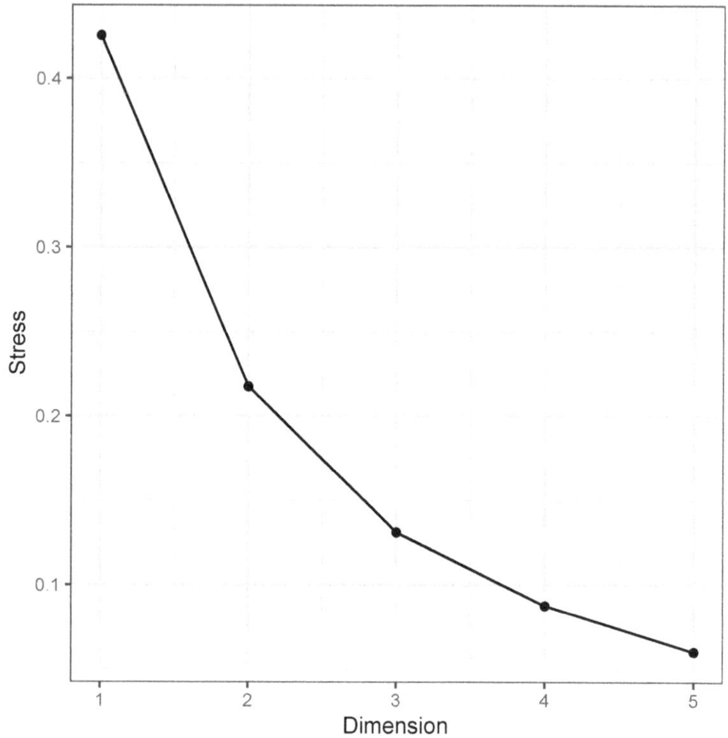

FIGURE 3.4: Stress-1 Values for SMACOF (Majorization) Metric Scaling of the Nations Similarities Matrix

3.1.4.1 Using the smacof Package to Analyze Matrices with Missing Data

One of the attractive features of the MDS functions included in the `smacof` package (de Leeuw and Mair, 2009) in `R` is that they can be easily altered to analyze similarities data with missing values. Below we replace the squared distance between Brazil (nation #1) and Congo (nation #2) in the nations dissimilarities matrix with missing values.

```
d[1,2] <- d[2,1] <- NA
```

To analyze the matrix of squared distances `d` with the `smacofSym()` function as before we create a matrix of weights (`weightmat`, composed of the w_{jm} terms from Equation 3.16) in which all values corresponding to non-missing values in the matrix d are set to 1. We then replace the cells in `weightmat` that correspond to the missing values in `d` to 0.

```
weightmat <- d
weightmat[!is.na(d)] <- 1
weightmat[is.na(d)] <- 0
```

We must also replace the `NA` values in `d` with non-missing values in order to be analyzed with the `smacofSym()` function. We replace the `NA` values with the mean value in the matrix. Of course, it doesn't matter which value we use since their weight term is 0.

```
d[is.na(d)] <- mean(d, na.rm=TRUE)
```

We then re-run the `smacofSym()` function on the nations similarities matrix with missing values for the squared distance between Brazil and Congo. The estimated configuration is shown in Figure 3.5. The results are virtually unchanged from the same analysis conducted on the full matrix shown in Figure 3.3. This is because the relationship between Brazil and Congo can be modeled by proxy using the observed dissimilarities between them and common objects (i.e., Brazil and Congo's mutual distance from China, Cuba, etc.).

3.2 Nonmetric Multidimensional Scaling

While metric MDS procedures assume that the similarities data represents ratio or interval-level information, nonmetric MDS procedures assume that the data represents only the ordinal relationships between the stimuli (Rabinowitz, 1975). Hence, nonmetric MDS methods estimate a point configuration in which the inter-point distances reproduce the rank ordering of the observed dissimilarities rather than reproduce the observed distances as closely as possible. For example, in the *nations* data, the difference between Israel and China 6.00 units and the difference between Israel and the US is 3.06 units. Metric MDS methods will attempt to estimate a configuration in which the distance between Israel and China is about twice the distance between Israel and the US. Nonmetric MDS methods will attempt to estimate a configuration in which the distance between Israel and China is greater than or equal to the distance between Israel and the US. More formally, the nonmetric MDS function, f, assumes only weak monotonicity between the observed and estimated distances (Kruskal, 1964a,b); that is

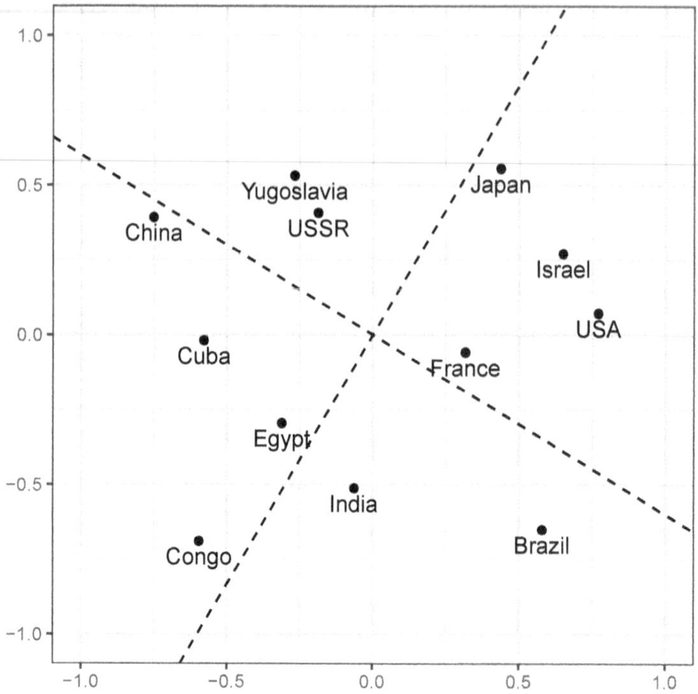

FIGURE 3.5: SMACOF (Majorization) Metric Scaling of the Nations Similarities Matrix with Missing Data

$$\text{if } \delta_{jm} < \delta_{kl}, \text{ then } d_{jm} \leq d_{kl} \tag{3.18}$$

Thus, nonmetric MDS is a more flexible means of estimating a configuration of points from proximities data. This is both a strength and a weakness. On the one hand, nonmetric MDS methods are likely to produce lower Stress values because there are less constraints on the solution. For example, Figure 3.6 plots the Stress values for metric and nonmetric (SMACOF) MDS of the `nations` data in 1-5 dimensions. Across dimensions, the nonmetric MDS Stress value is lower than metric MDS. However, because there is a wider range of configurations, nonmetric MDS methods are more vulnerable than their metric counterparts to locally optimal and degenerate solutions (Kruskal and Wish, 1978, p. 76). Locally optimal solutions are more undesirable the more they differ from a globally optimal solution (that is, the configuration that reduces the Stress to its minimum value). Degenerate solutions are problematic because they achieve an artificially low Stress value through an incoherent configuration of the stimuli, usually by clustering a few similar groups of stimuli together (Borg and Groenen, 2010, chap. 13). Nonmetric MDS methods are most likely to produce degenerate solutions when the number of dimensions is high relative to the number of stimuli (Rabinowitz (1975) recommends that the ratio of stimuli to dimensions be at least four).

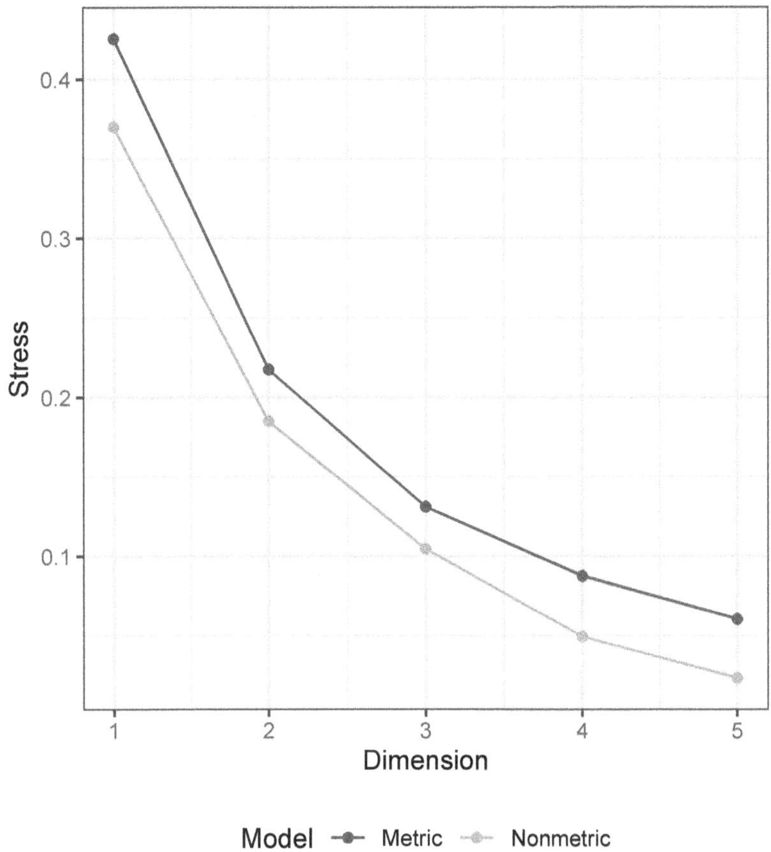

FIGURE 3.6: Stress Values of Metric and Nonmetric (SMACOF) Scaling of the Nations Similarities Matrix

3.2.1 Example 1: Nations Similarities Data

We return to the `smacof` package in `R` to perform nonmetric MDS on the `nations` data using the majorization algorithm. Indeed, we use the same function and syntax, with the exception that we change the argument `type`

= "interval" to type = "ordinal". The object smacof_nonmetric_result indicates that 75 iterations were required to achieve convergence. We would be most concerned about a degenerate solution if the Stress value is near zero and the points are arranged in a few tight clusters. However, an examination of the Stress-1 value (result$stress.nm) and the estimated point configuration (result$conf) indicates that it is very unlikely that this result is degenerate.

```
smacof_nonmetric_result <- smacofSym(delta=d, ndim=2,
                           type = "ordinal")
nm.conf <- smacof_nonmetric_result$conf

print(smacof_nonmetric_result$niter)
```

```
[1] 70
```

```
print(smacof_nonmetric_result$stress)
```

```
[1] 0.1850412
```

We then compare the two-dimensional results from metric and nonmetric MDS in Figure 3.7. The two configurations are highly correlated (Pearson's $r = 0.991$ for the first dimension coordinates and $r = 0.938$ for the second dimension; Kendall's $\tau = 0.878$ for the first dimension coordinates and $\tau = 0.848$ for the second dimension).

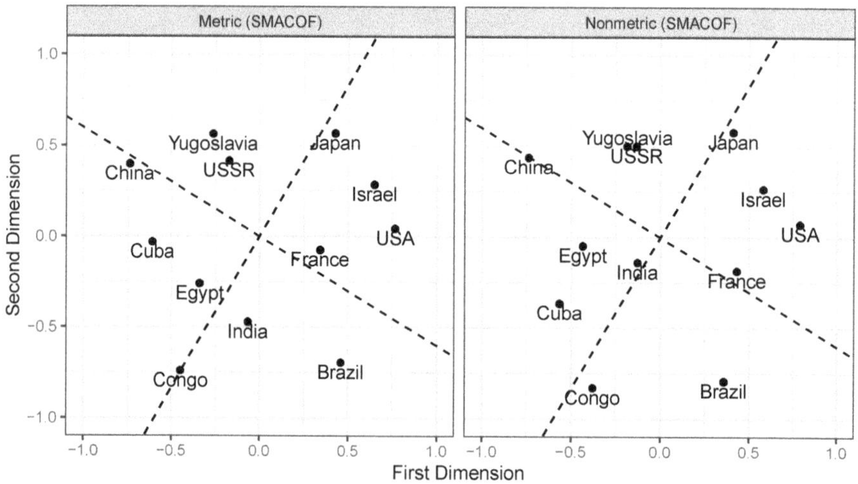

FIGURE 3.7: Metric and Nonmetric (SMACOF) Multidimensional Scaling of Nations Similarities Data

3.2.1.1 Rotating a Solution

We next demonstrate how to rotate a point configuration either by degrees or to match a target configuration as closely as possible. In either case, the point configuration is transformed by multiplying the points by a rotation matrix A that takes the form:

$$A = \begin{bmatrix} \cos\theta & -\sin\theta \\ \sin\theta & \cos\theta \end{bmatrix} \tag{3.19}$$

where θ is the angle by which the point configuration should be rotated. Positive angles produce counter-clockwise rotations and negative angles are used for clockwise rotations.

For instance, to rotate the metric (SMACOF) solution for the `nations` data so that the substantive dimensions (the dotted lines) match the geometric axes, we rotate the point configuration 30° clockwise so that the first dimension represents the undeveloped-developed dimension and the second dimension represents the Communist-West dimension. Hence we use the matrix:

$$A = \begin{bmatrix} \cos(-30^\circ) & -\sin(-30^\circ) \\ \sin(-30^\circ) & \cos(-30^\circ) \end{bmatrix} \tag{3.20}$$

Matrix multiplication is performed in `R` with the command \%*\%. The metric solution is rotated by multiplying the `A` matrix by the point coordinates as below. We then plot the rotated coordinates in Figure 3.8. The two substantive dimensions now match the geometric axes.

```
A <- matrix(c(cos(-30), -sin(-30), sin(-30), cos(-30)), nrow=2, ncol=2,
    byrow=TRUE)
rot.mds <- smacof_metric_result$conf %*% A
```

In addition, a Procrustes rotation procedure can be performed to transform a point configuration so that it is maximally comparable with another (target) configuration (Borg and Groenen, 2010, chap. 20). To rotate the nonmetric MDS configuration of the `nations` data to the target matrix of the rotated metric MDS configuration shown in the right panel of Figure 3.8, we use the `procrustes()` function in the `MCMCpack` package in `R` (Martin, Quinn and Park, 2011). In the `procrustes()` function, the `X` argument denotes the matrix to be rotated, the `Xstar` argument denotes the target matrix and the `translation` and `dilation` arguments denote whether the transformed matrix should be translated or dilated (by default, `FALSE`). Translation shifts the origin and dilation stretches or contracts the space by weighting the dimensions. The rotated configuration `proc$X.new` is then stored as the object `nonmetric.conf.rotated`. The target metric result and rotated nonmetric configurations are plotted in Figure 3.9.

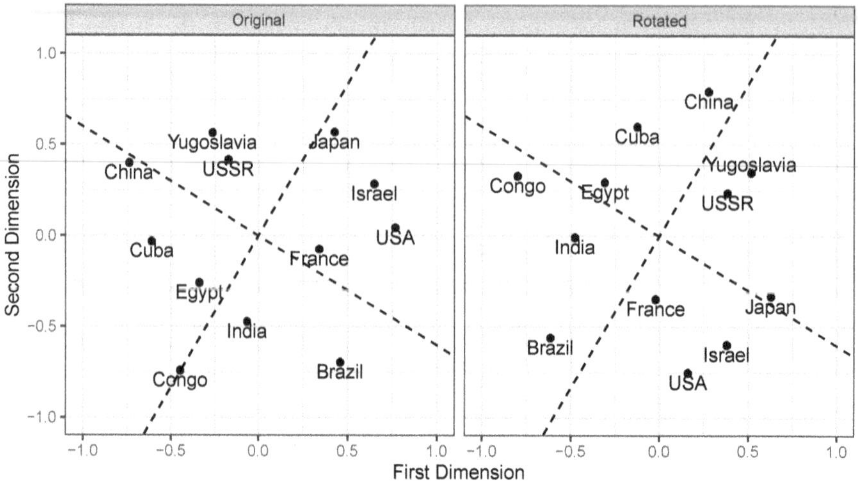

FIGURE 3.8: Original and Rotated Metric (SMACOF) Multidimensional Scaling of Nations Similarities Data

```
library(MCMCpack)
metric.conf <- rot.mds
nonmetric.conf <- smacof_nonmetric_result$conf
proc <- procrustes(X=nonmetric.conf, Xstar=metric.conf,
    translation=FALSE, dilation=FALSE)
nonmetric.conf.rotated <- proc$X.new
```

3.2.2 Example 2: 90th US Senate Agreement Scores

Legislative roll call data can be used to compute a symmetric matrix of agreement scores between each pair of legislators. The values denote how often the legislators vote together and range from 0 (no voting agreement) to 1 (perfect voting agreement).

In this section, we use the agreement score matrix of the 90th US Senate (1967-68) since there are two dimensions (liberal-conservative and region/civil-rights) underlying roll call voting during this period (Poole and Rosenthal, 1997). We show a section from the `senate.90` matrix below. There are 102 legislators included in the matrix: 100 Senators plus President Lyndon Johnson (who "voted" on select bills by announcing a position) and Senator Charles Goodell (R-NY) (who replaced Senator Robert F. Kennedy after Kennedy's assassination in June 1968).

```
data(senate.90)
print(senate.90[1:6,7:12])
```

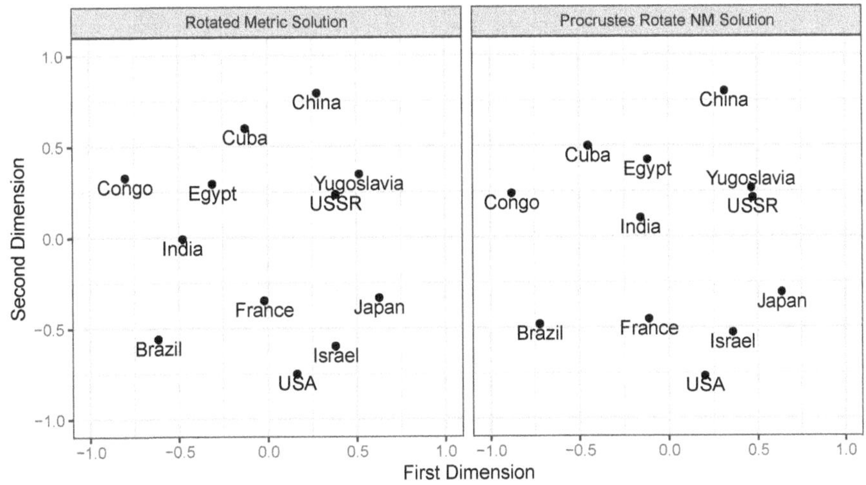

FIGURE 3.9: Procrustes Rotation of Nonmetric (SMACOF) Multidimensional Scaling of Nations Similarities Data

```
      name johnson sparkman  hill gruening bartlett
1  JOHNSON   1.000    0.611 0.510    0.524    0.651
2 SPARKMAN   0.611    1.000 0.899    0.510    0.650
3     HILL   0.510    0.899 1.000    0.530    0.628
4 GRUENING   0.524    0.510 0.530    1.000    0.762
5 BARTLETT   0.651    0.650 0.628    0.762    1.000
6   HAYDEN   0.700    0.850 0.786    0.583    0.704
```

The agreement scores themselves are stored in columns 8-109. The first seven columns are legislator information (e.g., state and party). We transform the agreement scores into distances (`d`) by subtracting them from 1 and then squaring the values. We then use the `smacofSym()` function to perform metric and nonmetric MDS.

```
d <- (1-senate.90[,8:109])^2
metric.result <- smacofSym(d, ndim=2, type = "interval")
metric.conf <- metric.result$conf
metric.stress <- metric.result$stress
nonmetric.result <- smacofSym(d, ndim=2, type = "ordinal")
nonmetric.conf <- nonmetric.result$conf
nonmetric.stress <- nonmetric.result$stress
```

Since these results are identified only up to a rotation and choice of origin, we reflect the recovered space about the x-axis so that liberal legislators are on the left side of each dimension. The code below flips the legislator coordinates (multiplying them by -1) if President Lyndon Johnson (Legislator #1) has a positive score on the dimension.

```
if (metric.conf[1,1] > 0) metric.conf[,1] <- -metric.conf[,1]
if (metric.conf[1,2] > 0) metric.conf[,2] <- -metric.conf[,2]
if (nonmetric.conf[1,1] > 0) nonmetric.conf[,1] <- -nonmetric.conf[,1]
if (nonmetric.conf[1,2] > 0) nonmetric.conf[,2] <- -nonmetric.conf[,2]
```

Figure 3.10 plots the results of the metric and nonmetric MDS of the `senate.90` data. We include the code used to produce the top panel of Figure 3.10 to show how conditions can be used to separately plot legislators by party and state (we want to distinctly label Southern and Northern Democrats, since the two groups split on both dimensions - especially civil rights - during this period). The two configurations are clearly very similar. Legislators' metric and nonmetric first dimension coordinates are correlated at $r = 0.998$ and $\tau = 0.979$. Their second dimensions are correlated at $r = 0.996$ and $\tau = 0.946$. Generally, metric and nonmetric results will be more similar as the number of stimuli increases because the ordinal constraints strongly limit the possible configurations of points (Borg and Groenen, 2010, p. 28).

```
senate.df <- data.frame(
  dim1 = c(metric.conf[,1], nonmetric.conf[,1]),
  dim2 = c(metric.conf[,2], nonmetric.conf[,2]),
  party = rep(senate.90$party, 2),
  statecode = rep(senate.90$statecode, 2),
  model = factor(rep(1:2, each=nrow(metric.conf)),
                 labels=c("Metric", "Nonmetric")))
senate.df$label <- NA
senate.df$label[which(senate.df$party == 100 &
        senate.df$statecode %in% c(40:49,51,53,56))] <- "S"
senate.df$label[which(senate.df$party == 100 &
        !(senate.df$statecode %in% c(99,40:49,51,53,56)))] <- "D"
senate.df$label[which(senate.df$party == 200)] <- "R"
senate.df$label[which(senate.df$statecode == 99)] <- "LBJ"
senate.df$label <- factor(senate.df$label,
                          levels=c("D", "S", "R", "LBJ"))
ggplot() +
  geom_text(data=subset(senate.df, label!="LBJ"),
            aes(x=dim1, y=dim2, color=label,
                      label=label), show.legend=FALSE) +
  geom_point(data=subset(senate.df, label!="LBJ"),
            aes(x=dim1, y=dim2, color=label), size=-1) +
  theme_bw() +
  facet_wrap(~model) +
  scale_color_manual(values=rev(gray.palette(4)), name="Party Group",
    labels=c("Democrat", "Southern Dem", "Republican", "LBJ")) +
  guides(colour = guide_legend(override.aes =
                      list(size = 4, pch=c("D", "S", "R")))) +
  geom_text(data=subset(senate.df, label=="LBJ"),
            aes(x=dim1, y=dim2, label=label),
            show.legend=FALSE, size=6) +
```

```
xlab("Liberal-Conservative") +
xlim(-1.25, 1.25) +
ylab("Region/Civil Rights") +
ylim(-1.25, 1.25) +
theme(aspect.ratio=1, legend.position="bottom")
```

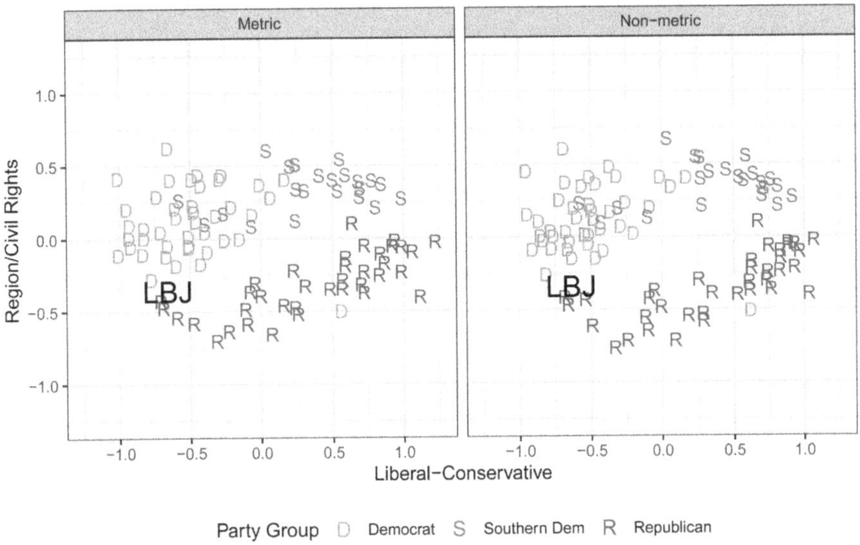

FIGURE 3.10: Metric and Nonmetric (SMACOF) Multidimensional Scaling of the 90th Senate Agreement Scores

Figure 3.11 shows what is known as a Shepard diagram for the `senate.90` MDS. Shepard diagrams plot the observed dissimilarities (δ_{jm}, stored in the objects `metric/nonmetric.obsdiss`) against the estimated distances (d_{jm}, stored in the objects `metric/nonmetric.dhat`), allowing for an assessment of the transformation of proximities data into inter-point distances. We include a lowess smoother (the gray line) to illustrate the relationship between the observed and estimated distances for each pair of legislators (Cleveland, 1981). Figure 3.11 indicates a good fit for both the metric and nonmetric MDS results, since the lowess smoother closely approximates the 45° line and the points are reasonably tightly clustered around the lowess line. The lowess line in the metric MDS solution is nearly linear, although as the deviation from linearity in the relationship between the observed and reproduced distances grows, we become more concerned about the appropriateness of metric MDS in modeling the data.

Figure 3.11 also provides a useful illustration of the consequences of the difference in how metric and nonmetric MDS treat ties in the similarities data. Note that the points in the right (nonmetric) panel are clustered in vertical bars, meaning that several groups of legislator pairings with the same observed dissimilarities are assigned different distances in the scaling procedure. This is because ties can take on different distances in nonmetric MDS because only a weak monotone transformation of the observed distances is being used (Equation 3.18). Conversely, metric MDS tries to reproduce tied observed dissimilarity values with identical interpoint distances. While this means that there is greater variation between the observed and estimated distances in nonmetric MDS than in metric MDS, it also helps to illustrate why nonmetric MDS has greater flexibility to produce a point configuration with lower Stress values. That is, the estimated distances must only comply with the ordinal properties of the data, so any given point at a fixed position on the x-axis (δ_{jm}) can move more freely along the y-axis (d_{jm}) in a manner that reduces Stress.

```
metric.obsdiss <- metric.result$dhat
metric.dhat <- metric.result$confdist
nonmetric.obsdiss <- nonmetric.result$dhat
nonmetric.dhat <- nonmetric.result$confdist
```

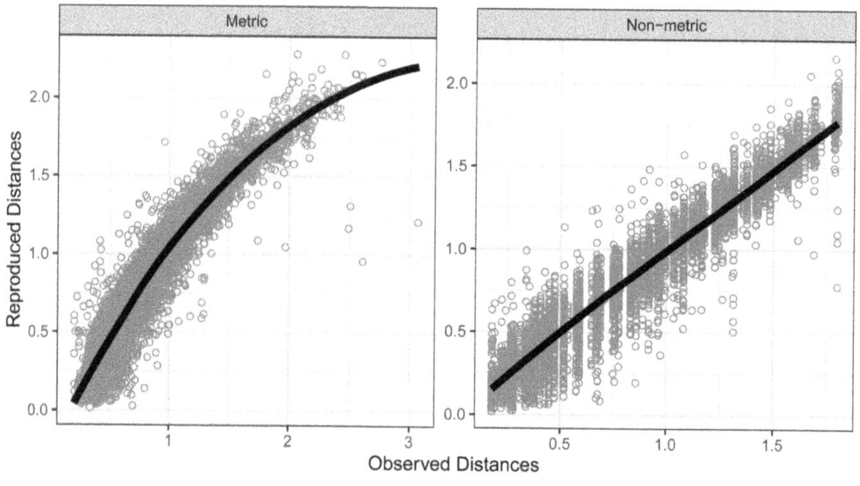

FIGURE 3.11: Shepard Diagrams of the 90th Senate Agreement Scores

3.2.2.1 Finding the Correlation of Two Sets of Interpoint Distances

We close by showing how to compute the correlation between a pair of interpoint distances, as with those from metric and nonmetric MDS of the `senate.90` data. The `dist()` function in the `stats` package in `R` automatically produces a (Euclidian) distance matrix between each of the objects in a given configuration. Hence, we can simply use the `cor()` function between the two distance matrices to find the correlation between the two sets of interpoint distances.

```
interpoint.d.metric <- dist(metric.conf)
interpoint.d.nonmetric <- dist(nonmetric.conf)
cor(interpoint.d.metric, interpoint.d.nonmetric)
```

```
[1] 0.9897796
```

3.3 Individual Differences Multidimensional Scaling

All of the previous methods covered in this chapter perform MDS on a single similarities or dissimilarities matrix. In many cases, these matrices are aggregated from choices or judgments from multiple sources. For instance, the `nations` data set from Wish (1971) consists of students' mean similarity ratings of each pair of nations. The similarities matrix from the `nations` data set could be broken down into 19 (one for each student) separate matrices. We can calculate a point configuration for each individual, and then find a matrix of weights that map these individual results onto a group configuration. This method is known as Individual Differences MDS.

This was an idea originally developed by Horan (1969) and forms the basic motivation underlying individual differences scaling. Individual differences scaling avoids the pitfall of treating respondents or sub-groups as if they use the same latent evaluative dimensions identically (as is the case when we average across respondents and recover a group configuration) (Bartholomew et al., 2008, pp. 59-60). By estimating separate weights for each dimension, more important (that is, more explanatory) dimensions are expanded and less important dimensions are compressed. For this reason, individual differences scaling is also referred to as Weighted MDS (Jacoby, 1986, 2009).

In this section we discuss Carroll and Chang's (1970) INDSCAL (INdividual Difference SCALing) method. In the INDSCAL problem, we are given n $(i = 1,...,n)$ matrices of the form

$$D_{zi}^* = D_{zi} + E = (diagZW_iZ')J_q' - 2ZW_iZ' + J_q(diagZW_iZ')' + E_i \tag{3.21}$$

where q $(j = 1,...,q)$ is the number of stimuli and we are asked to find the q by s $(k = 1,...,s)$ (where s is the number of dimensions) matrix Z and the n diagonal matrices W_i. There are n q by q symmetric matrices of distances between the q stimuli as judged by each of the n individuals.

Carroll and Chang (1970) begin by double centering each matrix; that is

$$Y_i^* = ZW_iZ' + E_i^* \tag{3.22}$$

where Y_i^* is q by q, Z is q by s, W_i is s by s, and E_i^*, the matrix of error, is q by q. Carroll and Chang (1970) distinguish between the two Z matrices in Y_i^*, calling one Z-left and the other Z-right. They do this in order to develop their alternating least-squares procedure shown below. Hence

$$Y_i^* = Z^{(L)}W_iZ'^{(R)} + E_i^* \tag{3.23}$$

It turns out that, at convergence, $Z^{(L)}$ and $Z^{(R)}$ will differ by only a diagonal transformation.

Starting values for $Z^{(L)}$ and $Z^{(R)}$ can be obtained by an eigenvalue-eigenvector decomposition of the average of the Y_i^*; that is set:

$$\bar{Y}^* = \frac{\sum_{i=1}^{n} Y_i^*}{n} = U\Lambda U' \tag{3.24}$$

and use the Eckart-Young Theorem (Eckart and Young, 1936) to get starting estimates of $Z^{(L)}$ and $Z^{(R)}$:

$$\hat{Z}^{(L)} = \hat{Z}^{(R)} = U_s\Lambda_s^{1/2} \tag{3.25}$$

where U_s is the n by s matrix consisting of the first s eigenvectors of U and Λ_s is the s by s diagonal matrix of the first s eigenvalues (singular values).

Carroll and Chang (1970) construct the linear system:

$$\Psi^* = \tilde{W}G' + E_1 \tag{3.26}$$

where Ψ^* is n by q^2, $\tilde{W}$ is n by s, G is q^2 by s, and E_1 is n by q^2. All of the elements of the Y_i^* are used in Equation 3.26. In particular, note that *each row of Ψ^* contains an entire Y_i^* matrix.* The rows of $\tilde{W}$ are the diagonals of the W_i, namely:

$$\tilde{W} = \begin{bmatrix} w_{11} & \cdots & w_{1s} \\ w_{21} & \cdots & w_{2s} \\ \cdot & & \cdot \\ \cdot & & \cdot \\ \cdot & & \cdot \\ w_{n1} & & w_{ns} \end{bmatrix} \tag{3.27}$$

Finally, G contains all of the cross-products of $Z^{(L)}$ and $Z^{(R)}$:

$$G = \begin{bmatrix} \hat{z}_{11}^{(L)}\hat{z}_{11}^{(R)} & \cdots & \hat{z}_{1s}^{(L)}\hat{z}_{1s}^{(R)} \\ \hat{z}_{11}^{(L)}\hat{z}_{21}^{(R)} & \cdots & \hat{z}_{1s}^{(L)}\hat{z}_{2s}^{(R)} \\ \vdots & \ddots & \vdots \\ \hat{z}_{11}^{(L)}\hat{z}_{q1}^{(R)} & \cdots & \hat{z}_{1s}^{(L)}\hat{z}_{qs}^{(R)} \\ \hat{z}_{21}^{(L)}\hat{z}_{11}^{(R)} & \cdots & \hat{z}_{2s}^{(L)}\hat{z}_{1s}^{(R)} \\ \vdots & \ddots & \vdots \\ \hat{z}_{21}^{(L)}\hat{z}_{q1}^{(R)} & \cdots & \hat{z}_{2s}^{(L)}\hat{z}_{qs}^{(R)} \\ \vdots & \ddots & \vdots \\ \hat{z}_{q1}^{(L)}\hat{z}_{11}^{(R)} & \cdots & \hat{z}_{qs}^{(L)}\hat{z}_{1s}^{(R)} \\ \hat{z}_{q1}^{(L)}\hat{z}_{21}^{(R)} & \cdots & \hat{z}_{qs}^{(L)}\hat{z}_{2s}^{(R)} \\ \vdots & \ddots & \vdots \\ \hat{z}_{q1}^{(L)}\hat{z}_{q1}^{(R)} & \cdots & \hat{z}_{qs}^{(L)}\hat{z}_{qs}^{(R)} \end{bmatrix} \tag{3.28}$$

Note that the first row of $\tilde{W}$ multiplied by the first column of G' is:

$$y_{111}^* = \sum_{k=1}^{s} w_{1k}\hat{z}_{1k}^{(L)}\hat{z}_{1k}^{(R)} \tag{3.29}$$

Similarly, the first row of $\tilde{W}$ multiplied by the first column of G' is:

$$y_{121}^* = y_{112}^* = \sum_{k=1}^{s} w_{1k}\hat{z}_{1k}^{(L)}\hat{z}_{2k}^{(R)} \tag{3.30}$$

because $\hat{Z^{(L)}} = \hat{Z^{(R)}}$ at the first step of the algorithm.

Now perform OLS to get an estimate of $\tilde{W}$; that is

$$\hat{\tilde{W}} = \Psi^* G(G'G)^{-1} \tag{3.31}$$

The $\hat{\tilde{W}}$ from Equation 3.31 and the starting estimate of $Z^{(R)}$ from Equation 3.25 are used to construct the linear system:

$$\Psi^{**} = \hat{Z}^{(L)} H_1' + E_2 \tag{3.32}$$

where Ψ^{**} is q by nq, $\hat{Z^{(L)}}$ is q by s, H_1 is nq by s, and E_2 is q by nq. All the elements of all the Y_i^* are used in Equation 3.32. In particular, note that *each row of* Ψ^{**} *contains the corresponding row of each of the n* Y_i^* *matrices.* The matrix H_1 is the partitioned matrix:

$$H_1 = \begin{bmatrix} \hat{Z}^{(R)}\hat{W}_1 \\ \hline \hat{Z}^{(R)}\hat{W}_2 \\ \hline \vdots \\ \hline \hat{Z}^{(R)}\hat{W}_n \end{bmatrix} \tag{3.33}$$

Each of the $\hat{Z^{(R)}}\hat{W}_i$ matrices in each partition is q by s so that the total number of rows is nq. Carroll and Chang (1970) now perform OLS to get an estimate of $\hat{Z^{(L)}}$; namely

$$\hat{Z}^{(L)} = \Psi^{**}H_1\left(H'_1H_1\right)^{-1} \tag{3.34}$$

The $\hat{W}_i$ from Equation 3.31 and $\hat{Z^{(L)}}$ from Equation 3.34 are used to construct:

$$\Psi^{**} = \hat{Z}^{(R)}H_2' + E_3 \tag{3.35}$$

where Ψ^{**} is q by nq and is the same matrix used in Equations 3.32 and 3.34, $\hat{Z^{(R)}}$ is q by s, H_2 is nq by s, and E_3 is q by nq.

H_2 has the same form as H_1, only now $\hat{Z^{(L)}}$ is used to construct it:

$$H_2 = \begin{bmatrix} \hat{Z}^{(L)}\hat{W}_1 \\ \hline \hat{Z}^{(L)}\hat{W}_2 \\ \hline \vdots \\ \hline \hat{Z}^{(L)}\hat{W}_n \end{bmatrix} \tag{3.36}$$

Carroll and Chang (1970) now perform OLS to get an estimate of $\hat{Z^{(R)}}$; namely

$$\hat{Z}^{(R)} = \Psi^{**}H_2\left(H'_2H_2\right)^{-1} \tag{3.37}$$

The estimates of $\hat{Z^{(R)}}$ from Equation 3.37 and $\hat{Z^{(L)}}$ from Equation 3.34 can now be used to construct G in Equation 3.26 thereby getting new estimates of the $\hat{W}_i$ in Equation 3.31. This process can be repeated as many times as desired. The sum of squared error will always decrease and the algorithm will always converge.

In summary, the Carroll and Chang (1970) INDSCAL procedure consists of the following steps:

1. Double center the n q by q symmetric matrices of squared distances to obtain the n Y_i^* matrices.

2. Obtain starting estimates of $\hat{Z^{(L)}}$ and $\hat{Z^{(R)}}$ from Equation 3.25.

3. Use $\hat{Z^{(L)}}$ and $\hat{Z^{(R)}}$ to construct G in Equation 3.26.

4. Run OLS to obtain estimates of the $\hat{W}_i$ from Equation 3.31.

5. The $\hat{W}_i$ from Equation 3.31 and $\hat{Z^{(R)}}$ from Equation 3.25 on the first pass and then from Equation 3.37 on later passes are used to construct Equation 3.32.

6. The $\hat{W}_i$ from Equation 3.31 and $\hat{Z^{(L)}}$ from Equation 3.34 are used to construct Equation 3.35.

7. Go to Step 3.

8. Repeat Steps 3-5 until convergence.

3.3.1 Example 1: 2009 European Election Study (French Module)

In this example, we revisit the party placement data from the French module of the 2009 European Election Study (EES) used in Chapter 2. This allows for a direct comparison of the results obtained by scaling two different forms of data from the same set of observations. Recall that the French module of the 2009 EES asked 1,000 respondents to place 8 major political parties on a ten-point left-right ideological scale (saved in the matrix `french.parties.individuals`). In Chapter 2, respondents' placements were scaled directly. In this case, we arrange each set of placements into a similarities matrix for each respondent.

To perform INDSCAL on the `french.parties.individuals` data, we first remove all rows with missing values with the `na.omit()` function.

```
data(french.parties.individuals)
french.parties.individuals <- na.omit(french.parties.individuals)
```

We then construct a dataframe (`parties`) for the n $q \times q$ dissimilarities matrices D^*_{zi}. For each cell of `parties`, we calculate the squared difference between individual i's rankings of the two parties. We also add a small positive constant (0.001) to each value, since INDSCAL cannot handle observed distances of 0 between non-identical parties on the off-diagonal. A couple of matrix commands makes the calculation of all of these symmetric dissimilarities matrices D^*_{zi} easy.

```
p <- as.matrix(french.parties.individuals)
parties <- lapply(1:nrow(french.parties.individuals),
        function(x)outer(p[x,], p[x,], "-")^2+.001)
for(i in 1:length(parties)){diag(parties[[i]]) <- 0}
```

The final product is a set of n $q \times q$ symmetric dissimilarity matrices stored in the dataframe `parties`. Values range from 0 (on the diagonal) and 0.001 (most similar stimuli) to 100.001 (most dissimilar stimuli). We print Respondent #1's matrix below.

```
                  1       2     3      4      5      6
Extreme Left   0.000   0.001  1.001 25.001 25.001 81.001
Communist      0.001   0.000  1.001 25.001 25.001 81.001
```

```
Socialist          1.001   1.001  0.000 16.001 16.001 64.001
Greens            25.001  25.001 16.001  0.000  0.001 16.001
UDF (Bayrou)      25.001  25.001 16.001  0.001  0.000 16.001
UMP (Sarkozy)     81.001  81.001 64.001 16.001 16.001  0.000
National Front 100.001 100.001 81.001 25.001 25.001  1.001
Left Party         1.001   1.001  0.001 16.001 16.001 64.001
                       7       8
Extreme Left   100.001   1.001
Communist      100.001   1.001
Socialist       81.001   0.001
Greens          25.001  16.001
UDF (Bayrou)    25.001  16.001
UMP (Sarkozy)    1.001  64.001
National Front   0.000  81.001
Left Party      81.001   0.000
```

We then perform INDSCAL on the **parties** dataframe using the **smacofIndDiff()** function in the **smacof** package in **R** (de Leeuw and Mair, 2009). We set three arguments in the **smacofIndDiff()** function. The dissimilarity matrices are assigned to the **delta** argument. The number of dimensions to be estimated is set in **ndim**. Finally, the **constraint** argument can be set to one of three options (**indscal**, **idioscal**, **identity**) which modify constraints on the configuration weights (W_i) obtained from INDSCAL. The **indscal** option follows the Carroll and Chang (1970) method in restricting W_i to a diagonal matrix.[‡] The **idioscal** routine was developed by Carroll and Chang (1972) as a generalization of the INDSCAL model that places no constraints on the W_i matrix (Cox and Cox, 2001, p. 211). Finally, the **identity** option constrains all individual configurations to be identical. We use the **indscal** option because **idioscal** produces a configuration that is more difficult to interpret (since each individual can rotate the space differently (Arabie, Carroll and DeSarbo, 1987, p. 45)) and **identity** produces a set of identical configurations.

```
indscal.result.2dim <- smacofIndDiff(delta=parties,
ndim=2, type="ratio",constraint="indscal",eps=0.00001)
```

The **smacofIndDiff()** function returns eleven objects stored in the dataframe **indscal.result.2dim**:

delta A list of dissimilarity matrices or a list objects of class dist.

ndim Number of dimensions.

weightmat Optional matrix with dissimilarity weights.

[‡]Note that the **smacofIndDiff()** function in **R** uses the SMACOF algorithm to minimize Stress across the dissimilarity matrices rather than the original Carroll-Chang CANDECOMP (canonical decomposition) algorithm.

init Matrix with starting values for configurations (optional).

ties Tie specification for nonmetric MDS.

constraint Either "indscal", "idioscal, or "identity" (see details).

verbose If TRUE, intermediate stress is printed out.

modulus Number of smacof iterations per monotone regression call.

itmax Maximum number of iterations.

eps Convergence criterion.

spline.degree Degree of the spline for "mspline" MDS type.

spline.intKnots Number of interior knots of the spline for "mspline" MDS type.

The `summary()` command can be used to display the estimated point configuration for the group space and the stress per point. The latter statistic indicates that the seventh and sixth stimuli (the National Front and UMP parties) contribute about 37% of the total Stress. The lowest amount of Stress is associated with the second stimuli (the Communist Party).

```
summary(indscal.result.2dim)
```

```
Group Stimulus Space (Joint Configurations):
                   D1      D2
Extreme Left    0.6969 -0.0698
Communist       0.4574 -0.0518
Socialist       0.1288 -0.0049
Greens          0.0873  0.0447
UDF (Bayrou)   -0.1112  0.0208
UMP (Sarkozy)  -0.5513  0.1333
National Front -1.1010 -0.0319
Left Party      0.3932 -0.0404


Stress per point:
  Extreme Left      Communist       Socialist         Greens
         13.58           8.79            9.39          10.34
  UDF (Bayrou)  UMP (Sarkozy) National Front     Left Party
         11.83          16.38           20.62           9.06
```

Figure 3.12 displays the group space recovered by INDSCAL using the SMACOF algorithm. There is a clearly dominant first dimension to this data, which is not surprising given that the concept being measured (left-right ideology) is unidimensional. The first-dimension INDSCAL coordinates (representing left-right placement) also share the same rank ordering as the results from ML Aldrich-McKelvey scaling and are correlated at 0.989.

```
parts <- rownames(indscal.result.2dim$gspace)
ind.df <- data.frame(
  x = indscal.result.2dim$gspace[,1],
  y = indscal.result.2dim$gspace[,2])
ind.df$party <- factor(parts, levels=parts[order(ind.df$x)])
ggplot(ind.df, aes(x=x, y=y, shape=party)) +
  geom_point(size=3) +
  scale_shape_manual(values=1:8, name="Party") +
  xlim(-1.15,1.15) +
  ylim(-1.15,1.15) +
  theme_bw() +
  xlab("First Dimension") +
  ylab("Second Dimension")+
  theme(legend.position="bottom", aspect.ratio=1) +
  guides(shape = guide_legend(nrow=3))
```

Finally, Figure 3.13 shows two respondents' point configurations: one with a low Stress per Subject (SPS) value (Respondent #176) and one with a high SPS value (Respondent #65). We can also investigate the weight terms for each dimension that map the group space onto the individual spaces. Weight terms close to 1 indicate that the individual space closely approximates the group configuration. This is the case for Respondent #176, especially on the first dimension where the weights are (1.06 and 1.29) for dimensions 1 and 2. In contrast, Respondent #65's weight terms (0.53 and 3.44) indicate that this individual views the parties as being compressed along the first (left-right) dimension but pushed apart on another (the second) dimension.

```
confs <- indscal.result.2dim$conf
weights <- rbind(diag(indscal.result.2dim$cweights[[176]]),
                 diag(indscal.result.2dim$cweights[[65]]))
iw.df <- data.frame(
    x = c(confs[[176]][,1], confs[[65]][,1]),
    y = c(confs[[176]][,2], confs[[65]][,2]),
    party = ind.df$party,
    respondent = factor(rep(c(1,2), each=nrow(confs[[1]])),
                        labels = c("# 176", "# 65"))
    )
ggplot(iw.df, aes(x=x, y=y, shape=party)) +
  geom_point(size=3) +
  scale_shape_manual(values=1:8, name="Party") +
  theme_bw() +
  xlab("First Dimension") +
  ylab("Second Dimension")+
  facet_wrap(~respondent) +
  xlim(-1.15,1.15) +
  ylim(-1.15,1.15) +
  theme(legend.position="bottom", aspect.ratio=1)
```

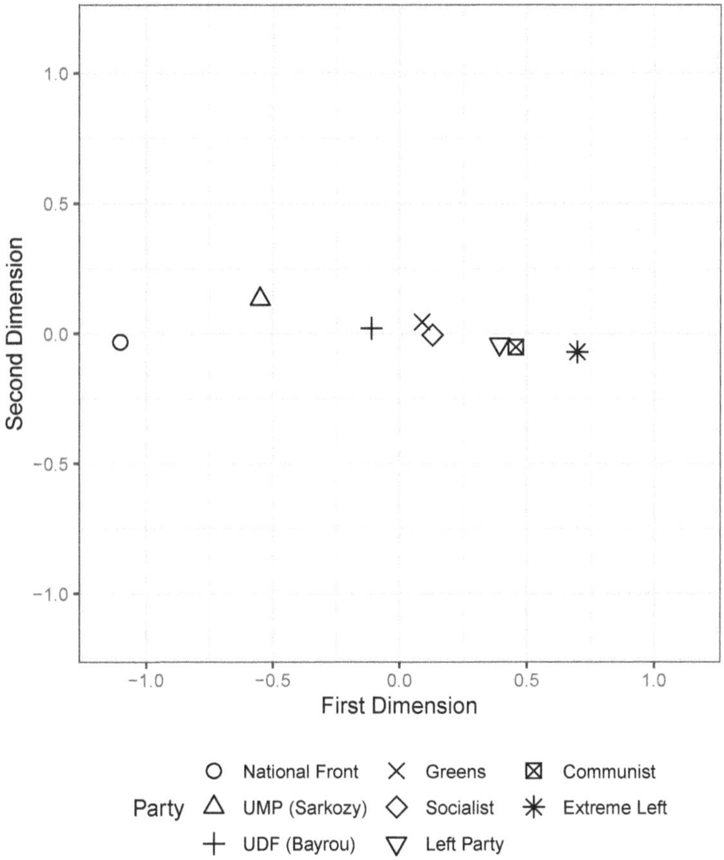

FIGURE 3.12: Individual Differences Scaling of Ideological Placements of French Political Parties (2009 European Election Study)

3.4 Conclusion

This chapter has detailed a suite of multidimensional scaling (MDS) methods used for the analysis of similarities or dissimilarities data. The goal of each of these techniques is to produce a geometric configuration of points that reproduces the observed distances data. However, as can be seen with the results produced from applying each of the MDS methods to the `nations` data, a distinct set of assumptions underlie each of these techniques. Consequently, the estimated `nations` configurations from metric and nonmetric MDS differed

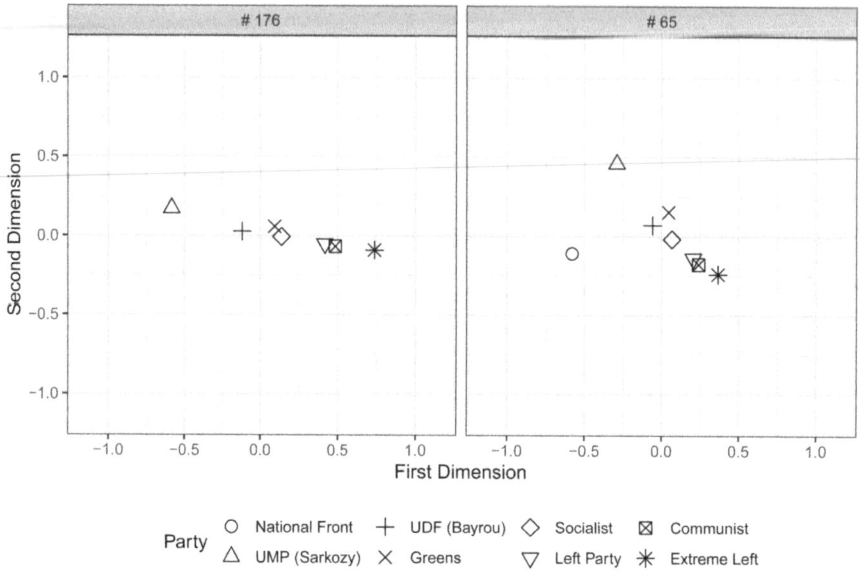

FIGURE 3.13: Individual Differences Scaling of Ideological Placements of French Political Parties (2009 European Election Study), Individual Configurations

slightly. This example emphasizes the importance of carefully considering the properties of the proximities data when deciding upon rival MDS procedures. In instances where the observed relationships between stimuli are better characterized as ordinal or rank-ordered, nonmetric MDS is more appropriate than metric MDS.

Moreover, within each method, more effort may be needed to estimate an optimal configuration. For metric MDS, numerical optimization techniques (particularly the SMACOF algorithm) should be used to minimize the Stress value of the estimated configuration. In nonmetric MDS, researchers must guard against degenerate solutions. Finally, within Bayesian MDS, additional constraints may be necessary to prevent sign-flips and sufficient iterations are of course needed to achieve convergence.

In addition, interpreting the dimensionality of MDS solutions can be something of an art. Scree plots are invaluable diagnostic tools when deciding upon the number of dimensions to estimate. However, careful consideration of the substantive meaning of the recovered dimensions is also critical, since it is quite common for dimensions (for example, the third dimension in the `nations` data) to be substantively meaningless but nonetheless provide an appreciable boost in fit. We caution against the inclusion of dimensions which appear to primarily be fitting noise in the data.

Finally, we note that although estimating a point configuration through the Torgerson solution of double-centering is something of a "primitive" method, it has proven to provide very accurate estimates of stimuli locations. Compare, for example, the high fidelity between the Torgerson solution for the `nations` data in Figure 3.1 with the results from other MDS methods. This is why double-centering provides such a versatile and quick means of estimating quality starting values for scaling methods, a point we discuss further in Chapters 5 and 6.

3.5 Exercises

1. Double-center the agreement score matrix of US Supreme Court Justices during the 2011 term stored in the data set `SupremeCourt.2011.Rda` following the steps below. The agreement scores in the matrix `SupremeCourt.2011` denote the proportion of cases that the Justices voted on the same side.[§]

   (a) Transform the agreement scores into squared distances and provide the code used.

   (b) Use the `doubleCenter()` function to double-center the squared distances and provide the code used.

   (c) Extract the eigenvalues and eigenvectors from the double-centered matrix and provide the code used.

      i. What are the first five eigenvalues?

   (d) Calculate the first and second dimension coordinates of the Justices using the eigenvalues and eigenvectors of the double-centered matrix (the Torgerson solution) and provide the code used.

   (e) Plot the first and second dimension Justice coordinates from the Torgerson solution, clearly labeling the Justice names.

2. Use the `smacofSym()` function to perform metric and nonmetric MDS on the `SupremeCourt.2011` agreement score matrix.

   (a) Plot the metric and nonmetric configurations of the Justices side-by-side, clearly labeling the Justices.

      i. Next, perform a Procrustes rotation of the nonmetric configuration using the metric configuration as the target. Provide the code used.

[§]Data from the *Harvard Law Review* 126(1): 390.

   ii. Plot the metric and rotated nonmetric configurations side-by-side.
   iii. Why does nonmetric MDS place Scalia and Thomas at virtually identical locations while metric MDS does not?

(b) Report the stress (Stress-1) values from metric and nonmetric scaling in one and two dimensions.

(c) Based upon your interpretation of the substantive meaning of the point configurations and the stress values, what does the first dimension represent and is there a meaningful second dimension to this data?

3. The 2010 Cooperative Congressional Election Study asked all respondents whether they favored or opposed eight pieces of legislation voted on by the 111th Congress. The data set `CCES2010.GA10.Rda` includes an agreement score matrix of how often nine respondents from the 10th Congressional District of Georgia shared the same positions on these eight issues. `CCES2010.GA10` is a list that stores the full agreement score matrix (`CCES2010.GA10$agreement`), the agreement score matrix with missing data (`CCES2010.GA10$missing.agreement`), respondents' answers to the eight policy questions (`CCES2010.GA10$votes`), and respondents' party identification (`CCES2010.GA10$party`). Missing values are coded as `NA`.

   (a) Convert the agreement score matrix entries to squared distances and perform nonmetric MDS in two dimensions using the `smacofSym()` function.
      i. How many iterations were required to reach convergence?
      ii. Plot the estimated point configuration and label the respondents by their letter identifier (A, B, C, etc.) and color them by their party identification (blue = Democrat, black = Independent, and red = Republican).

   (b) Repeat the steps from Exercise 3(a) to analyze the agreement score matrix with missing data (`CCES2010.GA10$missing.agreement`).
      i. Create a matrix of weights in which values of 1 correspond to non-missing values in the agreement score matrix and values of 0 correspond to missing values in the agreement score matrix. Provide the code used.
      ii. Use the weight matrix to perform nonmetric MDS on the data using the `smacofSym()` function. Plot the point configuration with the respondents' letter identifiers.
      iii. All of the missing values in the agreement score matrix involve Respondent D. How does the placement of Respondent D change from the configuration estimated from the full data set in Exercise 3(a)?

4. The 2010 Chapel Hill Expert Survey (CHES) asked expert informants to place European parties on three 11-point scales: general left-right (`leftright`), economic left-right (`econlr`), and social/cultural left-right (`galtan`). `CHES2010.France` is a list that includes the raw placements of six French parties by seven experts (`CHES2010.France$lr.placements`) and dissimilarity matrices for each expert constructed from the sum of the absolute distances between the parties across the three scales (`CHES2010.France$dissimilarity.matrices`).

   (a) Use the `smacofIndDiff()` function to run Individual Differences Scaling (INDSCAL) on `CHES2010.France$dissimilarity.matrices` in two dimensions with the `indscal` constraint.

   (b) Which respondent has the most stress? Which respondent has the least stress?

   (c) Which party has the most stress? Which party has the least stress?

   (d) Plot the group configuration, clearly labeling the party names.

5. Continuing with the 2010 CHES data, examine the raw placements of the parties on the three scales stored in `CHES2010.France$lr.placements`.

   (a) Print a table of the mean placements of each party on the general, economic, and social/cultural left-right scales.

   (b) The Communist, Green and Socialist Parties are clustered together on the first dimension but pushed apart on the second. On which scale(s) are they similar? On which scale(s) are they different? Does this help explain why they diverge on the second dimension?

6. Provide a side-by-side plot of the configurations of French parties for Expert 2 and Expert 6, setting `xlim` and `ylim` equal to `c(-1,1)` in both plots.

   (a) What are the configuration weights for Expert 2 and Expert 6 on the first and second dimensions?

   (b) Examining their raw placements of the parties, why are the Communist, Socialist, and Green Parties pushed apart on the second dimension in Expert 2's configuration and pushed together in Expert 6's configuration?

4

Unfolding Analysis of Rating Scale Data

In this chapter we discuss how unfolding methods can be used to analyze rating scale preferential choice data.* These types of data are arranged as rectangular matrices with the individuals on the rows and the stimuli on the columns. Common examples of this type of data in political science include feeling thermometers and propensity to vote measures. Feeling thermometer questions (first used in the 1964 American National Election Study (ANES)) ask respondents to rate their affect towards various political stimuli (parties, candidates, and sociopolitical groups) on a 0 to 100-point scale. Propensity to vote questions are most commonly included in European political surveys (e.g., the European Election Study (EES)), and ask respondents to rate how likely they are to vote for a given party or candidate on a similar (although usually 10 or 11-point) scale.

Data such as feeling thermometers and propensity to vote ratings are the best available measures of voters' expected utility from political alternatives. These measures capture the bundle of considerations that voters use to make political choices: policy positions, party labels, valence or personal affect, etc. This explains the extensive literature that these data (especially feeling thermometers) have spawned in political science (e.g., Weisberg and Rusk, 1970; Rabinowitz, 1978; Poole and Rosenthal, 1984; Jacoby and Armstrong, 2014).

We use Coombs' (1950; 1952; 1958; 1964) unfolding model to analyze this type of preferential choice data. Clyde Coombs developed unfolding analysis to deal with data where individuals ranked stimuli in order of preference. Coombs came up with the idea of a most preferred point (ideal point) and a single-peaked preference function to account for the observed rank orderings. The aim of an unfolding analysis is to arrange the individuals' ideal points and points representing the stimuli on a common evaluative scale (the J scale) so that the distances between the ideal points and the stimuli points reproduce the observed rank orderings. An individual's rank ordering is computed from her ideal point so that the reported ordering is akin to picking up the dimension (as if it were a piece of string) at the ideal point so that both sides of the dimension fold together to form a line with the individual's ideal point at the

*In Section 5.4.3 we demonstrate how rating or rank order scale data can be transformed into a series of pairwise comparisons and analyzed using unfolding methods for binary choice data.

end. Coombs called this an unfolding analysis because the researcher must take the rank orderings and "unfold" them. Because the unfolding model is an ideal point model, it is entirely consistent with the spatial (geometric) model of choice and exceptionally well-suited to the analysis of political choice data.

We begin with a general discussion of the thermometers problem and proceed to an exposition of the MLSMU6 and SMACOF least squares metric unfolding methods. The `smacofRect()` function in the `smacof` package (de Leeuw and Mair, 2009) uses the SMACOF optimization procedure introduced in Chapter 3 to perform metric unfolding on rectangular matrices in `R`. We then discuss and demonstrate the use of Bakker and Poole's (2013) Bayesian metric multidimensional unfolding method to analyze rectangular matrices of preferential choice data.

4.1 Solving the Thermometers Problem

Let T be the n by q matrix of thermometer scores (customarily ranging from 0 to 100) where $i = 1,...,n$ is the number of respondents and $j = 1,...,q$ is the number of political/social stimuli with ratings. T can be regarded as a matrix of inverse distances between the respondents and the stimuli. Specifically, apply the linear transformation:

$$d^*_{ij} = \left(\frac{100 - T}{50}\right) = (2 - 0.02T) = d_{ij} + \varepsilon_{ij} \tag{4.1}$$

Where the observed data are now *noisy* distances that range from zero to two, that is, $0 \leq d^*_{ij} \leq 2$, which are assumed to be equal to some true distance plus a random error term $(d_{ij} + \varepsilon_{ij})$. This transformation is convenient because it has the practical effect of confining the estimated respondent and stimuli points to a unit hypersphere.

Recall that our n by s matrix of individual (respondent) coordinates is:

$$X = \begin{bmatrix} x_{11} & x_{12} & \cdots & x_{1s} \\ x_{21} & x_{22} & \cdots & x_{2s} \\ \cdot & \cdot & \cdot & \cdot \\ \cdot & \cdot & \cdot & \cdot \\ \cdot & \cdot & \cdot & \cdot \\ x_{n1} & x_{n2} & \cdots & x_{ns} \end{bmatrix} \tag{4.2}$$

and our q by s matrix of stimuli coordinates is:

$$Z = \begin{bmatrix} z_{11} & z_{12} & \cdots & z_{1s} \\ z_{21} & z_{22} & \cdots & z_{2s} \\ \cdot & \cdot & \cdot & \cdot \\ \cdot & \cdot & \cdot & \cdot \\ \cdot & \cdot & \cdot & \cdot \\ z_{q1} & z_{q2} & \cdots & z_{qs} \end{bmatrix} \tag{4.3}$$

The n by q matrix of squared distances between X and Z (individuals and stimuli) is:

$$D = \begin{bmatrix} \sum_{k=1}^{s}(x_{1k}-z_{1k})^2 & \sum_{k=1}^{s}(x_{1k}-z_{2k})^2 & \cdots & \sum_{k=1}^{s}(x_{1k}-z_{qk})^2 \\ \sum_{k=1}^{s}(x_{2k}-z_{1k})^2 & \sum_{k=1}^{s}(x_{2k}-z_{2k})^2 & \cdots & \sum_{k=1}^{s}(x_{2k}-z_{qk})^2 \\ \cdot & \cdot & \cdot & \cdot \\ \cdot & \cdot & \cdot & \cdot \\ \cdot & \cdot & \cdot & \cdot \\ \sum_{k=1}^{s}(x_{nk}-z_{1k})^2 & \sum_{k=1}^{s}(x_{nk}-z_{2k})^2 & \cdots & \sum_{k=1}^{s}(x_{nk}-z_{qk})^2 \end{bmatrix} \tag{4.4}$$

As shown by Equations 3.3 - 3.7, this can be written in matrix algebra as the product of two partitioned matrices:

$$[diag(XX')| - 2X|J_n] \begin{bmatrix} J_q' \\ \overline{Z'} \\ \overline{diag(ZZ')'} \end{bmatrix} \tag{4.5}$$

Note that the rank of D, $\rho(D)$, must be less than or equal to $s+2$; i.e., $\rho(D) \leq s+2$.

If there was no error, then Equation 4.4 can be solved using the method of Schönemann (1970). Part of Schönemann's solution is to work with the double-centered matrix. Recall:

$$Y = X^*Z^{*\prime} = \begin{bmatrix} x_{11}-\bar{x}_1 & x_{12}-\bar{x}_2 & \cdots & x_{1s}-\bar{x}_s \\ x_{21}-\bar{x}_1 & x_{22}-\bar{x}_2 & \cdots & x_{2s}-\bar{x}_s \\ \cdot & \cdot & \cdot & \cdot \\ \cdot & \cdot & \cdot & \cdot \\ \cdot & \cdot & \cdot & \cdot \\ x_{n1}-\bar{x}_1 & x_{n2}-\bar{x}_2 & \cdots & x_{ns}-\bar{x}_s \end{bmatrix} \begin{bmatrix} z_{11}-\bar{z}_1 & z_{12}-\bar{z}_2 & \cdots & z_{1s}-\bar{z}_s \\ z_{21}-\bar{z}_1 & z_{22}-\bar{z}_2 & \cdots & z_{2s}-\bar{z}_s \\ \cdot & \cdot & \cdot & \cdot \\ \cdot & \cdot & \cdot & \cdot \\ \cdot & \cdot & \cdot & \cdot \\ z_{q1}-\bar{z}_1 & z_{q2}-\bar{z}_2 & \cdots & z_{qs}-\bar{z}_s \end{bmatrix}' \tag{4.6}$$

Y is an n by q matrix which is equal to the product of an n by s matrix X^* and a q by s matrix Z^*. It is *double-centered* because X^* is defined with respect to the coordinate system where X is *centered* at the origin $\bar{x}$ and Z^* is defined with respect to the coordinate system where Z is *centered* at the origin $\bar{z}$.

Where the double-centered matrix comes into play in the thermometers problem is that we can use singular value decomposition to get *starting coordinates* for either X or Z to use in a gradient-style solution. For example, let the singular value decomposition of Y be $U\lambda V'$ and the let the starting coordinates for X be $U\lambda^{\frac{1}{2}}$ and the starting coordinates for Z be $V\lambda^{\frac{1}{2}}$. Given these starting coordinates we can compute the d_{ij} for our standard squared error loss function:

$$\mu = \sum_{i=1}^{n}\sum_{j=1}^{q}\varepsilon_{ij}^2 = \sum_{i=1}^{n}\sum_{j=1}^{q}\left(d_{ij}^* - d_{ij}\right)^2 \tag{4.7}$$

Where, from above,

$$d_{ij} = \sqrt{\sum_{k=1}^{s}(x_{ik} - \mathrm{z_{jk}})^2} \tag{4.8}$$

4.2 Metric Unfolding Using the MLSMU6 Procedure

The MLSMU6 (Multidimensional Least-Squares Metric Unfolding) procedure was developed by Poole (1984, 1990).

Recall that x_{ik} represents the position of individual i ($i = 1,...,n$) on the kth ($k = 1,...,s$) dimension and z_{jk} represents the position of stimulus j ($j = 1,...,q$) on the kth dimension. The first derivatives of the loss function in Equation 4.7 are:

$$\frac{\partial\mu}{\partial z_{jk}} = 2\sum_{i=1}^{n}\left\{\left(\frac{d_{ij}^*}{d_{ij}} - 1\right)(x_{ik} - \mathrm{z_{jk}})\right\} \tag{4.9}$$

$$\frac{\partial\mu}{\partial x_{ik}} = -2\sum_{j=1}^{q}\left\{\left(\frac{d_{ij}^*}{d_{ij}} - 1\right)(x_{ik} - \mathrm{z_{jk}})\right\} \tag{4.10}$$

Setting Equation 4.9 equal to zero and solving for z_{jk}:

$$\hat{z}_{jk} = \frac{1}{n}\sum_{i=1}^{n}\left[x_{ik} + \frac{d_{ij}^*}{d_{ij}}(z_{jk} - \mathrm{x_{ik}})\right] \tag{4.11}$$

Continuing, setting Equation 4.10 equal to zero and solving for x_{ik}:

$$\hat{x}_{ik} = \frac{1}{q}\sum_{i=1}^{q}\left[z_{jk} + \frac{d_{ij}^*}{d_{ij}}(x_{ik} - \mathrm{z_{jk}})\right] \tag{4.12}$$

Note that the solution (Equations 4.11 and 4.12) is in the form $z = f(x,z)$ and $x = g(x,z)$. That is, the solutions for z and x are values such that when they are plugged into $f(x,z)$ and $g(x,z)$ they reproduce themselves!

Define:

$$z_{jki} = x_{ik} + \frac{d_{ij}^*}{d_{ij}}\left(z_{jk} - x_{ik}\right) \tag{4.13}$$

$$x_{ikj} = z_{jk} + \frac{d_{ij}^*}{d_{ij}}\left(x_{ik} - z_{jk}\right) \tag{4.14}$$

So that Equations 4.11 and 4.12 can be rewritten as:

$$\hat{z}_{jk} = \frac{1}{n}\sum_{i=1}^{n} z_{jki} \tag{4.15}$$

$$\hat{x}_{ik} = \frac{1}{q}\sum_{j=1}^{q} x_{ikj} \tag{4.16}$$

Using Equation 4.13, note that the point $z_{j,i}$ is:

$$z_{j,i} = \begin{bmatrix} x_{i1} + \frac{d_{ij}^*}{d_{ij}}\left(z_{j1} - x_{i1}\right) \\ x_{i2} + \frac{d_{ij}^*}{d_{ij}}\left(z_{j2} - x_{i2}\right) \\ \cdot \\ \cdot \\ \cdot \\ x_{is} + \frac{d_{ij}^*}{d_{ij}}\left(z_{js} - x_{is}\right) \end{bmatrix} = x_i + \frac{d_{ij}^*}{d_{ij}}\left(z_j - x_i\right) \tag{4.17}$$

Similarly:

$$x_{i,j} = \begin{bmatrix} z_{j1} + \frac{d_{ij}^*}{d_{ij}}\left(x_{i1} - z_{j1}\right) \\ z_{j2} + \frac{d_{ij}^*}{d_{ij}}\left(x_{i2} - z_{j2}\right) \\ \cdot \\ \cdot \\ \cdot \\ z_{js} + \frac{d_{ij}^*}{d_{ij}}\left(x_{is} - z_{js}\right) \end{bmatrix} = z_j + \frac{d_{ij}^*}{d_{ij}}\left(x_i - z_j\right) \tag{4.18}$$

Where $z_j = \begin{bmatrix} z_{j1} \\ z_{j2} \\ \cdot \\ \cdot \\ \cdot \\ z_{js} \end{bmatrix}$ and $x_i = \begin{bmatrix} x_{i1} \\ x_{i2} \\ \cdot \\ \cdot \\ \cdot \\ x_{is} \end{bmatrix}$ are points and $\frac{d_{ij}^*}{d_{ij}}$ is a scalar.

Equations 4.17 and 4.18 are basic equations of a line that passes through z_j and x_i. The general formula for a line equation is:

$$Y(t) = A + t(B - A) \tag{4.19}$$

Where A and B are points and t is a scalar. Note that if $0 < t < 1$, then Equation 4.19 defines a line that runs between points A and B.

Once specific values are plugged into Equations 4.15 and 4.16 then the solution for the point $\hat{z}_j = \begin{bmatrix} \hat{z}_{j1} \\ \hat{z}_{j2} \\ \cdot \\ \cdot \\ \cdot \\ \hat{z}_{js} \end{bmatrix}$ is simply the centroid of the n $z_{j,i}$ points and $\hat{x}_i = \begin{bmatrix} \hat{x}_{i1} \\ \hat{x}_{i2} \\ \cdot \\ \cdot \\ \cdot \\ \hat{x}_{is} \end{bmatrix}$ is simply the centroid of the q $x_{i,j}$ points!

Finally, note that the squared distance between the points z_j and $z_{j,m}$ is:

$$\begin{aligned}
\sum_{k=1}^{s} \left(z_{jk} - z_{jki}\right)^2 &= \sum_{k=1}^{s} \left[z_{jk} - \left(x_{ik} + \frac{d_{ij}^*}{d_{ij}}\left(z_{jk} - x_{ik}\right)\right)\right]^2 \\
&= \sum_{k=1}^{s} \left[\left(z_{jk} - x_{ik}\right)\left(1 - \frac{d_{ij}^*}{d_{ij}}\right)\right]^2 \\
&= \frac{\left(d_{ij} - d_{ij}^*\right)^2}{d_{ij}^2}\left(\sum_{k=1}^{s} \left(z_{jk} - x_{ik}\right)^2\right) \\
&= \varepsilon_{ij}^2
\end{aligned} \tag{4.20}$$

So that the squared error is represented directly on the surface of the s-dimensional hyperplane.

Similarly, the squared distance between the points x_i and $x_{i,j}$ is:

$$\begin{aligned}
\sum_{k=1}^{s} \left(x_{ik} - x_{ikj}\right)^2 &= \sum_{k=1}^{s} \left[x_{ik} - \left(z_{jk} + \frac{d_{ij}^*}{d_{ij}}\left(x_{ik} - z_{jk}\right)\right)\right]^2 \\
&= \sum_{k=1}^{s} \left[\left(x_{ik} - z_{jk}\right)\left(1 - \frac{d_{ij}^*}{d_{ij}}\right)\right]^2 \\
&= \frac{\left(d_{ij} - d_{ij}^*\right)^2}{d_{ij}^2}\left(\sum_{k=1}^{s} \left(x_{ik} - z_{jk}\right)^2\right) \\
&= \varepsilon_{ij}^2
\end{aligned} \tag{4.21}$$

Another interesting property is the following identities:

$$\begin{aligned} d_{ij}^{*2} &= \sum_{k=1}^{s} \left(x_{ik} - z_{jki}\right)^2 \\ &= \sum_{k=1}^{s} \left\{ x_{ik} - \left[x_{ik} + \frac{d_{ij}^*}{d_{ij}} \left(z_{jk} - x_{ik}\right) \right] \right\}^2 \\ &= \frac{d_{ij}^{*2}}{d_{ij}^2} \sum_{k=1}^{s} \left(z_{jk} - x_{ik}\right)^2 \\ &= d_{ij}^{*2} \end{aligned} \tag{4.22}$$

$$\begin{aligned} d_{ij}^{*2} &= \sum_{k=1}^{s} \left(z_{jk} - x_{ikj}\right)^2 \\ &= \sum_{k=1}^{s} \left\{ z_{jk} - \left[z_{jk} + \frac{d_{ij}^*}{d_{ij}} \left(x_{ik} - z_{jk}\right) \right] \right\}^2 \\ &= \frac{d_{ij}^{*2}}{d_{ij}^2} \sum_{k=1}^{s} \left(x_{ik} - z_{jk}\right)^2 \\ &= d_{ij}^{*2} \end{aligned} \tag{4.23}$$

Intuitively, the observed distances, the d_{ij}^*, are the lengths of the vectors *attached* to the x_i that produce the $z_{j,i}$ points. Similarly, the d_{ij}^* are the lengths of the vectors *attached* to the z_j that produce the $x_{i,j}$ points.

It is a relatively simple process to iterate back and forth between Equations 4.11 and 4.12 (equivalently, Equations 4.15 and 4.16) until the $\hat{x}$'s and $\hat{z}$'s reproduce each other.

Note that this process is *strictly descending* (Poole, 1984); that is:

$$\begin{aligned} \sum_{i=1}^{n} \varepsilon_{ij}^{(h+1)^2} &= \sum_{i=1}^{n} \sum_{k=1}^{s} \left(z_{jk}^{(h+1)} - z_{jki}^{(h+1)} \right)^2 \leq \sum_{i=1}^{n} \sum_{k=1}^{s} \left(z_{jk}^{(h+1)} - z_{jki}^{(h)} \right)^2 \\ &\leq \sum_{i=1}^{n} \sum_{k=1}^{s} \left(z_{jk}^{(h)} - z_{jki}^{(h)} \right)^2 = \sum_{i=1}^{n} \varepsilon_{ij}^{(h)^2} \end{aligned} \tag{4.24}$$

where h is the iteration number. Poole (1990) shows that the procedure almost always converges to the global minimum.

This is a useful method with interesting mathematical properties, but as a *statistical* model it is not realistic for two reasons. First, the ratio of observed distances to reproduced distances $(\frac{d_{ij}^*}{d_{ij}})$ in Equations 4.11-4.12 is undefined when $x_i = z_j$ so that $d_{ij} = 0$. Second, because distances cannot be negative, the errors cannot follow standard Gauss-Markov assumptions. As a result, the statistical properties of the error process are not clear and any assumptions about the errors are dicey at best. Nonetheless, finding the values of x_{ik} and z_{jk} that minimize the loss function (i.e., the sum of squared errors) is relatively easy with the MLSMU6 or, as we discuss shortly, SMACOF metric unfolding procedures.

4.2.1 Example 1: 1981 Interest Group Ratings of US Senators Data

We demonstrate use of the MLSMU6 procedure using interest group ratings of US Senators in 1981. Interest groups rate members of Congress on a scale based on their votes on a select number of roll calls of concern to the organization. These scales are customarily either ratio level (0-100) or based on an ordinal grading system (e.g., the National Rifle Association rates candidates and elected officials using letter grades between "A" and "F"). For our purposes, we use ratings collected from 30 interest groups that used a 0-100 scale to rate Senators in 1981 (the first year of the 97th Congress).

The data we use (stored in the object `interest1981.Rda`) are taken from a much larger data set compiled by Keith Poole containing nearly 200,000 interest group ratings of members of Congress between 1959 and 1981.† These data have been analyzed by Poole (1981, 1984, 1990) and Poole and Daniels (1985). Following this work, we perform the MLSMU6 unfolding procedure in two dimensions.

```
library(asmcjr)
data("interest1981")
```

The data `interest1981` are arranged such that the individuals (Senators) are on the rows and the stimuli (the interest groups) are on the columns. The interest group ratings themselves are stored in columns 9-38, which we assign to the matrix `input`. We delete Senators with less than five interest group rankings with the command `input <-input[rowSums(!is.na(input))>= cutoff,]`. We then transform the ratings into distances that range between 0 and 2 and square the transformed values (see Section 4.1), storing the resulting matrix in the object `input2`. Finally, we replace missing ratings in the object `input2` with the squared mean value of `input`.

```
input <- as.matrix(interest1981[,9:38])
cutoff <- 5
input <- input[rowSums(!is.na(input))>=cutoff,]
input <- (100-input)/50
input2 <- input*input
input2[is.na(input)] <- (mean(input,na.rm=TRUE))^2
```

†Data available for download from: `https://legacy.voteview.com/dwnl_5.htm`.

As in Chapter 3, we use double-centering to get good starting values for the MLSMU6 algorithm. This first requires us to program the function `doubleCenterRect()` as below. We then process the matrix of squared ratings through the function and store the double-centered matrix in the object `inputDC`.

```
inputDC <- doubleCenterRect(input2)
```

In order to get starting coordinates for the stimuli (`stims`) and individuals (`inds`), we perform singular value decomposition on the double-centered matrix with the base function `svd`. We are unfolding in two dimensions. Consequently, the first two columns of the right singular vectors multiplied by the square root of the first and second singular values (`xsvd$d`) are used as the starting values of `stims`; the first two left singular vectors (`xsvd$u`) multiplied by the square root of the first and second singular values (`xsvd$d`) are used as the starting values of `inds`.

```
xsvd <- svd(inputDC)
ndim <- 2
stims <- xsvd$v[,1:ndim]
inds <- xsvd$u[, 1:ndim]# matrix(0, nrow=n, ncol=ndim)
for (i in 1:ndim){
stims[,i] <- stims[,i] * sqrt(xsvd$d[i])
inds[,i] <- inds[,i] * sqrt(xsvd$d[i])
}
```

We provide a function in the `asmcjr` package to run the MLSMU6 procedure, including the preparatory steps identified above. As detailed above, the algorithm alternates between estimation of the x_{ik} and z_{jk} (Equations 4.11-4.12) values until convergence (that is, when the improvement in the loss function (the sum of squared errors) between subsequent iterations is less than some pre-determined tolerance, here 0.005). MLSMU6 converges on the result for the 1981 interest group data in 12 iterations, reducing the sum of squared errors to 194.132 (this information is stored in the `iter` element of the returned results from the `mlsmu6` procedure).

```
out <- mlsmu6(input = interest1981[,9:38], ndim=2, cutoff=5,
    id=factor(interest1981$party, labels=c("D", "R")))
tail(out$iter)
```

We plot the estimated configuration of Senators and 12 selected interest groups in Figure 4.1. Note that most of the interest groups are located at the edges of the space and are exterior to the Senators. The first dimension represents the liberal-conservative continuum, with the very liberal groups American Federation of Teachers and Americans for Democratic Action and the very conservative groups Conservative Coalition and the National Taxpayers' Union furthest apart on this dimension. Senators' first dimension coordinates also correlate with their first dimension (liberal-conservative) DW-NOMINATE (see Chapter 5) scores at $r = 0.937$. The results allow us to assess the relative ideological positions of interest groups; for example, that the National Federation of Independent Business is more moderate than the Chamber of Commerce. An alternative method to place legislators and interest groups in the same space is to allow the interest groups to "vote" on the roll calls used in their ratings and then to use NOMINATE scaling of the roll calls, as in Poole and Rosenthal (2007, chap. 7).

Consistent with Poole and Daniels (1985), we find another cleavage running diagonally from the top-left to the bottom-right of the space that separates Democrats from Republicans. We can imagine rotating the configuration approximately 45° counter-clockwise so that nearly all Democratic Senators are to the left of the Senate Republicans. The interest groups National Farmers Union and the Ripon Society (a liberal-Republican organization) anchor the ends of this dimension.

4.3 Metric Unfolding Using Majorization (SMACOF)

As discussed in Chapter 3, SMACOF (Scaling by MAjorizing a COmplicated Function) is an iterative technique that constructs a quadratic function that *always lies above* the loss function in Equation 4.7 (de Leeuw, 1977; de Leeuw and Heiser, 1977; de Leeuw, 1988). To see the logic, expand Equation 4.7:

$$\sum_{i=1}^{n}\sum_{j=1}^{q}\left(d_{ij}^{*}-d_{ij}\right)^{2}=\sum_{i=1}^{n}\sum_{j=1}^{q}d_{ij}^{*2}+\sum_{i=1}^{n}\sum_{j=1}^{q}d_{ij}^{2}-2\sum_{i=1}^{n}\sum_{j=1}^{q}d_{ij}^{*}d_{ij} \tag{4.25}$$

Now, let X^h and Z^h be the coordinates at the hth iteration and let X^{h+1} and Z^{h+1} be the coordinates at the $h+1^{th}$ iteration. The first term, $\sum_{i=1}^{n}\sum_{j=1}^{q}d_{ij}^{*2}$ is always a constant, at iteration $h+1$ the second term is $\sum_{i=1}^{n}\sum_{j=1}^{q}d_{ij}^{(h+1)^2}$. Because the first and second terms are fixed at iteration $h+1$, bounding the loss function boils down to finding a bound such that $d_{ij} \geq \Delta_{ij}$ for all i, j where Δ_{ij} is easily constructed. De Leeuw's (1977) solution is:

$$\Delta_{ij}^{(h+1)}=\frac{\sum_{k=1}^{s}\left(x_{ik}^{(h+1)}-z_{jk}^{(h+1)}\right)\left(x_{ik}^{(h)}-z_{jk}^{(h)}\right)}{d_{ij}^{(h)}} \tag{4.26}$$

The numerator comes from the Cauchy-Schwarz inequality, which in this case:

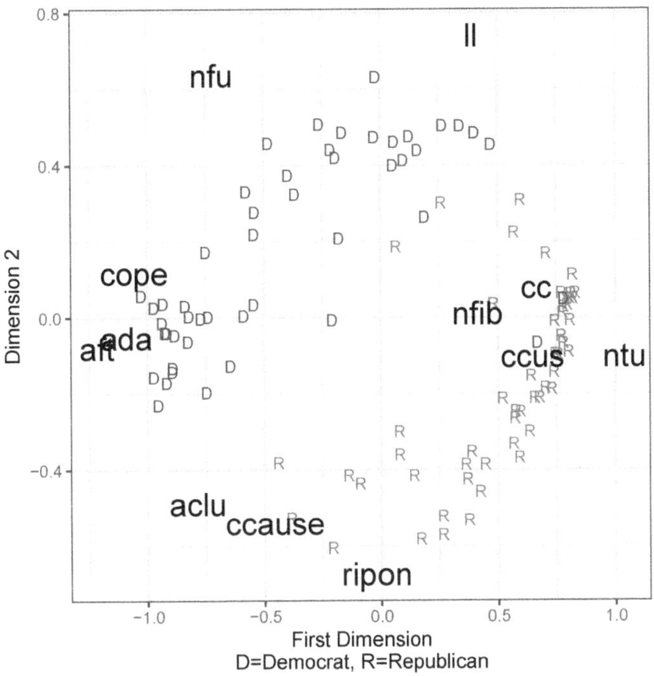

Code	Interest Group Name
ACLU	American Civil Liberties Union
ADA	Americans for Democratic Action
AFT	American Federation of Teachers
CC	Conservative Coalition (CQ)
CCAUSE	Common Cause
CCUS	Chamber of Commerce of the United States
COPE	Comittee on Political Education AFL-CIO
LL	Liberty Lobby
NFIB	National Federation of Independent Business
NFU	National Farmers Union
NTU	National Taxpayers' Union
RIPON	Ripon Society

FIGURE 4.1: Metric Unfolding of 1981 Interest Group Ratings Data Using the MLSMU6 Procedure

$$\sum_{k=1}^{s}\left(x_{ik}^{(h+1)}-z_{jk}^{(h+1)}\right)\left(x_{ik}^{(h)}-z_{jk}^{(h)}\right)\leq\left(\sum_{k=1}^{s}\left(x_{ik}^{(h+1)}-z_{jk}^{(h+1)}\right)^{2}\right)^{1/2} \tag{4.27}$$

$$\left(\sum_{k=1}^{s}\left(x_{ik}^{(h)}-z_{jk}^{(h)}\right)^{2}\right)^{1/2}=d_{ij}^{(h+1)}d_{ij}^{(h)}$$

This implies that:

$$\Delta_{ij}^{(h+1)} \leq d_{ij}^{(h+1)} \tag{4.28}$$

Note that when $X^h = X^{h+1}$ and $Z^h = Z^{h+1}$ then $\Delta_{ij}^{h+1} = d_{ij}^h$. Hence:

$$\sum_{i=1}^{n}\sum_{j=1}^{q} d_{ij}^* d_{ij}^{(h+1)} \geq \sum_{i=1}^{n}\sum_{j=1}^{q} d_{ij}^* \Delta_{ij}^{(h+1)} = \tag{4.29}$$
$$\sum_{i=1}^{n}\sum_{j=1}^{q} d_{ij}^* \frac{\sum_{k=1}^{s}\left(x_{ik}^{(h+1)} - z_{jk}^{(h+1)}\right)\left(x_{ik}^{(h)} - z_{jk}^{(h)}\right)}{d_{ij}^{(h)}}$$

This gives us the bounding function we seek:

$$\sum_{i=1}^{n}\sum_{j=1}^{q} d_{ij}^{*2} + \sum_{i=1}^{n}\sum_{j=1}^{q} d_{ij}^{(h+1)^2} - 2\sum_{i=1}^{n}\sum_{j=1}^{q} d_{ij}^* d_{ij}^{(h+1)} \leq \tag{4.30}$$
$$\sum_{i=1}^{n}\sum_{j=1}^{q} d_{ij}^{*2} + \sum_{i=1}^{n}\sum_{j=1}^{q} d_{ij}^{(h+1)^2} - 2\sum_{i=1}^{n}\sum_{j=1}^{q} d_{ij}^* \Delta_{ij}^{(h+1)}$$

Taking derivatives of the right-hand side:

$$\frac{\delta rhs}{\delta x_{ik}^{(h+1)}} = 2\sum_{j=1}^{q}\left(x_{ik}^{(h+1)} - z_{jk}^{(h+1)}\right) - 2\sum_{j=1}^{q} d_{ij}^* \frac{\left(x_{ik}^{(h)} - z_{jk}^{(h)}\right)}{d_{ij}^{(h)}} \tag{4.31}$$
$$= qx_{ik}^{(h+1)} - \sum_{j=1}^{q} z_{jk}^{(h+1)} - \sum_{j=1}^{q} d_{ij}^* \frac{\left(x_{ik}^{(h)} - z_{jk}^{(h)}\right)}{d_{ij}^{(h)}}$$

$$\frac{\delta rhs}{\delta z_{jk}^{(h+1)}} = -2\sum_{i=1}^{n}\left(x_{ik}^{(h+1)} - z_{jk}^{(h+1)}\right) + 2\sum_{i=1}^{n} d_{ij}^* \frac{\left(x_{ik}^{(h)} - z_{jk}^{(h)}\right)}{d_{ij}^{(h)}} \tag{4.32}$$
$$= nz_{jk}^{(h+1)} - \sum_{i=1}^{n} x_{ik}^{(h+1)} + \sum_{i=1}^{n} d_{ij}^* \frac{\left(x_{ik}^{(h)} - z_{jk}^{(h)}\right)}{d_{ij}^{(h)}}$$

We have $s(n+q)$ equations with $s(n+q)$ unknowns so we can solve for the minimum of the bounding function. Note that:

$$x_{ik}^{(h+1)} = \bar{z}_k^{(h+1)} + \frac{1}{q}\sum_{j=1}^{q} d_{ij}^* \frac{\left(x_{ik}^{(h)} - z_{jk}^{(h)}\right)}{d_{ij}^{(h)}} \tag{4.33}$$

$$z_{jk}^{(h+1)} = \bar{x}_k^{(h+1)} - \frac{1}{n}\sum_{i=1}^{n} d_{ij}^* \frac{\left(x_{ik}^{(h)} - z_{jk}^{(h)}\right)}{d_{ij}^{(h)}} \tag{4.34}$$

Hence, if we assume that Z^{h+1} is centered at the origin, $\bar{z}_1^{h+1} = \bar{z}_2^{h+1} = ... = \bar{z}_s^{h+1} = 0$, then we have a solution for all the x_{ik}^{h+1}. Given the x_{ik}^{h+1} we can compute the $\bar{x}_k^{h+1}$ and then the z_{jk}^{h+1}.

Another solution would be:

$$z_{jk}^{h+1} = \bar{z}_k^{h+1} + \frac{1}{nq}\sum_{i=1}^{n}\sum_{j=1}^{q} d_{ij}^* - \frac{1}{n}\sum_{i=1}^{n} d_{ij}^* \frac{(x_{ik}^{(h)} - z_{jk}^{(h)})}{d_{ij}^{(h)}} \tag{4.35}$$

4.3.1 Example 1: 2009 European Election Study (Danish Module)

Metric unfolding using the SMACOF procedure can be performed in R with the `smacofRect()` (or `unfolding()`) function in the `smacof` package (de Leeuw and Mair, 2009). In this example, we use data from the Danish module of the 2009 European Election Study (EES) to demonstrate use of the `smacofRect()` function. The EES asked 1,000 Danes to rate their propensity (how likely they would be) to vote for each of eight parties on a 0-10 point scale, with 0 denoting "not at all possible" and 10 denoting "very probable." There are only 61 missing ratings in the 1000×8 matrix.

The full data set is stored in the object `denmarkEES2009.Rda` and we assemble the propensity to vote ratings in the matrix `T`. Missing values in `T` are replaced with `NA` and the ratings in `T` are transformed to distances that range between 0 and 2. Note that in this case the ratings are on a 0-10 point scale and hence the ratings are subtracted from 10 rather than 100 and divided by 5 rather than 50. We then delete respondents from `T` who provided less than 5 party ratings.

```
library(smacof)
data(denmarkEES2009)
input.den <- as.matrix(denmarkEES2009[,c("q39_p1","q39_p2","q39_p3",
    "q39_p4","q39_p5","q39_p6","q39_p7","q39_p8")])
colnames(input.den) <- c("Social Democrats",
    "Danish Social Liberal Party", "Conservative Peoples Party",
    "Socialist Peoples Party", "Danish Peoples Party",
    "Liberal Party", "Liberal Alliance", "June Movement")
input.den[input.den == 77 | input.den == 88 |
    input.den == 89] <- NA
input.den <- (10-input.den)/5
cutoff <- 5
input.den <- input.den[rowSums(!is.na(input.den))>=cutoff,]
```

As we discussed in Section 3.1.4.1, the scaling procedures in the `smacof` package can handle missing data by manipulating the weight matrix (the `weightmat=` parameter). The weight matrix has the same dimensions as the data matrix, and values of 1 and 0, respectively, are assigned to cells in the weight matrix that correspond to non-missing and missing values in the data matrix `T`. We then replace missing values in `T` with the mean rating.

```
weightmat <- input.den
weightmat[!is.na(input.den)] <- 1
weightmat[is.na(input.den)] <- 0
input.den[is.na(input.den)] <- mean(input.den,na.rm=TRUE)
```

The `smacofRect()` function requires several arguments. First is the matrix of distances `delta=input.den`. The number of dimensions to be estimated is specified with the `ndim` parameter. The `weightmat` argument is used to assign the weight matrix. Starting values for the parameters can be provided with the argument `init` and `verbose` can be used to print out stress values during estimation. The `itmax` parameter sets the maximum number of iterations to be executed. Finally, the argument `reg` specifies a regularization factor that prevents 0 distances and `eps` specifies the convergence criterion: the minimum difference in values of the loss function (Stress) between successive iterations that constitutes convergence on the solution. The default values of the `reg` and `eps` parameters are 0.00001.

```
result <- smacofRect(delta=input.den, ndim=2, itmax=1000,
    weightmat=weightmat, init=NULL, verbose=FALSE)
```

The `smacofRect()` function returns 12 objects, which we store in the dataframe `result` and detail below. The estimated locations of the voters and parties are stored in the objects `result$conf.row` and `result$conf.col`, respectively. We flip the signs on the first dimension coordinates of both if the left-wing Social Democratic Party (the first party) has a positive first dimension score.

obsdiss Observed distances.

confdiss Estimated distances.

conf.row Estimated configuration of individuals (rows).

conf.col Estimated configuration of stimuli (columns).

stress Total stress value.

spp.row Stress per point, individuals (rows).

spp.col Stress per point, stimuli (columns).

ndim Number of dimensions estimated.

model Type of SMACOF model used.

niter Number of iterations until convergence.

nind Number of individuals (rows).

nobj Number of stimuli (columns).

```
voters <- result$conf.row
parties <- result$conf.col
if (parties[1,1] > 0){
parties[,1] <- -1 * parties[,1]
voters[,1] <- -1 * voters[,1]
}
```

Figure 4.2 shows the estimated two-dimensional configuration of Danish voters and parties from the propensity to vote ratings. What is readily apparent is that the first dimension taps into the main axis of competition between the two largest parties in Denmark: the center-left Social Democrats and the center-right Liberal Party. Indeed, it appears that the first dimension closely corresponds to the left-right ideological positions of the parties, with the left-wing Social Democrats, Socialist Peoples Party, and June Movement in a cluster on the left- and the right-wing Liberal Party, Conservative Peoples Party, and Danish Peoples Party in a cluster on the right. However, the June Movement and Danish Peoples Party are the most ideologically extreme of these parties but are not located to the far-left and far-right, respectively, on the first dimension. Instead, these minor parties are pushed outward on the second dimension such that they are exterior to nearly all of the voters. The same pattern occurs for the centrist Danish Social Liberal Party and Liberal Alliance: both are (correctly) estimated at the center of the first dimension, but are also pushed outwards towards the top of the second dimension. This result leads us to suspect that the second dimension is tapping into a viability dimension.

```
voters.dat <- data.frame(
  dim1 = voters[,1],
  dim2 = voters[,2]
)
parties.dat <- data.frame(
  dim1 = parties[,1],
  dim2 = parties[,2],
  party = factor(1:8, labels=rownames(parties))
)
g <- ggplot() +
  geom_point(data = voters.dat, aes(x=dim1, y=dim2), pch=1,
             color="gray65") +
  geom_point(data=parties.dat, aes(x=dim1, y=dim2), size=2.5) +
  geom_text(data=parties.dat, aes(x=dim1, y=dim2, label = party),
            nudge_y=.1, size=5) +
  theme_bw() +
  labs(x="First Dimension", y="Second Dimension") +
  theme(aspect.ratio=1)
g
```

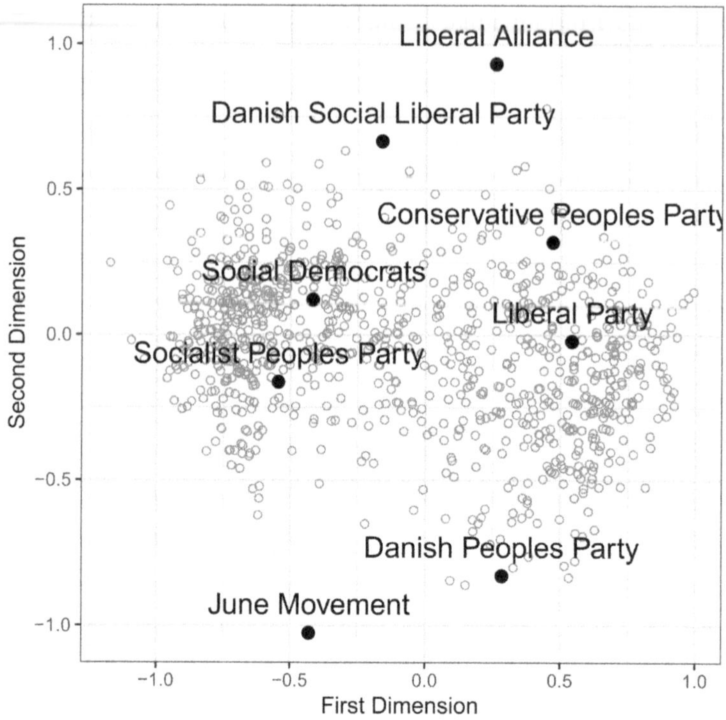

FIGURE 4.2: SMACOF (Majorization) Metric Unfolding of Propensity to Vote Ratings of Danish Political Parties (2009 European Election Study)

To investigate the substantive meaning of the dimensions, we next compare external measures of left-right ideology and viability with the scaling results. To measure left-right ideology, we run Aldrich-McKelvey scaling (Chapter 2) on respondents' left-right placements of the parties and for our viability measure we calculate the sum of respondents' propensity to vote ratings for each party. The simplest way to use these measures is to compute their correlations with the party coordinates on each dimension. Doing so with the Danish example, we find that first dimension scores and parties' left-right positions correlate at $r = 0.943$ and the propensity to vote totals correlate with the *absolute value* of the second dimension coordinates at $r = 0.952$.

As we showed with Basic Space scaling in Section 2.2.1, we can also map these external measures onto the recovered space using a method described by Poole (2005, pp. 152-155). In this procedure, a regression is performed using the external measure as the response variable and the recovered dimension(s) as the independent variable(s), and the coefficient values of the dimensions are used to calculate the normal vector of the external measure.

First we use OLS (via the `lm()` base function in `R`) to regress the Aldrich-McKelvey estimates of the parties' left-right positions on their first and second dimension scores. The R^2 is 0.894. As expected, the first dimension is the workhorse in explaining left-right variation as $\beta_1 = 0.571$ (standard error = 0.089), $\beta_2 = -0.031$ (0.057) and the intercept term is approximately 0.

```
ols <- lm(AM.result$stimuli ~ parties[,1] + parties[,2])
printCoefmat(summary(ols)$coefficients)
```

```
                Estimate  Std. Error t value  Pr(>|t|)
(Intercept)    1.4655e-14  4.8313e-02  0.0000 1.0000000
parties[, 1]   8.3691e-01  1.2175e-01  6.8743 0.0009964 ***
parties[, 2] -4.8008e-02  7.8946e-02 -0.6081 0.5696750
---
Signif. codes:
0 '***' 0.001 '**' 0.01 '*' 0.05 '.' 0.1 ' ' 1
```

We next use Equation 4.36 to calculate the normal vector of the Aldrich-McKelvey scores in the recovered two-dimensional space from the regression coefficients above (recall that the intercept term plays no role, see Poole (2005, pp. 37-40)). `N1` is 0.999 and `N2` is −0.054, meaning that the normal vector runs between the origin and the point (0.999, −0.054). Its reflection (−`N1`, −`N2`) runs between the origin and the point (−0.999, 0.054).

$$NV_k = \frac{\beta_k}{\sqrt{\beta_1^2 + \ldots + \beta_s^2}} \tag{4.36}$$

```
N1 <- ols$coefficients[2] /
    (sqrt((ols$coefficients[2]^2) + (ols$coefficients[3]^2)))
N2 <- ols$coefficients[3] /
    (sqrt((ols$coefficients[2]^2) + (ols$coefficients[3]^2)))
```

We then plot the normal vector (and its reflection) of the Aldrich-McKelvey left-right scores in Figure 4.3 with the commands below. The `exp.factor` value is simply used to expand the normal vector and its reflection beyond unit length for illustration purposes. We can now see that the projection of our external measure of parties' left-right positions closely corresponds to the first recovered dimension. This method is tremendously useful as a way to interpret and verify the substantive meaning of scaling results.

```
exp.factor <- 1.1
g +
  geom_segment(aes(x=exp.factor*-N1, y=exp.factor*-N2,
    xend=exp.factor*N1, yend=exp.factor*N2), lty=2, size=1)
```

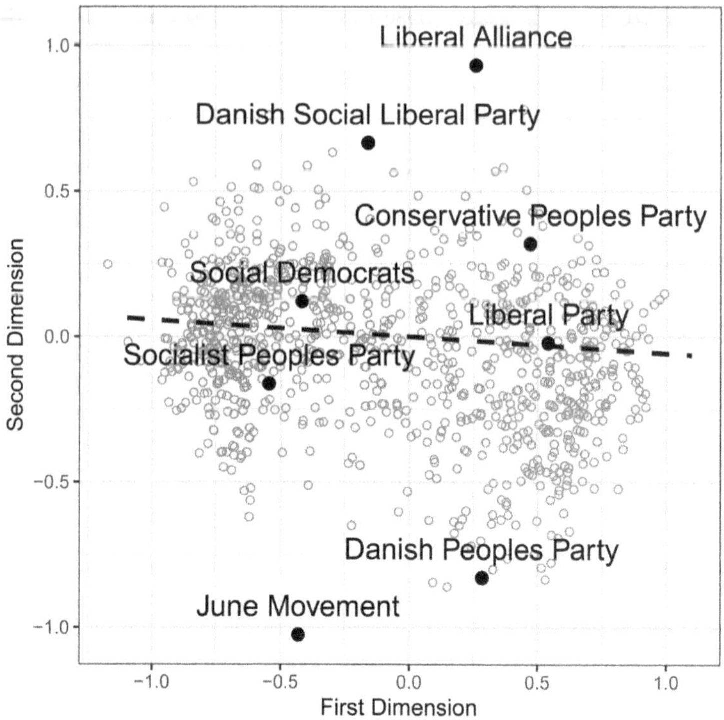

FIGURE 4.3: SMACOF (Majorization) Metric Unfolding of Propensity to Vote Ratings of Danish Political Parties (2009 European Election Study) with Normal Vector of Left-Right Scores

4.3.2 Comparing the MLSMU6 and SMACOF Metric Unfolding Procedures

Since the MLSMU6 and SMACOF procedures for metric unfolding are both least squares optimization methods, it is appropriate to ask whether they produce meaningfully different point estimates. In order to compare the two procedures, we conducted Monte Carlo tests in which the two-dimensional ideal points of 12 interest groups and 100 legislators were randomly drawn from a uniform distribution between −1 and 1. Ratings data was then generated by treating the ratings as a function of the distance between each group and legislator plus random normal error. We ran MLSMU6 and SMACOF metric unfolding on the simulated data and computed the correlations between the true and estimated interest group and legislator scores on each dimension. This process was repeated 1,000 times using four different specifications (0%, 5%, 10% and 20%) for the amount of missing data.

Both procedures perform identically well in the recovery of the true locations of the interest groups and legislators when there is no missing data. The mean correlation of the true and estimated legislator ideal points across the 1,000 trials for both dimensions is $r = 0.867$ for SMACOF and $r = 0.869$ for MLSMU6 (for the interest groups the correlations were $r = 0.848$ for SMACOF and $r = 0.849$ for MLSMU6). However, the disparity between SMACOF and MLSMU6 grows as the level of missing data increases such that when 20% of the ratings are missing, SMACOF produces mean correlations of $r = 0.853$ for the legislators and $r = 0.840$ for the interest groups, while for MLSMU6 the mean correlations are $r = 0.826$ for the legislators and $r = 0.829$ for the interest groups. With 20% missing data, the legislator first and second dimension correlations from SMACOF are higher than those from MLSMU6 in 79-80% of the trials. For the interest groups, the SMACOF correlations are higher than the MLSMU6 correlations in 67-69% of the trials.

These results make sense when considering the different ways the two procedures handle missing data. In its implementation in R, SMACOF mutes the influence of missing data by assigning 0 weights. MLSMU6, on the other hand, imputes missing data with the mean of the matrix to obtain starting coordinates. While the imputation process could be made more sophisticated, it won't be able to match the advantage that the SMACOF algorithm has in simply ignoring missing values. When there is no missing data, MLSMU6 and SMACOF will produce virtually identical point estimates because they are both effective optimization procedures and in this case will be working with the same data.

4.4 Conclusion

The least squares unfolding methods discussed in this chapter (MLSMU6 and SMACOF) are efficient and produce reasonable results. However, they have three important drawbacks as statistical methods: two methodological and the other substantive. Methodologically, these methods rely on an incorrect statistical model. Because distances are inherently positive and smaller distances should have smaller variances, standard Gauss-Markov assumptions about the error process are violated. In addition, there is no good way to get convincing standard errors from these methods. Substantively, because there are usually so many minimum and maximum ratings of the stimuli (e.g., 0 and 100 thermometer scores), least squares unfolding methods exhibit a strong tendency to push the stimuli to the edge of the space and locate them exterior to most of the individuals. This contradicts classical spatial voting theory and seems to lack face validity in many cases. For example, 1968 presidential candidate George Wallace was ideologically extreme, but do we really believe that he was more extreme than nearly *all* of his supporters in the electorate?

Among the least squares methods discussed, our Monte Carlo simulations showed that both methods utilize effective optimization procedures. However, SMACOF outpaced MLSMU6 in its recovery of the true individual and stimuli ideal points when missing data was present. For this reason and its implementation in the `smacof` package in `R`, we recommend use of the SMACOF procedure when performing least squares metric unfolding.

4.5 Exercises

1. Use the `smacofRect()`/`unfolding()` function to perform metric unfolding using SMACOF on the feeling thermometers data from the 2012 American National Election Study (ANES) stored in `ANES2012$thermometers` in the data set `ANES2012.Rda`. Respondents to the 2012 ANES were asked to rate eight stimuli (Barack Obama, Mitt Romney, Joe Biden, Paul Ryan, Hillary Clinton, George W. Bush, and the Democratic and Republican Parties) on 0-100 feeling thermometers. Missing values are coded as 999.

   (a) Write and provide code used to remove respondents who provide less than five valid thermometer ratings (or, equivalently, have more than three missing ratings). Save the new matrix as the object `thermometers`.

   (b) Write and provide code to create a weight matrix comprised of 0's that correspond to missing values in `thermometers` and 1's that correspond to non-missing values in `thermometers`. Then replace missing values in `thermometers` with values of 50.

   (c) Write and provide code that rescales the values in `thermometers` to be distances that range between 0 and 2.

   (d) Run the `smacofRect()`/`unfolding()` function on the thermometers data in two dimensions using the constructed weight matrix.

      i. How many iterations were required to reach convergence?

      ii. What is the total stress?

      iii. Which stimuli has the most stress? The least stress?

2. Plot the stimuli configuration estimated from the 2012 ANES feeling thermometers data in Exercise 1, clearly labeling the candidate/party names.

   (a) What is your interpretation of the first and second dimensions?

(b) Add labels for the estimated locations of Romney and Obama voters stored in `result$conf.row`. Note that this will require you to select only elements from `ANES2012$presvote` that correspond to respondents who provided at least five valid thermometer ratings (from Exercise 1(a)).

3. Analyze 2011 Canadian Election Study (CES) respondents' propensity to vote ratings of five parties (the Conservative Party, the Liberal Party, the NDP, the Bloc Quebecois, and the Green Party) stored in the matrix `CES2011$vote.propensity` using the `smacofRect()/unfolding()` function. The propensity ratings are on a 0 (no chance that he or she would ever vote for the party) to 10 (absolutely certain to vote for the party) point scale.

   (a) Delete respondents who provide less than three party ratings and transform the ratings into distances that range between 0 and 2. Then, create a weight matrix corresponding to the matrix of propensity ratings in which values of 1 correspond to non-missing ratings and values of 0 correspond to missing ratings.

      i. How many respondents provided at least two party ratings?

   (b) Run `smacofRect()/unfolding()` on the 2011 CES propensity ratings.

      i. Which two parties have the greatest stress?
      ii. Plot the estimated configuration of parties (`result$conf.col`) and respondents (`result$conf.row`), labeling respondents by their preferred party (note that this will require selecting only elements of `CES2011$party` that correspond to respondents who provided at least two propensity ratings).
      iii. Provide the code used to produce this plot.

4. Run `smacofRect()/unfolding()` on the 2016 ANES thermometers data (stored in the object `ANES2016$thermometers`). These data include feeling thermometer ratings of 27 stimuli (political candidates and groups) by 4,271 respondents.

   (a) First, you will need to delete rows with all missing data (i.e., respondents who did not provide any ratings). You can do so with the command: `therms[!rowSums(is.na(therms))==27,]`.

   (b) Now run the same function after adjusting the weight matrix so that extreme ratings (those of 0 or 100) have greater influence (i.e., a weight value of 10), while ambivalent ratings (those of 50) have much less influence (i.e., a weight value of 0.1).

   (c) Use a Procrustes rotation (discussed in Chapter 3) to make the two estimated configurations maximally comparable to each other.

(d) Compare and contrast the two (rotated and target) configurations. What appears to be the influence of the alternate weighting scheme?

5

Unfolding Analysis of Binary Choice Data

The spatial model of choice has been enormously successful in its application to the study of legislative behavior. In this chapter, we use the term "legislative voting," but the methods can be used to analyze behavior in other settings (e.g., judicial bodies) in which members cast a series of binary* (i.e., "Yea" or "Nay") votes.† This type of analysis begins with the assumption that legislators or other political actors such as judges or voters have ideal points in a latent (i.e., abstract) policy space and vote for the policy alternative closest to their ideal point (subject to some random error). A roll call vote, then, is understood as the product of the distance between a legislator's ideal point and a policy proposal. And, because we observe roll call votes, we can use this data to recover the locations of the legislators' ideal points and the policy alternatives in the latent space. This is the basic idea underlying spatial analyses of binary choice data.

Beginning with MacRae's (1958; 1970) path-breaking work on the structure of Congressional voting, the fields of ideal point estimation and legislative studies in particular have grown closely intertwined and have benefited from this symbiotic relationship. Scaling methods have revealed numerous insights into the structure and dynamics of legislative and judicial voting (Poole and Rosenthal, 1997, 2007; Martin and Quinn, 2007; Epstein et al., 2007), voting in historical legislative bodies like the Constitutional Convention (Dougherty and Heckelman, 2006) and the French Fourth Republic (Rosenthal and Voeten, 2004), and contemporary political phenomena like partisan polarization (McCarty, Poole and Rosenthal, 2006).‡ Likewise, legislative voting has proven a fertile ground for the development of scaling methods like NOMINATE (Poole and Rosenthal, 1985, 1997), MCMC or α-NOMINATE (Carroll et al., 2013), and Optimal Classification (Poole, 2000, 2005).

*This data can also encompass situations in which ratio-scale or rank order data is converted into a series of pairwise comparisons (i.e., the more preferred of every set of two alternatives) in order to analyze them using the unfolding methods discussed in this chapter.

†Abstention represents a third choice. True abstentions are now very rare in legislatures in the United States, but common in other chambers. We treat them as missing data in this chapter, but they may be modeled or treated as a form of Nay vote. For example, Voeten (2000) treats abstentions as equivalent to a Nay vote in the United Nations General Assembly.

‡For a broad review of legislative applications, see Carroll and Poole (2014).

Part of the reason that legislative roll call data is so widely used in the scaling community owes to its wide availability and relative abundance as a source of choice data. But more importantly, the roll call setting is seen as well-suited to the assumptions of the standard spatial model (and, more generally, the rational choice model); namely, that individuals have ordered preferences over an abstract policy space and act so as to maximize their utility (Riker and Ordeshook, 1973). Presumably, legislators are able to cast votes consistent with their instrumental goals because legislative institutions disseminate information and structure choices in an organized manner (Shepsle, 2010, chap. 12).

This chapter begins by demonstrating how to process binary choice data in `R`. We then discuss two methods to perform unfolding analysis of binary choices: NOMINATE, and Optimal Classification (OC). These methods differ in their assumptions about legislators' utility functions and the error process. NOMINATE and other parametric unfolding methods employ the random utility model developed in economics (McFadden, 1976) and treat voting probabilistically. That is, vote choices are assumed to be the product of separate deterministic and stochastic components. The deterministic portion represents the difference in utilities between the policy alternatives, wherein utility is based on the distance between the legislator's ideal point and the policy alternatives specified by a probability distribution. NOMINATE assumes Gaussian (normal) utility. The stochastic portion introduces random error and is modeled using some probability distribution. Conversely, nonparametric methods (OC) do not treat voting probabilistically and avoid parametric assumptions about the utility functions and the error process.

These methods are included in the **`wnominate`** (Poole et al., 2011), **`anominate`** (Lo et al., 2013), **`pscl`** (Jackman, 2012), **`MCMCpack`** (Martin, Quinn and Park, 2011), and **`oc`** (Poole et al., 2012) packages in `R`. The chapter closes by theoretically and practically comparing and contrasting these methods.

5.1 The Geometry of Legislative Voting

Let the two outcomes corresponding to Yea and Nay on the jth $(j = 1, ..., q)$ roll call be represented by O_{jy} and O_{jn}, respectively. In most cases it is more convenient to work with the midpoint of the two outcomes:

$$Z_j = \frac{O_{jy} + O_{jn}}{2} \tag{5.1}$$

In one dimension Z_j is known as a *cutting point* that divides the Yeas from the Nays. With perfect (errorless) spatial voting, all the legislators to the left of Z_j vote for one outcome and all the legislators to the right of Z_j vote for the opposite outcome. In two dimensions, a *cutting line* will divide the Yeas and Nays.

In one-dimensional perfect (no error) roll call voting, both the legislator and the roll call midpoints are represented by points—X_i and Z_j, respectively—and a joint rank ordering of the legislators and roll call midpoints can be found that exactly reproduces the roll call votes (Poole, 2005). In two- or more dimensional perfect voting, a legislator is still represented by a point—the s by 1 vector X_i where s is the number of dimensions—but a roll call is now represented by a plane that is perpendicular to a line joining the Yea and Nay policy points—the s by 1 vectors O_{jy} and O_{jn}—and passes through the midpoint, the s by 1 vector Z_j.

The *normal vector* to this cutting plane is unit-length vector that starts at the origin of the space and is parallel to the line joining the Yea and Nay policy points. The normal vector indicates the direction of the issue in multi-dimensional space; i.e., which dimension or dimensions are most important in modeling voting patterns on the roll call vote. Imagine a normal vector that runs along the first dimension (between the origin and the coordinate (1,0)). This means that the cutting plane, which is perpendicular to the normal vector, will divide legislators along the first dimension. Hence, we would say that the first dimension explains roll call voting in this instance, as indicated by the direction of its normal vector.

Figure 5.1 illustrates the normal vector and cutting plane of a roll call vote with perfect voting in two dimensions. In two dimensions, if a variety of voting coalitions form amongst the legislators, then the q cutting planes will criss-cross one another in several directions. The configuration of all cutting planes is known as the *Coombs mesh* (Poole, 2005, p. 32). The Coombs mesh displays the bounded regions—known as *polytopes*—created by the intersections of cutting planes. With perfect voting, a legislator is only defined up to a polytope. That is, each legislator could be anywhere in the polytope that correspondents to her roll call choices (Poole, 2005).

If error is present (which, as a practical matter, it usually is), then the problem of estimating the cutting planes is equivalent to a probit or logit analysis depending upon the assumptions made about the error. That is, legislators cast a series of binary (Yea/Nay) votes based on the distance between the latent spatial locations of their own ideal points and the Yea/Nay outcomes. The parametric methods of roll call analysis, like generalized linear models in their specification of a link function, make assumptions about the distribution of the errors in order to obtain interval-level information for the legislator and roll call parameters.

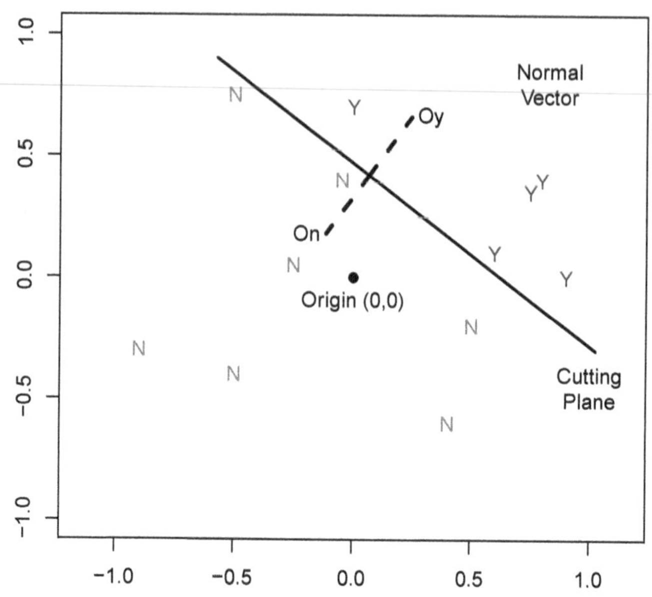

FIGURE 5.1: Normal Vector and Cutting Plane of Vote with 12 Legislators in Two Dimensions

5.2 Reading Legislative Roll Call Data into R with the pscl Package

Most of the `R` packages for the analysis of legislative roll call data discussed in this chapter require the data to be formatted as an object of class `rollcall`. `rollcall` objects are lists that separately store several objects; namely, the matrix of roll call votes (`$votes`), the legislator data (`$legis.data`), and the values that correspond to Yea, Nay, Not in Legislature, and Missing roll call votes (`$codes`). This can be done using one of two functions in the `pscl` package in `R` (Jackman, 2012). The first function, `readKH()`, is used specifically to read Congressional roll call matrices in a format (`.ord`) created by Keith T. Poole and Howard Rosenthal and convert them to `rollcall` objects. Poole and Rosenthal have compiled House and Senate roll call data sets in this format spanning the history of the US Congress which are maintained at `http://www.voteview.com`. The second function, `rollcall()`, is used to create `rollcall` objects based on numeric roll call matrices read in from various formats (e.g., Stata, Excel, or text files). See the help files for `hr108` or `rc_ep` in the `asmcjr` package to see how we used these functions to prepare the data.

Most of the `R` packages for the analysis of legislative roll call data discussed in this chapter require the data to be formatted as an object of class `rollcall`. `rollcall` objects are lists that separately store several objects; namely, the matrix of roll call votes (`$votes`), the legislator data (`$legis.data`), and the values that correspond to Yea, Nay, Not in Legislature, and Missing roll call votes (`$codes`). This can be done using one of two functions in the `pscl` package in `R` (Jackman, 2012). The first function, `readKH()`, is used specifically to read Congressional roll call matrices in a format (`.ord`) created by Keith T. Poole and Howard Rosenthal and convert them to `rollcall` objects. Poole and Rosenthal have compiled House and Senate roll call data sets in this format spanning the history of the US Congress which are maintained at `http://www.voteview.com`. The second function, `rollcall()`, is used to create `rollcall` objects based on numeric roll call matrices read in from various formats (e.g., Stata, Excel, or text files). See the help files for `hr108` or `rc_ep` in the `asmcjr` package to see how we used these functions to prepare the data.

5.3 Parametric Methods - NOMINATE

In the 1980s, Poole and Rosenthal used the random utility model developed in economics (McFadden, 1976), the spatial theory of voting, and alternating estimation methods developed in psychometrics (Chang and Carroll, 1969; Carroll and Chang, 1970; Young, de Leeuw and Takane, 1976; Takane, Young and de Leeuw, 1977) to develop NOMINATE, an unfolding method for parliamentary roll call data (Poole and Rosenthal, 1985, 1991, 1997; Poole, 2005).[§] NOMINATE is an acronym for **Nominal** **T**hree-Step **E**stimation. "Nominal" refers to the nominal (binary) nature of the data, and "three-step estimation" describes the alternating estimation procedure used to estimate the legislator ideal points, the roll call parameters, and a signal-to-noise parameter that captures the extent to which voting appears probabilistic. It is helpful to discuss each component of the NOMINATE model in turn.

[§]Their student Ladha (1991) found that ideal point estimation for roll call scaling was analogous to the item response models (IRT) used in educational testing, which are now extensively applied in political contexts. Chapter 6 provides an extensive discussion of IRT methods as they relate to the types of spatial models covered in this book.

First, NOMINATE is an unfolding method (see the discussion of unfolding in Chapter 4). Second, the NOMINATE model is based explicitly on the spatial theory of voting. Legislators have ideal points in an abstract policy space and vote for the policy alternative closest to their ideal point. Each roll call vote has two policy points - one corresponding to Yea and one to Nay. Each legislator's utility function is treated as having (1) a *deterministic* component that is a function of the distance between the legislator and a roll call outcome; and (2) a *stochastic* component that represents the idiosyncratic component of utility. The deterministic portion of the utility function is assumed to have a normal distribution. A legislator's overall utility for voting Yea is the sum of a deterministic utility and a random error. The same is true for the utility of voting Nay.

Suppose there are n legislators, q roll calls, and s dimensions indexed by $i = 1,..,n$, $j = 1,...,q$, and $k = 1,...,s$, respectively. Legislator i's utility for the Yea outcome on roll call j is:

$$U_{ijy} = u_{ijy} + \varepsilon_{ijy} \tag{5.2}$$

where u_{ijy} is the deterministic portion of the utility function and ε_{ijy} is the stochastic or random portion of the utility function. If there is no error, then the legislator votes Yea if $U_{ijy} > U_{ijn}$. Equivalently, if the difference, $U_{ijy} - U_{ijn}$, is positive, the legislator votes Yea. With random error the utility difference is:

$$U_{ijy} - U_{ijn} = u_{ijy} - u_{ijn} + \varepsilon_{ijy} - \varepsilon_{ijn} \tag{5.3}$$

So that the legislator votes Yea if:

$$u_{ijy} - u_{ijn} > \varepsilon_{ijn} - \varepsilon_{ijy} \tag{5.4}$$

That is, the legislator votes Yea if the difference in the deterministic utilities is greater than the difference between the two random errors. Since the errors are unobserved, we must make an assumption about the error distribution from which they are drawn. We can calculate the *probability* that the legislator will vote Yea or Nay. That is,

$$\begin{aligned} \mathbf{P}(\text{Legislator } i \text{ Votes Yea}) &= \mathbf{P}(U_{ijy} - U_{ijn} > 0) \\ &= \mathbf{P}(\varepsilon_{ijn} - \varepsilon_{ijy} < u_{ijy} - u_{ijn}) \\ \mathbf{P}(\text{Legislator } i \text{ Votes Nay}) &= \mathbf{P}(U_{ijy} - U_{ijn} < 0) \\ &= \mathbf{P}(\varepsilon_{ijn} - \varepsilon_{ijy} > u_{ijy} - u_{ijn}) \end{aligned} \tag{5.5}$$

So that $\mathbf{P}(\text{Yea}) + \mathbf{P}(\text{Nay}) = 1$.

The squared distance of the *ith* legislator (X_{ik}) to the Yea outcome for roll call j on the *kth* dimension (O_{jky}) is:

$$d_{ijky}^2 = (X_{ik} - O_{jky})^2 \tag{5.6}$$

In the NOMINATE model, the deterministic utility function is Gaussian (normal). The normal distribution assigns high utility to outcomes near the individual's ideal point, but utility approaches zero as the choices become more and more distant. Legislators are roughly indifferent between two extreme alternatives, but are sensitive to shifts in alternatives close to their ideal point.

With the normal distribution for the deterministic portion of utility, legislator i's utility for the Yea outcome on roll call j is:

$$U_{ijy} = u_{ijy} + \varepsilon_{ijy} = \beta e^{\left(-\frac{1}{2}\sum_{k=1}^{s} w_k^2 d_{ijky}^2\right)} + \varepsilon_{ijy} \tag{5.7}$$

where w_k are salience weights ($w_k > 0$); and because there is no natural metric β "adjusts" for the overall noise level and is proportional to the variance of the error distribution.[¶] The w_k allow the indifference curves of the utility function to be ellipses rather than circles.

In earlier versions of NOMINATE (D-NOMINATE and W-NOMINATE), the stochastic component is a random draw from the logit distribution. The logit distribution was selected because of computational limitations in the 1980s. DW-NOMINATE, which was developed in the 1990s, models the error term with the normal distribution. The normal distribution has the advantage that it guarantees that ε_{ijy} and ε_{ijn} are a random sample from a known distribution (i.e., they are independent and identically distributed (iid) errors) and that their distributions are symmetric and unimodal (Poole, 2005, p. 98).

If the stochastic portion of the utility function is normally distributed with common variance, β is proportional to $\frac{1}{\sigma^2}$, where

$$\varepsilon \sim N(0, \sigma^2) \tag{5.8}$$

Hence the probability that legislator i votes Yea on the jth roll call is:

$$\begin{aligned} P_{ijy} &= P(U_{ijy} > U_{ijn}) \qquad\qquad (5.9)\\ &= P(\varepsilon_{ijn} - \varepsilon_{ijy} < u_{ijy} - u_{ijn}) \\ &= \Phi(u_{ijy} - u_{ijn}) \\ &= \Phi\left[\beta\left(e^{\left(-\frac{1}{2}\sum_{k=1}^{s} w_k^2 d_{ijky}^2\right)} - e^{\left(-\frac{1}{2}\sum_{k=1}^{s} w_k^2 d_{ijkn}^2\right)}\right)\right] \end{aligned}$$

Let Y be the $n \times q$ matrix of observed roll call choices (y_{ij}). Given Y, NOMINATE estimates the combination of parameters for legislator ideal points

[¶]The deterministic utility function in D-NOMINATE—the original version of NOMINATE—did not include the w_k term, but all subsequent versions of NOMINATE include the w_k term and hence allow the weights to vary by dimension.

and roll call outcomes that maximizes the joint probability of the observed data (choices) (King, 1989). The likelihood function is computed over the legislators and roll calls:

$$L(u_{ijy} - u_{ijn}|Y) = \prod_{i=1}^{n}\prod_{j=1}^{q}\prod_{\tau=1}^{2} P_{ij\tau}^{C_{ij\tau}} \quad (5.10)$$

where τ is the index for Yea and Nay, $P_{ij\tau}$ is the probability of voting for choice τ as given by Equation 5.9, and $C_{ij\tau} = 1$ if the legislator's actual choice is τ, and 0 otherwise. The natural log of the likelihood function in Equation 5.10 is:

$$L = \sum_{i=1}^{n}\sum_{j=1}^{q}\sum_{\tau=1}^{2} C_{ij\tau}\, ln\, P_{ij\tau} \quad (5.11)$$

The likelihood function to be optimized, then, is a continuous distribution over the $ns + 2qs + s$ hyperplane, where s is the number of estimated dimensions. ns is the number of legislator coordinates (X_{ik}), $2qs$ is the number of roll call parameters (the locations of the Yea and Nay alternatives O_{jky} and O_{jkn}), and s is the number of the β and $w_2,...,w_s$ terms. As an example, for a hypothetical US Senate session with 100 legislators and 500 roll call votes estimated in two dimensions, this amounts to finding the maximum of a likelihood function over a 2202-dimensional hyperplane. Given the scale of such an optimization problem, the NOMINATE method uses an alternating maximum likelihood procedure. Initially, starting values for the legislator ideal points are calculated from the agreement score matrix between legislators and provisional values for the scaling constants β and w are assigned.

With these starting values, the three-step iterative estimation algorithm (hence NOMINAl Three-step Estimation) begins. First, holding the legislator ideal points and the β and w terms fixed, NOMINATE finds the values for the roll call parameters that maximize the likelihood function (the BHHH hill-climbing procedure (Berndt et al., 1974) is used in all three steps to optimize the likelihood function). Then, the likelihood function is maximized over the β and w terms while holding fixed the legislator ideal points and roll call parameters. Finally, the roll call parameters and the β and w terms are held fixed while searching for the values of the legislator ideal points that maximize the likelihood function. The process is repeated until convergence.

5.3.1 Obtaining Uncertainty Estimates with the Parametric Bootstrap

Lewis and Poole (2004) develop a method to estimate uncertainty measures for NOMINATE scores using the parametric bootstrap (Efron and Tibshirani, 1993, pp. 53-56). Recall from Equation 5.9 that NOMINATE calculates the probabilities associated with legislators' observed roll call vote choices. These values form a $n \times q$ matrix of probabilities ($\hat{P}_{ijc}$) of each legislator's vote on each roll call. The parametric bootstrap then generates a series of new roll call matrices by reproducing each observed choice (c_{ij}) with probability ($\hat{P}_{ijc}$), and inserting the opposite choice with probability (1 - ($\hat{P}_{ijc}$)). For example, if a legislator voted Yea and her probability of doing so was 0.75 (that is, 75%), then the parametric bootstrap "flips" a weighted coin that has a $\frac{3}{4}$ chance of producing a Yea vote for the sampled choice ($\hat{c}_{ij}$) and a $\frac{1}{4}$ chance of producing a Nay vote.

The parametric bootstrap repeats this process for each choice, producing a specified number (usually 1,000) of bootstrapped roll call matrices. NOMINATE is then run separately on each matrix, producing a distribution of NOMINATE scores for each legislator. The variances of these distributions are used to calculate the standard errors of the NOMINATE scores. The parametric (rather than the non-parametric) bootstrap is used so as to express the uncertainty associated with the choices themselves. So as $\hat{P}_{ijc} \to 1$, the $\hat{c}_{ij} \to c_{ij}$, so that the standard errors of the point estimates will be small for legislators who have high probabilities associated with their observed choices, and will be large for legislators whose observed choices are less likely given their position in the spatial model (Lewis and Poole, 2004, p. 109).

5.3.2 Types of NOMINATE Scores

The NOMINATE procedure has evolved over its history, and quite understandably there is often confusion about which type of estimates should be used in particular research applications (e.g., Carson et al., 2004). Below we detail the different categories of NOMINATE scores. All scores can be accessed from: `https://voteview.com/`.

1. **D-NOMINATE scores**: These are the original, two-dimensional NOMINATE scores used in *Congress: A Political-Economic History of Roll Call Voting* (Poole and Rosenthal, 1997). Each legislator's point is dynamic and is allowed to move as a linear function of time as measured by the Congress number (higher polynomials in time did not appreciably increase the fit). That is, a legislator's point is constant within a Congress but moves along a linear path between Congresses. Because of the "overlapping generations" nature of the estimation, scores in one Congress are directly comparable with scores in another Congress, although comparisons are most meaningful between Congresses occurring during one of the stable two-party periods of American history.

2. **W-NOMINATE scores**: W-NOMINATE is a static (i.e., meant to be applied to only one or possibly a few Congresses) version of D-NOMINATE, with a number of improvements being designed to increase the efficiency of the algorithm so that it can be run on a desktop personal computer. W-NOMINATE differs from D-NOMINATE in two

ways. First, it uses a slightly different deterministic utility function ($U_{ijy} = \beta e^{(-\frac{1}{2}\sum_{k=1}^{s} w_k^2 d_{ijky}^2)}$) that allows the weights to vary by dimension (the first dimension weight is set to 0.5 but then is adjusted to cancel any expansion of the space due to constraints, see the Appendix of Poole and Rosenthal (1997)). Second, because it is a static algorithm, it constrains the legislators and roll call midpoints to lie within an s-dimensional hypersphere of radius one (in contrast to the rather flexible constraint structure necessitated by the dynamic model).

3. **DW-NOMINATE scores**: DW-NOMINATE is a dynamic version of W-NOMINATE and is very similar to the original D-NOMINATE procedure. The only differences are that DW-NOMINATE is based on normally distributed errors rather than on logit errors and that each dimension has a distinct (salience) weight (the weight of the first dimension is always 1.0). DW-NOMINATE scores in one Congress are directly comparable with scores in another Congress. However, as with D-NOMINATE scores, cross-Congress comparisons are most meaningful between Congresses occurring during one of the stable two-party periods of American history. Also, the DW-NOMINATE scores cannot be compared across chambers. Finally, because these coordinates were estimated using a *weighted* utility model, if distances are computed between legislators, this weighting must be taken into account. This is not true of the original D-NOMINATE coordinates. DW-NOMINATE scores were introduced in McCarty, Poole and Rosenthal (1997) and analyzed in *Polarized America* (McCarty, Poole and Rosenthal, 2006), *Ideology and Congress* (Poole and Rosenthal, 2007), and *Political Bubbles* (McCarty, Poole and Rosenthal, 2013).[‖]

4. **Common Space DW-NOMINATE scores**: The DW-NOMINATE procedure can be applied to data covering multiple chambers, provided there is overlap between the roll call matrices. To produce "common space" scores for the US Congress, the House and Senate are treated as if they were one legislature by using the legislators who served in both chambers as "bridge" observations.[**] That is, a single ideal point is esti-

[‖]See also Nokken and Poole poole scores (Nokken and Poole, 2004) which are designed to allow the maximum amount of movement of legislators from Congress to Congress. First, the DW-NOMINATE two-dimensional constant coordinate model, that is, every legislator has the same ideal point throughout his or her career in the two-dimensional space, is run to convergence. Second, holding the roll call outcomes from the two-dimensional constant model fixed, an ideal point for every legislator in every Congress is estimated. Changes in ideal points of members from Congress to Congress can then be analyzed against the background of the fixed cutting lines.

[**]Another possible source of bridge observations is to treat interest groups as voters in both chambers because they rate members on select roll call votes. See Poole and Rosenthal (2007, chap. 8) and Gerber and Lewis (2004). See also Shor and McCarty (2011), which

mated for each member of Congress based upon his or her entire record of service in Congress. Both dimensions have equal salience weights and the scores are constrained to lie within the unit hypersphere. Presidential ideal points (beginning with President Eisenhower) are estimated by treating presidents' announced positions on legislation before Congress (using CQ Presidential Support Scores) as roll call votes (see McCarty and Poole, 1995).

Poole and Rosenthal (2007, chap. 10) survey many applications in the literature that have used scores from the NOMINATE family. While extensive, the survey is less than complete and totally ignores the many applications from more recent years.

5.3.3 Accessing DW-NOMINATE Scores

In this section, we offer an example that demonstrates both how to access DW-NOMINATE scores for the US Congress and plot smoothed histograms of legislator ideal points by party. It has been our experience that the smoothed histogram is one of the most effective ways to illustrate the ideological distribution of legislators. This is especially true when dealing with legislatures with many members, whose tokens tend to overstrike when plotting them individually. To do so, we will use the `densityplot()` function in the popular `R` graphics package `ggplot2`. DW-NOMINATE scores for the US Congress can be accessed from `http://www.voteview.com` in various formats.[††]

```
data("rcx", package="asmcjr")
```

Because DW-NOMINATE scores are comparable over time within chambers (and with Common Space scores, *across* chambers), we can assess the ideological positions of the parties over time. We examine the 90th, 100th, and 110th Houses, storing their members' first dimension scores, party, and the Congress identifier in `cong1`, `cong2`, and `cong3`. We then stack the Congresses by rows in the matrix `polarization`.

```
ncong <- c(90, 100, 110)
sub <- with(rcx, which(cong %in% ncong & dist != 0))
polarization <- cbind(rcx[sub, c("dwnom1", "party")],
    congress = factor(paste("House", rcx$cong[sub]),
    levels=c("House 90", "House 100", "House 110")))
```

---

uses the responses of state legislators to common survey items to bridge ideological estimates across states and Bailey (2007) which uses the announced policy positions of US Supreme Court Justices, members of Congress, and presidents to bridge across these institutions over time.

[††]Note that there is not an official package for the DW-NOMINATE procedure within `R`, but May, Poole and McCarty (2019) provide an interface for DW-NOMINATE that is available at https://github.com/wmay/dwnominate

```
polarization$party <- factor(polarization$party,
    levels=c(100, 200), labels=c("D", "R"))
```

The first dimension DW-NOMINATE scores (`~polarization[,1]`) are then plotted separately by Congress (`| polarization[,3]`) and within Congress by party affiliation (`groups=polarization[,2]`). Figure 5.2 shows the gradual but dramatic ideological divergence of the parties in the House of Representatives between the 90th House (1967 – 1969) and the 110th House (2007 – 2009).

```
library(ggplot2)
ggplot(polarization, aes(x=dwnom1, group=party,
    colour=party, fill=party, colour=party)) +
  geom_density(adjust=2.5, alpha=.2) +
  facet_wrap(~congress, ncol=2) +
  xlab("DW-Nominate\n(First Dimension)") +
  theme_bw() +
  scale_colour_manual(values=c("gray25", "gray75"), name="Party") +
  scale_fill_manual(values=c("gray25", "gray75"), name="Party") +
  theme(aspect.ratio=1, legend.position="bottom")
```

5.3.4 The wnominate Package in R

The W-NOMINATE procedure can be performed on legislative roll call data sets via the `wnominate` package in `R` (Poole et al., 2011). As discussed, W-NOMINATE assumes a legislator utility function ($U_{ijy} = \beta e^{(-\frac{1}{2}\sum_{k=1}^{s} w_k^2 d_{ijky}^2)}$) that allows different salience weights on the estimated dimensions. W-NOMINATE also constrains the legislators and roll call midpoints to lie within an s-dimensional hypersphere of radius one (in other words, between -1 and 1 in the one-dimensional model and within the unit circle in the two-dimensional model).

5.3.5 Example 1: The 108th US House

To demonstrate use of the `wnominate()` function, we use roll call data from the 108th US House of Representatives (2003-2005). The 108th House conducted 843 recorded roll call votes and 440 Representatives served in the chamber. We omit President George W. Bush from the roll call matrix. We showed how to read in the 108th House roll call data in Section 5.2, where it was saved as the object `hr108` in class `rollcall`. Once in this format, `hr108` can be analyzed by the `wnominate()` function.

The `wnominate()` function requires several arguments. First is the name of the `rollcall` object (`hr`). `ubeta` and `weight` are the initial values for the β and w_k terms. The default starting values of `ubeta` and `weight` are 15 and 0.5, respectively. `dim` is the number of dimensions to be estimated (following (Poole and Rosenthal, 1997), we fit two dimensions to the 108th House roll

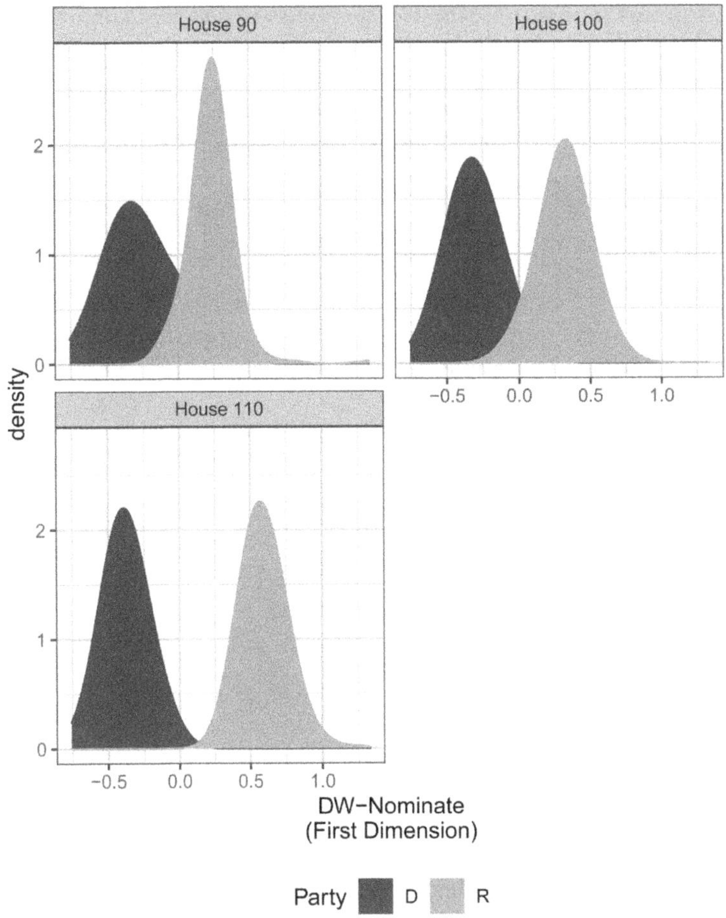

FIGURE 5.2: Partisan Polarization in the 90th, 100th, and 110th Houses

call data). `minvotes` specifies the minimum number of scalable votes that a legislator needs to cast in order to be included in the scaling (the default value is 20). `lop` is the "lopsided" threshold that excludes unanimous or lopsided roll call votes from the analysis. The default value for `lop` is 0.025 (used by Poole and Rosenthal), meaning that votes are omitted if less than 2.5% of voting legislators are in the minority.

The `wnominate()` function includes the parametric bootstrap procedure discussed in Section 5.3.2 to calculate standard errors for the point estimations. Bootstrap trials are run when `trials` is set above four (we recommend using 1,000 trials). The default value of `trials` is 3, so that the parametric bootstrap will not be run by default. Be advised that the computing time required to run the parametric bootstrap—especially with more than 1,000 trials—can be quite high.

We must also set the polarity of the estimated configuration by identifying right-wing legislators (i.e., holding positive scores) on both dimensions. To do so, we can look at `hr$legis.data` to identify the row numbers of the legislators to be constrained. In this case, we set Rep. Jo Bonner (R-AL) (#1) to have a positive score on the first dimension and Rep. Robert Cramer (D-AL) (#5), a Southern "Blue Dog" Democrat, to have a positive score on the second dimension. Of course, if uncertain about which legislators will correctly orient the space, one can always run the `wnominate` procedure to convergence and determine *post hoc* who should be constrained to hold positive scores in each of the s dimensions.

Finally, `verbose` specifies whether the names of legislators and roll calls that are omitted should be printed while the `wnominate()` function is being executed (either `TRUE` or `FALSE`). The default is `FALSE`.

```
#install.packages("wnominate")
library(wnominate)
result <- wnominate(hr108, ubeta=15, uweights=0.5, dims=2,
  minvotes=20, lop=0.025, trials=1, polarity=c(1,5), verbose=FALSE)
```

The `wnominate()` function returns eight objects that are stored in the dataframe `result`:

legislators Estimates of the legislators:

> **state, party, etc.** Legislator-specific variables stored in the subset `$legis.data` of the `rollcall` object passed to `wnominate`.
>
> **correctYea** Yea votes correctly predicted.
>
> **wrongYea** Nay votes incorrectly predicted as Yea votes.
>
> **wrongNay** Yea votes incorrectly predicted as Nay votes.
>
> **correctNay** Nay votes correctly predicted.
>
> **GMP** Geometric Mean Probability.
>
> **coord1D** First dimension W-NOMINATE score, with subsequent dimensions numbered accordingly.
>
> **se1D** If bootstrap trials run, the bootstrapped standard errors of the legislator ideal points, with subsequent dimensions numbered accordingly.
>
> **corr.1** If bootstrap trials runs, the correlation between the first and second dimension W-NOMINATE scores, with subsequent dimensions numbered accordingly.

rollcalls Estimates of the roll calls:

correctYea Yea votes correctly predicted.

wrongYea Nay votes incorrectly predicted as Yea votes.

wrongNay Yea votes incorrectly predicted as Nay votes.

correctNay Nay votes correctly predicted.

GMP Geometric Mean Probability.

PRE Proportional Reduction in Error.

spread1D First dimension spread, with subsequent dimensions numbered accordingly.

midpoint1D First dimension midpoint, with subsequent dimensions numbered accordingly (the roll call spread is added and subtracted from the roll call midpoint to calculate the Yea and Nay locations).

dimensions Number of dimensions estimated.

eigenvalues Eigenvalues of the double-centered agreement score matrix.

beta The estimated β value.

weights The estimated salience weight(s) (w_k).

fits In s dimensions, the correct classification rate, Aggregate Proportional Reduction in Error (APRE), and Geometric Mean Probability (GMP) fit statistics.

The command `summary(result)` offers an overview of the W-NOMINATE estimates. This is a useful way to quickly assess the face validity of the results; if not, there may be a problem with the format of the roll call matrix or an argument in the call to `wnominate`. Note that the first value listed for each of the fit statistics (`Correct Classification`, `APRE`, and `GMP`) is for the one-dimensional result, the second value is for the two-dimensional result.[‡‡] All three fit statistics indicate that there is considerable unidimensional structure to voting in the 108th House, with 92.1% of legislator vote choices correctly

[‡‡]The correct classification rate is simply the number of choices correctly classified divided by the total number of choices made. PRE (proportional reduction in error) is a fit statistic that measures how much improvement the model provides over classifying all votes for each roll call at the baseline (modal) category. For example, on a 65-35 vote in favor of some policy proposal, we could classify all choices as Yea votes and achieve a correct classification rate of 65%. PRE measures how much a model improves classification among the 35 Nay (minority) votes, and APRE simply aggregates across all roll calls. APRE is calculated as: $\frac{\sum_{j=1}^{q} (\text{Minority Vote} - \text{Classification Errors})_j}{\sum_{j=1}^{q} \text{Minority Vote}_j}$. Finally, W-NOMINATE calculates the probability (likelihood) associated with each observed choice. The GMP (Geometric Mean Probability) is the exponential of the average log-likelihood of all choices; that is: $e^{\frac{\ell \text{ of all observed choices}}{N}}$, where N is the total number of choices.

classified with the use of a single, ideological dimension. The addition of a second dimension provides only a minimal improvement in fit, increasing correct classification to 92.9% (the APRE values increase from 0.776 to 0.798 and the GMP values increase from 0.818 to 0.838 by moving from one dimension to two dimensions).

```
summary(result)

SUMMARY OF W-NOMINATE OBJECT
----------------------------

Number of Legislators:          440 (0 legislators deleted)
Number of Votes:      843 (375 votes deleted)
Number of Dimensions:      2
Predicted Yeas:             179490 of 191618 (93.7%) predictions correct
Predicted Nays:             145802 of 158655 (91.9%) predictions correct
Correct Classifiction:     92.09% 92.87%
APRE:                     0.776 0.798
GMP:                     0.819 0.838


The first 10 legislator estimates are:
                  coord1D coord2D
BONNER (R AL-1)     0.640  -0.134
EVERETT (R AL-2)    0.731   0.245
ROGERS (R AL-3)     0.576   0.019
ADERHOLT (R AL-4)   0.595   0.004
CRAMER (D AL-5)    -0.177   0.373
BACHUS (R AL-6)     0.606   0.062
DAVIS (D AL-7)     -0.417   0.099
YOUNG (R AK-1)      0.570  -0.057
RENZI (R AZ-1)      0.517   0.128
FRANKS (R AZ-2)     0.897   0.442
```

We next demonstrate how to graphically present the results from W-NOMINATE, first for the legislator ideal points and then for the roll calls.

5.3.5.1 Plotting Legislator Estimates

Recall from Section 5.3.2 that the W-NOMINATE procedure sets the first dimension weight equal to 0.5 (it then is adjusted to cancel any expansion of the space due to the constraints), but the weights on additional dimensions are allowed to vary. We calculate `weight` as the ratio of dimension weights from the W-Nominate routine $\left(\frac{w_2}{w_1}\right)$, which is applied to legislators' second dimension coordinates. The value of `weight` is about 0.8, meaning that the second dimension should be compressed slightly. We save the first and second-dimension legislator coordinates as objects `coord1D` and `coord2D` (multiplying the original second dimension estimates by `weight`).

In Figure 5.3 we plot legislators' first and second dimension W-NOMINATE scores. We plot the legislators separately by party and state using identifying tokens. We also divide Democrats based on whether they represent Northern or Southern (the 11 states of the Confederacy plus Kentucky and Oklahoma) states. Because legislators are constrained to lie within the unit hypersphere (forming a circle in two dimensions) in W-NOMINATE, several of the Republicans in the top-right quadrant are on the edge of the circle.

```
library(ggplot2)
group <- rep(NA, nrow(result$legislators))
group[with(result$legislators, which(partyCode == 100
    & icpsrState %in% c(40:51, 53, 54)))] <- 1
group[with(result$legislators, which(partyCode == 100
    & icpsrState %in% c(1:39, 52, 55:82)))] <- 2
group[with(result$legislators, which(partyCode == 200))] <- 3
group[with(result$legislators, which(partyCode == 328))] <- 4
group <- factor(group, levels=1:4,
    labels=c("Southern Dems", "Northern Dems",
             "Republicans", "Independents"))
plot_wnom_coords(result, shapeVar=group, dropNV=FALSE) +
  scale_color_manual(values=gray.colors(4, end=.75), name="Party Group",
  labels=c("Southern Dems", "Northern Dems",
           "Republicans", "Independents")) +
  scale_shape_manual(values=c("S", "N", "R", "I"), name="Party Group",
  labels=c("Southern Dems", "Northern Dems",
           "Republicans", "Independents")) +
  theme_bw() +
  theme(aspect.ratio=1, legend.position="bottom") +
  xlab("First Dimension") +
  ylab("Second Dimension") +
  guides(colour=guide_legend(nrow=2))
```

5.3.5.2 Plotting Roll Calls: The Partial-Birth Abortion Ban Act of 2003

We next demonstrate how to plot legislators' vote choices on an individual roll call vote. As an example, we use a vote from the 108th House of Representatives on the Partial-Birth Abortion Ban Act of 2003. The bill, signed into law by President George W. Bush in November 2003 and upheld by the Supreme Court in 2007 in *Gonzales v. Cahart*, criminalizes the class of D&X (dilation and extraction) procedures in which late-term abortions are performed by partially delivering the fetus, puncturing the skull and then removing its contents (Gordon, 2004). The final bill passed the House on October 2, 2003 in a $281-142$ vote and the Senate on October 21, 2003 in a $64-34$ vote. As suggested by the roll call margins, the issue divided Democrats and those with pro-choice views more strongly than it did Republicans (indeed, an October

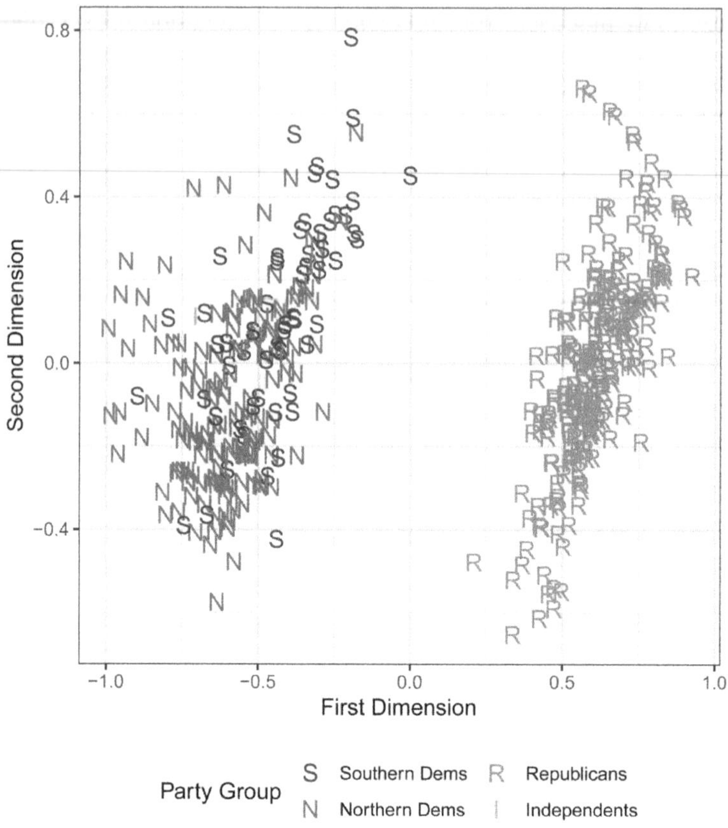

FIGURE 5.3: W-NOMINATE Ideal Point Estimates of Members of the 108th US House of Representatives

2003 Gallup poll indicated that 50% of self-identified pro-choice individuals supported a ban on late-term or "partial-birth" abortions (Saad, 2003)).

The House vote on the bill was Roll Call #528. We can use the `plot_rollcall` function from the `asmcjr` package to plot the cutting line for that particular rollcall vote.

```
weight <-  result$weights[2]/result$weights[1]
wnom.dat <- data.frame(
    coord1D = result$legislators$coord1D,
    coord2D = result$legislators$coord2D*weight,
    group=group)
pdf("spatial-rollcallplot.pdf", height=6.25, width=5)
plot_rollcall(result, hr108, wnom.dat, 528,
              wnom.dat$group, dropNV=TRUE) +
  theme_bw() +
```

```
  scale_shape_manual(values=c("S", "D", "R", "I"),
                     name="Party Group") +
  xlab("First Dimension") +
  ylab("Second Dimension") +
  theme(aspect.ratio=1, legend.position="bottom") +
  guides(shape = guide_legend(nrow=2))
dev.off()
```

For the specified roll call (`rcnum`), in the function, we capture the first and second dimension spreads and midpoints from the `result` object. The spreads are signed so that they can be subtracted from the midpoint to calculate the location of the Yea alternative and added to the midpoint to calculate the location of the Nay alternative (O_{jy} and O_{jn}, respectively, from Equation 5.1 and Figure 5.1). The coordinate (O_{1y}, O_{2y}) marks the starting point of the segment while (O_{1n}, O_{2n}) marks its endpoint. The actual values are (0.13,0.60) for the Yea outcome, (−0.92,−0.56) for the Nay outcome. Also note that the second dimension coordinates are weighted by the ratio of weights from the first and second dimensions from the `result` object.

The normal vector will be perpendicular to the cutting plane between O_{jy} and O_{jn}. However, one endpoint of the normal vector will be at the origin. To calculate the normal vector in the weighted metric, we first calculate the signed distances on each dimension by subtracting O_{jy} from O_{jn}. We then calculate the length of the normal vector as the square root of the sum of these distances squared. The end-point coordinates of the normal vector are the two distances divided by the length of the vector. Note that the normal vector is then reflected around the origin by multiplying the endpoint coordinates by −1.

With the normal vector of the cutting plane for the roll call vote in hand, we next calculate the point (x_0, y_0) on the normal vector (or its reflection) through which the cutting plane passes. These values are calculated as Mv_j, where M is the midpoint and v_j is the normal vector coordinate for dimension j. The overall midpoint is calculated by $M = \sum_{j=1}^{2} n_j m_j$ where m_j is the midpoint for each dimension. With the normal vector and cut point coordinates, it is easy to calculate the cutting plane since we know that the cut point is where the cutting plane intersects the normal vector perpendicularly. Specifically, the cutting plane is the line between the points $(x_0 + v_2, y_0 - v_1)$ and $(x_0 - v_2, y_0 + v_1)$.

Finally, we calculate and display the roll call vote counts to produce Figure 5.4, a "map" of voting patterns on this particular roll call. We can see that the vote divides the Democratic Caucus primarily along the second dimension, which represents some of the lingering regional divides in the parties. Historically, Southern Democrats and Northern and Midwestern Republicans have been more likely to defect from their party on social/cultural issues like abortion and gun control than on economic matters (although as the parties have grown more homogenous, these intra-party divisions—specifically on abortion—have faded in Congressional voting) (Poole and Rosenthal, 2007, pp. 143-144).

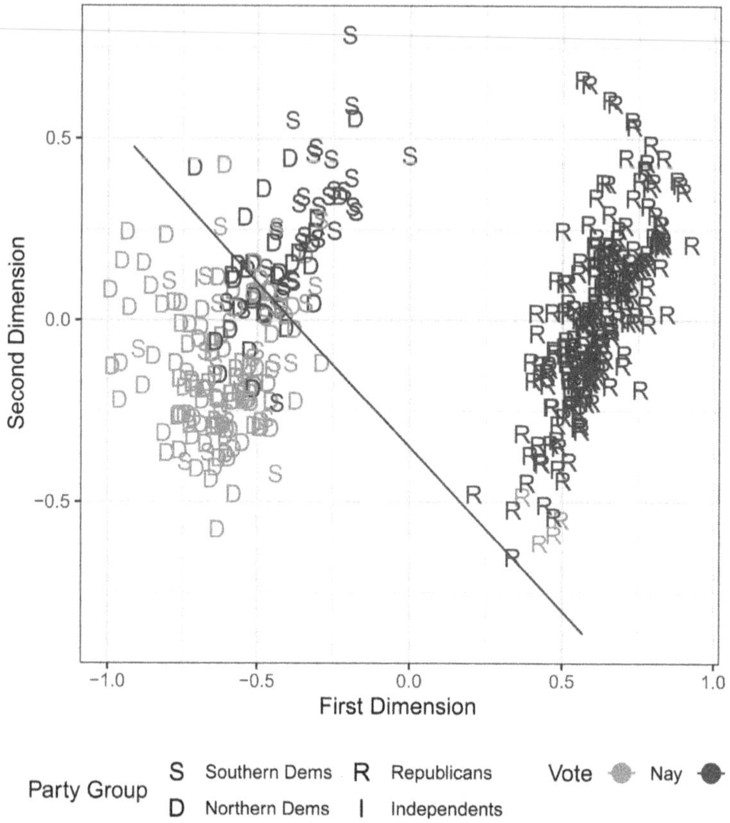

FIGURE 5.4: W-NOMINATE Analysis of the 108th House Vote on the Partial-Birth Abortion Ban Act of 2003, All Legislators

Figure 5.4 shows the roll call choices of all voting legislators, but we next want to identify those legislators who committed voting "errors" (that is, voted opposite of what they were predicted to vote based on their ideological position). We calculate `polarity` for each legislator on the roll call vote. `polarity` will be positive if the legislator's ideal point is on one side of the cutting plane, and negative if she is on the opposite side.

The identification of which sides of the cutting plane represent spatially predicted Yea and Nay votes depends on which configuration produces the least errors. From Figure 5.4, it is clear that the area to the right of the cutting line is the Yea side, and the area to the left is the Nay side.

We can calculate the PRE (Proportional Reduction in Error) statistic and display it in the plot below the number of errors as below:

```
rc.errors(result, hr108, 528)[c("tot.errors", "PRE")]
```

To plot only those legislators who committed voting errors, we can use the `plot_rollcall()` function from above. We simply specify the `onlyErrors=TRUE` option and only the errors will be plotted.

```
plot_rollcall(result, hr108, wnom.dat, 528,
              shapeVar=wnom.dat$group, onlyErrors=TRUE) +
  scale_shape_manual(values=c("S", "N", "R"), name="Party Group") +
  theme_bw() +
  theme(aspect.ratio=1, legend.position="bottom") +
  xlab("First Dimension") +
  ylab("Second Dimension") +
  guides(shape = guide_legend(nrow=2))
```

The voting errors are then isolated in Figure 5.5. This plot is valuable in showing the proximity of the errors to the cutting plane. The cutting plane, at the midpoint between the Yea and Nay alternatives, represents indifference between the two options. The closer a legislator is to the cutting line, the more ambivalent she should be about the issue. Accordingly, errors that are close to the cutting plane are treated by parametric methods like NOMINATE as less severe than those distant from the cutting plane. Figure 5.5 shows that most of the 36 errors are clustered around the cutting line, meaning that the spatial model performed well in capturing voting patterns on this roll call vote.* The PRE value (0.75) is also high.

5.3.5.3 The Coombs Mesh and Cutting Line Angles

As discussed in Section 5.1, the Coombs mesh displays the configuration of cutting planes created by a series of roll call votes. The Coombs mesh also illustrates how cutting planes intersect to form polytopes (bounded regions in the space). Every scaling method produces a Coombs mesh, although it is most central to the Optimal Classification procedure (which explicitly aims to place legislators in the polytope that maximizes the correct classification of their choices).

*When there are errors far from the cutting plane, the spatial model is not entirely appropriate for the roll call. This can happen when there is "both ends against the middle" voting. Such voting is rare in the US Congress (Poole and Rosenthal, 2007).

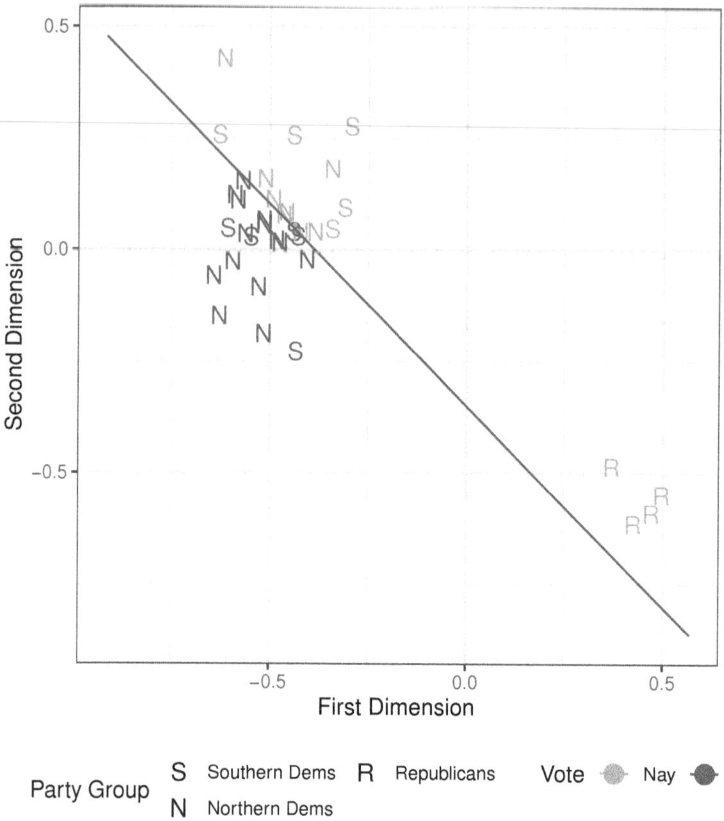

FIGURE 5.5: W-NOMINATE Analysis of the 108th House Vote on the Partial-Birth Abortion Ban Act of 2003, Errors Only

Nonetheless, the Coombs mesh is also a useful tool for assessing the dimensionality of roll call voting and identifying voting coalitions in the legislature. For example, if most cutting planes are perpendicular to the first dimension, then this would suggest that roll call voting is primarily one-dimensional. Likewise, cutting planes may seem to repeatedly divide a certain group of legislators, perhaps within a party. Cutting planes that are not purely horizontal or vertical indicate a mix of the dimensions.

To plot the Coombs mesh, we need to calculate the normal vectors as in Section 5.3.5.2. However, for the Coombs mesh, we need the normal vectors of *all* roll call votes. In cases where there are many (i.e., hundreds) of roll calls, it may be necessary to plot only a sample of cutting lines for graphical purposes. Below, we select and plot 100 random (without replacement) roll calls in the partial Coombs mesh in Figure 5.6. Figure 5.6 shows that most of the sampled cutting lines run vertically (near 90°), especially in a manner that divides the parties. Cutting lines that internally divide the parties along the first dimension (vertical) or the second dimension (horizontal) are very much in the minority in the 108th House. We can use the `plot.cutlines()` function from the `wnominate` package to generate the figure.

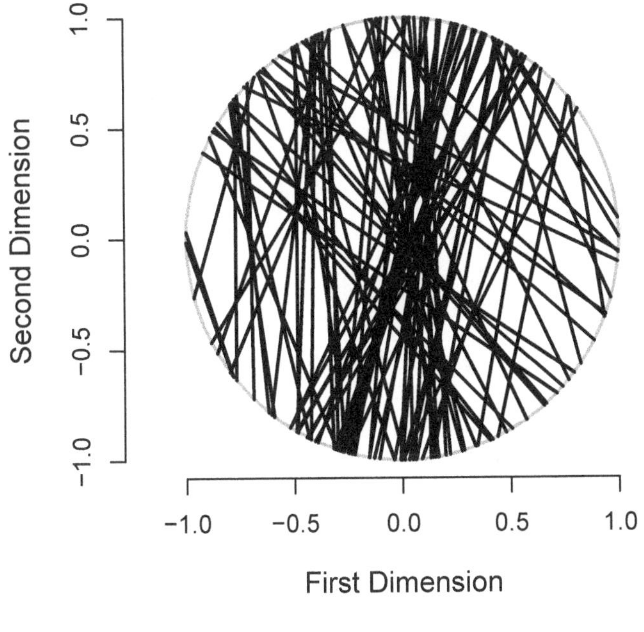

FIGURE 5.6: Partial Coombs Mesh of the 108th US House

Another way to quantify roll call cutting lines is to calculate their angle off the x-axis. In two dimensions, this means that the angle of a perfectly horizontal cutting line would be 0°, a perfectly vertical cutting line would be 90°, a diagonal cutting line running from the bottom left to the top right would be 45°, and a diagonal cutting line running from the top left to the bottom right would be −45°.

The commands below calculate the cutting line angles of all scaled roll calls in the 108th House. (`c1, c2`) is a point perpendicular to the normal vector. Its sign is flipped if `c1` is negative to ensure than the angle will range between $-90°$ and $90°$ (since it is calculated based on the location of the cutting line in the northeast and southeast quadrants). Positive angles indicate that the cutting line runs from the bottom left to the top right, while cutting lines with negative angles run from the top left to the bottom right. To calculate the cutting line angles, we initially set $c_{i1} = v_{i2}$ (the normal vector coordinate for the second dimension) and $c_{i2} = -v_{i1}$ (the negative of the normal vector coordinate for dimension 1), where i indexes the rollcall vote. Then, if $c_{i1} < 0$, both c_{i1} and c_{i2} are multiplied by -1. We then define $\theta = \text{atan2}(c_{i1}, c_{i2})$, the angle between the x-axis and the vector that emanates from the origin to (c_{i1}, c_{i2}). We convert from radian to degrees by multiplying θ by $\frac{180}{\pi}$. The function **`makeCutlineAngles()`** produces the cutline angles for all rollcall votes. This returns a data frame that adds the θ in degrees and the two normal vector coordinate values to the `rollcalls` element of the `results` object.

```
angles <- makeCutlineAngles(result)
head(angles)
```

```
     angle       N1W        N2W
1 69.68510 0.9377987 -0.3471796
2 69.61763 0.9373892 -0.3482837
3 71.37551 0.9476320 -0.3193645
4 70.77895 0.9442555 -0.3292135
5 87.45190 0.9990112 -0.0444581
6       NA        NA         NA
```

Below we print the cutting line angle and normal vectors for the roll call vote in Figures 5.4 and 5.5. The cutting line angle is about $-42°$, indicating that both dimensions are important in modeling legislator choices on this roll call vote.

```
print(angles[528,])
```

```
        angle       N1W       N2W
528 -42.06908 0.6700262 0.7423375
```

The average cutting line also provides a useful summary statistic of the dimensionality of roll call voting in a legislature. Below we find that 73.3% of cutting lines angles in the 108th House have an absolute value greater than $60°$, suggesting that voting was primarily one-dimensional across roll calls.

```
mean(abs(angles$angle), na.rm=T)
```

```
[1] 68.61649
```

```
mean(abs(angles$angle) > 60, na.rm=T)
```

```
[1] 0.7330961
```

Finally, we can also show the frequency of cutting line angles with a histogram, as in Figure 5.7 below. Figure 5.7 further supports the conclusion that voting in the 108th House was primarily one-dimensional, as most cutting line angles are near $-90°$ or $90°$. Very few roll call votes have cutting line angles around $0°$, which would indicate a split along the second dimension.

```
ggplot(angles, aes(x=angle)) +
    geom_histogram() +
    theme_bw() +
    theme(aspect.ratio=1) +
    xlab("Cutline Angles")
```

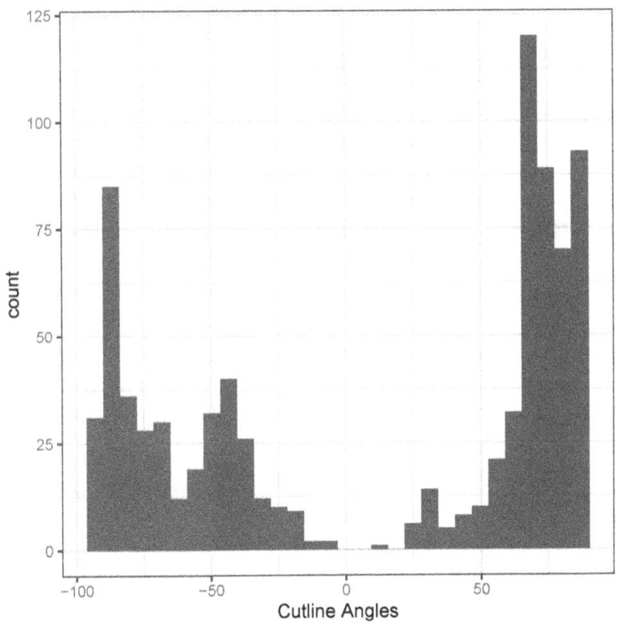

FIGURE 5.7: Histogram of Cutting Line Angles of Roll Call Votes in the 108th US House

5.3.6 Example 2: The First European Parliament (Using the Parametric Bootstrap)

We demonstrate use of the parametric bootstrap in the `wnominate()` function with roll call data from the First European Parliament (1979-1984) collected and analyzed by Hix, Noury and Roland (2006). This data set was loaded in Section 5.2 as an example of how to read non-`.ord` roll call files using the `rollcall()` function in the `pscl` package. The data is stored in the object `rc`.

Hix, Noury and Roland (2006) use W-NOMINATE to analyze roll call voting in the first five European Parliaments (spanning 1979 – 2001), finding two main dimensions underlying the data: the standard left-right dimension and a dimension representing positions on European integration. We use `wnominate` to recover and plot the ideal points of legislators (MEPs) who served in the First European Parliament in Figure 5.8. We use Neil Balfour (row #25), a member of the British Conservatives and allies party group, as the rightward stimuli on both dimensions. We plot only MEPs of five party groups—British Conservatives and allies, French Gaullists and allies, Liberals, the Radical Left, and Socialists—for purposes of space. Note that rimming occurs for several of the legislators, a consequence of the unit hypersphere (circle in two dimensions) constraint on legislator ideal points in W-NOMINATE.

The parametric bootstrap is executed in the `wnominate()` function by setting the argument `trials` to an integer greater than 3. We use 101 bootstrap trials to obtain standard errors for the ideal point estimates.

```
result <- wnominate(rc_ep, ubeta=15, uweights=0.5, dims=2, minvotes=20,
    lop=0.025, trials=101, polarity=c(25,25), verbose=FALSE)
```

We can then use the `plot_wnom_coords()` function to plot the coordinates.

One of Hix, Noury and Roland's (2006, p. 499) results is that the British Conservatives moved abruptly from the top (pro-) on the second (European integration) dimension in the First and Second European Parliaments to the bottom (anti-) on the second dimension in the Fifth European Parliament. This accords with British Conservatives' shift in position on European integration in this period. Suppose we wish to test whether the difference between British Conservatives and a consistently anti-European integration party group (e.g., French Gaullists and allies) on the second dimension is statistically significant in the First European Parliament. We can construct confidence intervals around the ideal point estimates using the bootstrapped standard errors to make these kinds of inferences.

The `plot_wnom_coords` allows us to optionally plot confidence ellipses. First, we want to identify those observations that are either Conservatives or Gaullists. Then, we set all of the other party labels to `NA`. Using `dropNA=TRUE` in our code then removes all of those parties from the plot. Setting `ci=TRUE` will plot confidence ellipses for all of the included legislators. The `level` argument allows you to control the confidence level of the ellipse. The ellipse uses a bivariate normal density to define the confidence regions.

We plot 95% confidence intervals around the ideal points of MEPs who are members of the British Conservative and French Gaullist party groups in

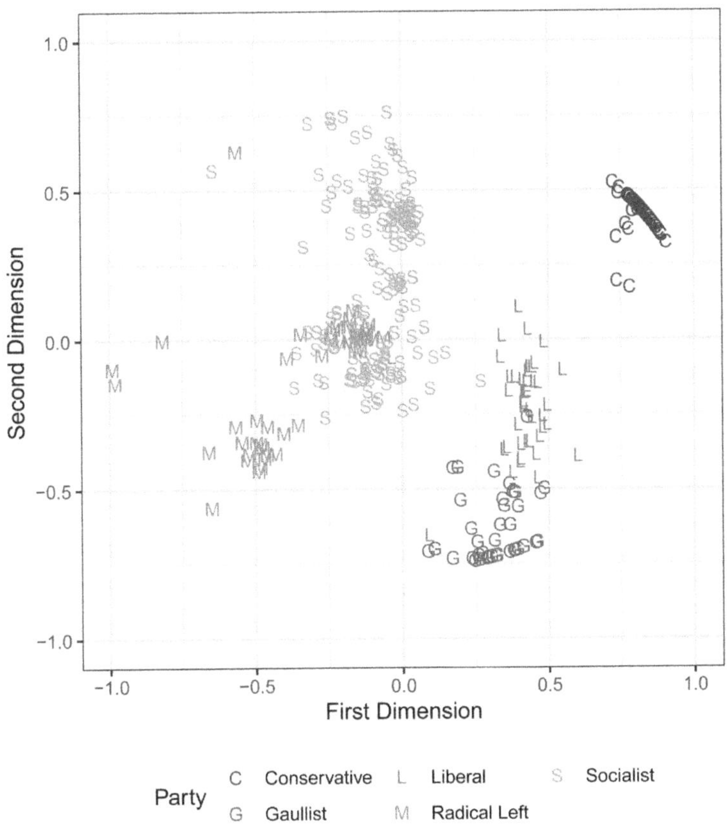

FIGURE 5.8: W-NOMINATE Ideal Point Estimates of Members of the First European Parliament

Figure 5.9. We do so by multiplying the bootstrapped standard errors by 1.96 on each dimension. We also draw an ellipse around the confidence intervals if the first and second dimension ideal point coordinates correlate at greater than 0.30. The results show no overlap between British Conservative and French Gaullist MEPs on the second dimension, indicating that the difference between the two parties on European integration was indeed statistically significant during the First European Parliament.

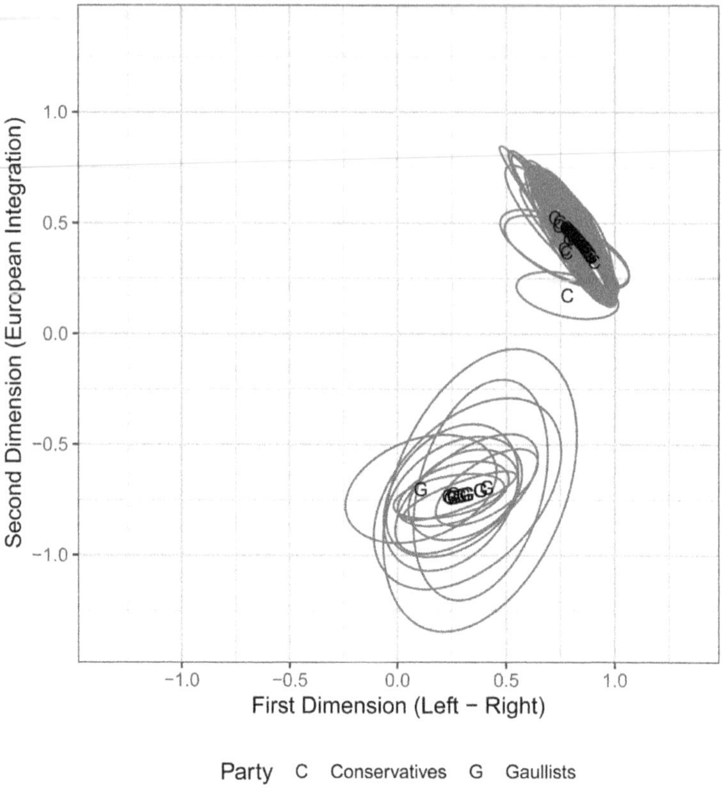

FIGURE 5.9: W-NOMINATE Ideal Point Estimates and Confidence Intervals of British Conservative and French Gaullist Members of the First European Parliament

5.4 Nonparametric Methods - Optimal Classification

Optimal Classification (OC) is a nonparametric unfolding technique developed by Poole (2000, 2005). OC is nonparametric in that it does not make strong parametric assumptions about the functional form of the error distribution or legislators' utility functions. Instead, it assumes only that the latent space is Euclidian, individuals vote sincerely and maintain the same policy positions throughout the duration of the legislature, and that legislators' utility functions are single-peaked and symmetric. As an unfolding procedure, OC uses legislative choice data to recover the locations of both legislators and policy alternatives in latent ideological space.

Unlike parametric methods such as NOMINATE, OC is not designed to find the parameters that maximize the joint likelihood of the observed legislative choice data. The goal of OC is to estimate the configuration of legislator ideal points and roll call cutting planes to maximize the correct classification of the choices themselves. Equivalently, OC seeks to *minimize* the number of classification errors, meaning that all errors are treated equally. The parametric and nonparametric approaches to ideal point estimation may—and often do—produce virtually identical results. But their results may also differ considerably as a consequence of the trade-off between stringent parametric assumptions and precise estimation of the parameters. For example, when the error rate is low, Rosenthal and Voeten (2004) show that there can be important differences between the results from OC and W-NOMINATE. We discuss the trade-offs between parametric and nonparametric methods in the conclusion of this chapter.

OC is executed in three steps. First, a starting configuration of the legislator coordinates is generated through an eigenvalue/eigenvector decomposition of the double-centered agreement score matrix. Second, from this configuration, the cutting plane procedure uses an iterative process to position cutting planes on each roll call vote such that the number of classification errors is minimized. Finally, the legislator procedure then locates the polytope for each voter which maximizes the correct classification of the choices. This produces the best available configuration of legislator ideal points and cutting planes in a space of specified dimensionality.

5.4.1 The oc Package in R

The `oc` package in `R` (Poole et al., 2012) facilitates the use of OC on legislative roll call data sets. It also contains several ancillary functions (`plot.OCcoords`, `plot.OCangles`, `plot.OCcutlines`, and `plot.OCscree`) for the analysis of the output from the `oc()` function.

5.4.2 Example 1: The French National Assembly during the Fourth Republic

We demonstrate use of the `oc()` function with roll call data from the National Assembly of the French Fourth Republic (1946-1958). The French Fourth Republic was established after World War II and lasted 12 turbulent years until its dissolution following the Algerian crisis in 1958. The roll call data from its three legislative sessions was assembled by Duncan MacRae, Jr., in the French Representation Study, 1946-1958 (ICPSR Study #52) and analyzed in MacRae (1967). The data set is not exhaustive: it includes all 739 roll calls in the *L'Annèe Politique* and a random sample of 50 roll calls from each of the three legislatures, producing 889 total votes.

Rosenthal and Voeten (2004) analyzed this roll call data using OC and have made the data and extensive documentation available online (`http://www9.georgetown.edu/faculty/ev42/france.htm`). We adopt their approach of estimating a party-switcher model in which a separate ideal point (1,416 in total) is estimated each time a legislator changes party affiliation. However, Rosenthal and Voeten (2004) also find that the latent ideological space is stable over the course of the French Fourth Republic. Hence, we do not segment the roll call data by legislative session.

We load the roll call data below. The first through fifth columns of `france4` are legislator-specific variables. In order, they are `NAME` (deputy name), `MID` (deputy ID, constant if the deputy switches party), `CASEID` (deputy ID, not duplicated if the deputy switches party), `PAR` (party affiliation), and `PARSEQ` (sequence of party affiliation for party-switching deputies). The sixth through final columns are the roll call votes. Because the names of party-switching deputies are included more than once, we append the `france4$CASEID` variable to `france4$NAME` for the `legis.names` argument in order to prevent duplicate legislator names.

```
data("france4", package="asmcjr")
rc <- rollcall(data=france4[,6:ncol(france4)],
    yea=1,
    nay=6,
    missing=7,
    notInLegis=c(8,9),
    legis.names=paste(france4$NAME,france4$CASEID,sep=""),
    vote.names=colnames(france4[6:ncol(france4)]),
    legis.data=france4[,2:5],
    vote.data=NULL,
    desc="National Assembly of the French Fourth Republic")
```

The `oc()` function requires six arguments also required by the `wnominate()` function. The first argument is the roll call matrix (`rc`) of class `rollcall`. The number of dimensions to be estimated is set with the `dims` argument. Legislators are not scaled if they cast fewer valid roll call votes than specified by the parameter `minvotes`. The default value of `lop` (0.025) omits roll call votes from the analysis if fewer than 2.5% of voting legislators are in the minority. `polarity` identifies the row number of the legislator(s) whose score is constrained to be positive on each dimension. We use Deputy Ahnne (#2), a member of the anti-tax Poujadist Party, as the right-leaning constraint on both dimensions. Finally, `verbose` specifies whether the identifiers of legislators and roll calls that are omitted should be printed while the `oc` function is being executed (either `TRUE` or `FALSE`).

```
#install.packages('oc')
library(oc)
result2 <- oc(rc, dims=2, minvotes=20, lop=0.025,
    polarity=c(2,2), verbose=FALSE)
```

The `oc()` function returns five objects that are stored in the dataframe `result`:

> **legislators** Estimates of the legislators:
>
> > **state, party, etc.** Legislator-specific variables stored in the subset `$legis.data` of the `rollcall` object passed to `oc`.
> >
> > **correctYea** Yea votes correctly predicted.
> >
> > **wrongYea** Nay votes incorrectly predicted as Yea votes.
> >
> > **wrongNay** Yea votes incorrectly predicted as Nay votes.
> >
> > **correctNay** Nay votes correctly predicted.
> >
> > **volume** Measure of the legislator's polytope size.
> >
> > **coord1D** First dimension OC score, with subsequent dimensions numbered accordingly.
> >
> > **rank** The rank-order of the legislators, only applicable if one dimension is estimated.
>
> **rollcalls** Estimates of the roll calls:
>
> > **correctYea** Yea votes correctly predicted.
> >
> > **wrongYea** Nay votes incorrectly predicted as Yea votes.
> >
> > **wrongNay** Yea votes incorrectly predicted as Nay votes.
> >
> > **correctNay** Nay votes correctly predicted.
> >
> > **PRE** Proportional Reduction in Error.
> >
> > **normvector1D** First dimension location of the unit normal vector, with subsequent dimensions numbered accordingly.
> >
> > **midpoints** The point on the normal vector through which the cutting plane passes.
>
> **dimensions** Number of dimensions estimated.
>
> **eigenvalues** Eigenvalues of the double-centered agreement score matrix.
>
> **fits** In s dimensions, the correct classification rate and the Aggregate Proportional Reduction in Error (APRE).

The command `summary(result)` offers a quick overview of the OC estimates. Note that unlike the `wnominate()` function, when multiple dimensions are estimated with `oc`, fit statistics are not provided for lower-dimensional solutions. To assess the dimensionality of roll call data with OC, one can run `oc` with different dimensional configurations and then compare their fit statistics. Below, after estimating one-dimensional (`result1`) and two-dimensional (`result2`) solutions, we evaluate their correct classification rates and APRE values stored in `$fits`. A one-dimensional model performs well, correctly classifying 93.3% of choices and producing an APRE of 0.812. The two-dimensional solution improves the correct classification rate to 96.6% and boosts the APRE value to 0.904. As do Rosenthal and Voeten (2004), we proceed with the two-dimensional configuration.

```
summary(result2)

SUMMARY OF OPTIMAL CLASSIFICATION OBJECT
----------------------------


Number of Legislators:   1454 (112 legislators deleted)
Number of Votes:        855 (34 votes deleted)
Number of Dimensions:        2
Predicted Yeas:              245634 of 252555 (97.3%) predictions correct
Predicted Nays:              204941 of 214120 (95.7%) predictions correct

The first 10 legislator estimates are:
            coord1D coord2D
Abelin1       0.179  -0.426
Ahnne2        0.282   0.571
Airoldi3      0.511  -0.010
Ait Ali4     -0.957   0.000
Aku5         -0.269  -0.226
Alduy6       -0.018   0.254
Allion7       0.511  -0.010
Alliot8       0.631   0.002
Alliot1197    0.103   0.037
Alliot1198       NA      NA

result1 <- oc(rc, dims=1, minvotes=20, lop=0.025,
    polarity=2, verbose=FALSE)

fits <- cbind(result1$fits, result2$fits)
colnames(fits) <- c("1 Dim", "2 Dim")
rownames(fits) <- c("% Correct", "APRE")
fits

              1 Dim     2 Dim
% Correct 0.9320169 0.9655006
APRE      0.8103793 0.9037732
```

5.4.2.1 Plotting Legislator Estimates

We plot the OC deputy ideal points (`result$legislators$coord1D` and `result$legislators$coord2D`) in Figure 5.10. Rosenthal and Voeten's (2004) substantive interpretation of the recovered space is that there are two dimensions that run diagonally: one running from the bottom left to the top right that represents the left-right continuum, and a dimension running from the top left to the bottom right that represents pro/anti-regime position. We isolate only deputies who are members of the four parties that occupy each of these quadrants: Communists (left, anti-regime), Socialists (left, pro-regime), Christian Democrats (MRP) (right, pro-regime), and Poujadists (right, anti-regime).

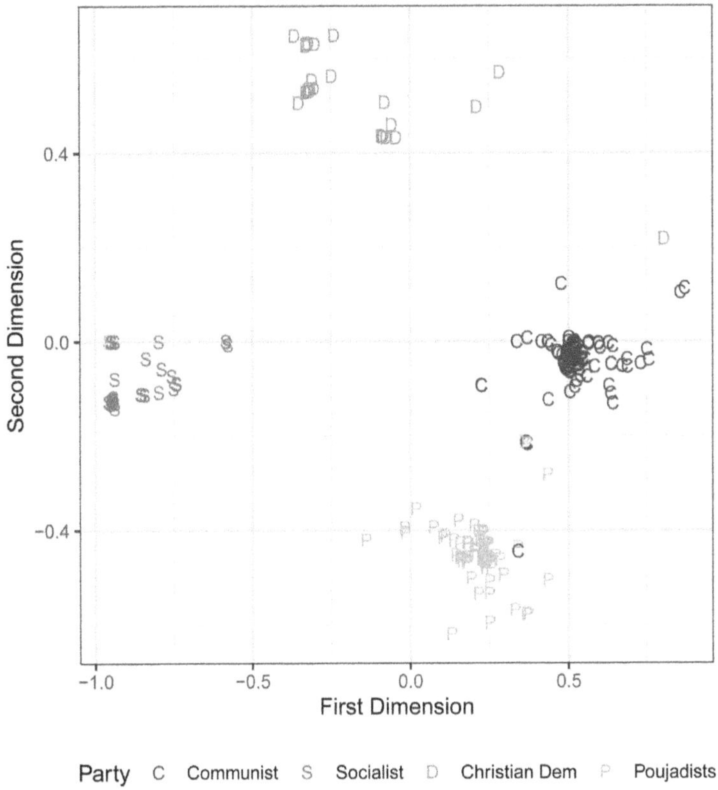

FIGURE 5.10: Optimal Classification Ideal Point Estimates of Deputies of the French Fourth Republic

In order to rotate the configuration 45 degrees clockwise so that the axes represent the two substantive dimensions, we apply a rotation matrix to the OC result. The rotation matrix A takes the form:

$$A = \begin{bmatrix} \cos\theta & -\sin\theta \\ \sin\theta & \cos\theta \end{bmatrix} \tag{5.12}$$

We apply the rotation matrix where $\theta = 45^{\circ}$ to the legislator ideal points and roll call normal vectors below. Matrix multiplication is performed in R with the command \%*\%. The OC solution is rotated by multiplying the A matrix by the two-dimensional coordinates of each ideal point and normal vector. The trigonometric functions in R arguments in radians. We can convert from degrees to radians by multiplying the angle in degrees by $\frac{\pi}{180}$. Below, we compute the rotation matrix A and pass that to the `plot_oc_coords` function, which uses it to rotate the coordinates and then plots them as in Figure 5.11. The two substantive dimensions now match the geometric axes.

```
deg2rad <- function(x)x*pi/180
rad45 <- deg2rad(45)
A <- matrix(c(cos(rad45), -sin(rad45),
              sin(rad45), cos(rad45)),
            nrow=2, ncol=2, byrow=TRUE)
```

To use some of the plotting functions, we create a copy of the object `result` and store that copy in `res`. We then replace the original variables `coord1D`, `coord2D` in `res$legislators`, and the 6th and 7th columns of the `res$rollcalls` element with their rotated counterparts calculated above. Then, passing `res` instead of `result` to the plotting functions will plot the rotated points.

```
pb <- rc$legis.data$PAR
pb <- car::recode(pb, '1="Communitsts"; 2="Socialists";
     5="Christian Dems"; 7 = "Poujadists"; else=NA',
     as.factor=TRUE)
plot_oc_coords(result, pb, dropNV=TRUE, ptSize=3, rotMat=A) +
    theme_bw() +
    coord_fixed() +
    scale_shape_manual(values=c("C", "S", "D", "P"),
        labels= c("Communist", "Socialist",
                  "Christian Dem", "Poujadists"),
        name = "Party") +
    scale_color_manual(values=gray.colors(5),
        labels= c("Communist", "Socialist",
                  "Christian Dem", "Poujadists"),
        name = "Party")+
  labs(x="First Dimension", y="Second Dimension") +
  theme(aspect.ratio=1, legend.position="bottom")
```

5.4.2.2 Plotting Roll Calls: Regulation of Labor Conflicts

To demonstrate how to plot roll call votes with the `oc()` function, we use as an example the Third Legislature's vote on a measure involving the regulation of labor conflicts in the public sector. The vote was held on 6 February 1957 and failed by a $222-306$ margin. The vote ID in the Rosenthal and Voeten (2004)

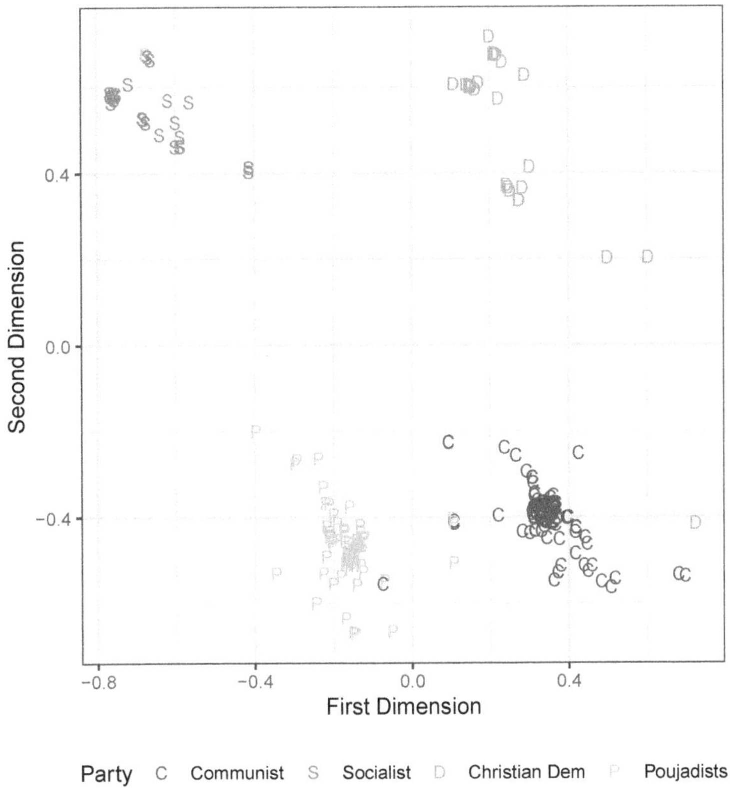

FIGURE 5.11: Optimal Classification Ideal Point Estimates of Deputies of the French Fourth Republic (with Rotation)

database is: *V*3090. However, because not all roll calls are included in the data set, we must find the number of the column in which the vote is stored. We do this by searching the column names in `rc$votes` for `V3090`, extracting the column number, and storing this value in the object `nrollcall`. We then store the deputy roll call votes in the object `vote`.

```
which(colnames(rc$votes)=="V3090")
```

```
[1] 807
```

We use the `plot_oc_rollcall` function to plot the normal vector and the votes. The cutting line is plotted in Figure 5.12. As would be expected for a roll call vote involving government intervention in the economy, Figure 5.12 indicates that the vote pitted deputies primarily along the first (left-right) dimension. Note the intra-party split among French Socialists that is captured by the cutting line.

```
vote <- rc$votes[, "V3090"]
plot_oc_rollcall(result, rc, shapeVar=pb, 807, dropNV=TRUE,
                 ptSize=3, onlyErrors=FALSE)  +
  theme_bw() +
  theme(aspect.ratio=1, legend.position="bottom") +
  xlim(-1.1,1.1) + ylim(-1.1,1.1) +
  scale_shape_manual(values=c("D", "C", "P", "S"),
      labels=c("Christian Dems", "Communists",
               "Poujadists", "Socialists"),
      name = "Party") +
  xlab("First Dimension (Left-Right)") +
  ylab("Second Dimension (Pro / Anti-Regime") +
  guides(shape=guide_legend(nrow=2)) +
  annotate("text", label=paste0("Yea = ",
    sum(vote %in% rc$codes$yea)), colour="gray33", x=-.9, y=1) +
  annotate("text", label=paste0("Nay = ",
    sum(vote %in% rc$codes$nay)), colour="gray67", x=-.9, y=.9) +
  annotate("text", label="Predicted\nYea",
    colour="gray33", x=.1, y=.2 ) +
  annotate("text", label="Predicted\nNay",
    colour="gray67", x=-.6, y=-.5 )
```

We next identify and plot the legislators who committed voting errors.

```
rc.errors(result, rc, 807)[c("tot.errors", "PRE")]
```

```
$tot.errors
yeaerror nayerror
       3       74

$PRE
[1] 0.6515837
```

Figure 5.13 isolates the voting errors on the roll call vote. OC incorrectly classifies all of the Christian Democrat (MRP) deputies, who were unanimously against the measure, but only misclassifies the Socialist Yea votes who are in close proximity to the cutting line. The vote's PRE value of 0.65, although low compared to many other roll call votes in the French Fourth Republic, is fairly high in absolute terms.

5.4.3 Example 2: 2008 American National Election Study Feeling Thermometers Data

Though our focus to this point has been on cases in which legislators or judges cast binary votes, the methods discussed in this chapter can also be used for

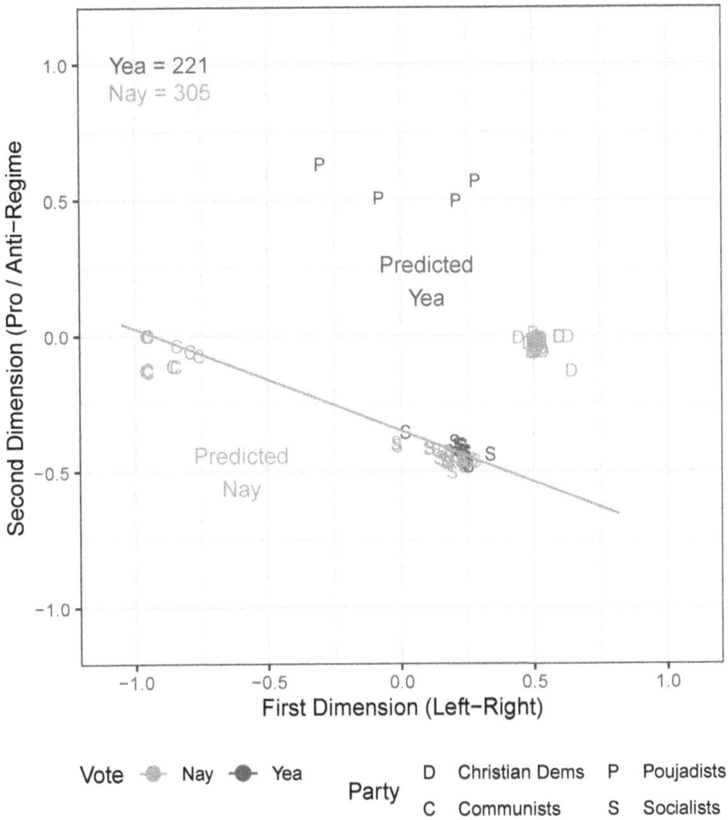

FIGURE 5.12: Optimal Classification Analysis of the French Fourth Republic Vote on the Regulation of Labor Conflicts, All Legislators

the analysis of public opinion survey data. OC is particularly attractive in this context because it avoids the pitfall of making parametric assumptions about the choices of survey respondents. As discussed in Chapter 2, respondents can vary considerably in their level of political knowledge, the structure of their preference functions, and how they understand and answer survey questions (Brady, 1985; Alvarez, 1997; King et al., 2004). One way to ameliorate this problem is to proceed with the use of nonparametric methods. Another is to treat the data as providing ordinal-level (*rankings*) information rather than interval-level (*ratings*) information Brady (1990).

In this section, we demonstrate how OC can be used to analyze feeling ther-

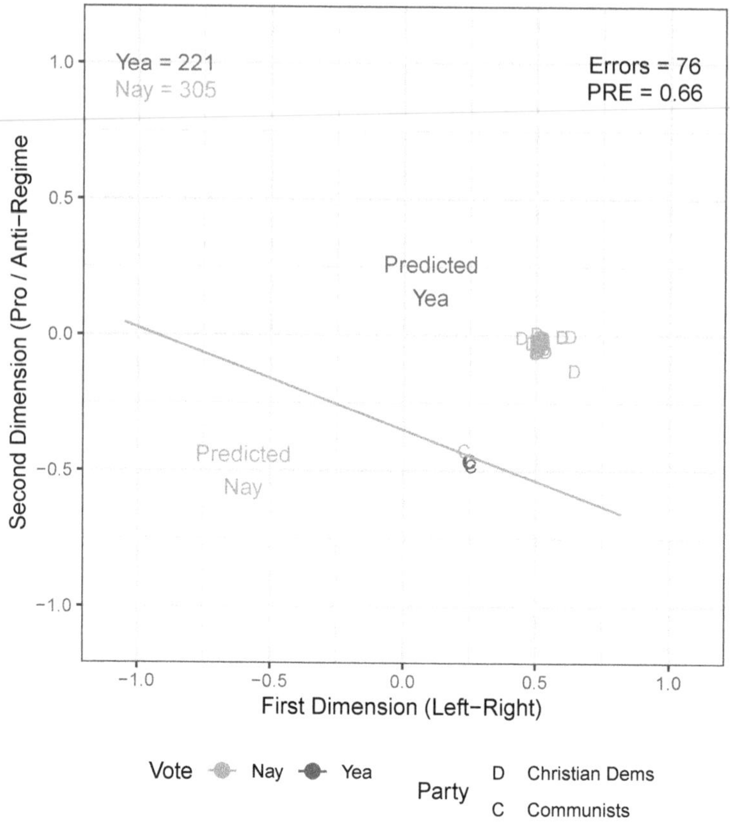

FIGURE 5.13: Optimal Classification Analysis of the French Fourth Republic Vote on the Regulation of Labor Conflicts, Errors Only

mometers data and estimate the latent positions of candidates and respondents in a common space. The feelings thermometers data are transformed into binary roll calls by creating a series of pairwise comparisons between each pair of stimuli (candidates). This produces $\frac{q(q-1)}{2}$ pairwise comparisons, where q is the number of stimuli. Respondents "vote" on each comparison by rating one candidate more favorably than the other on the thermometer scale. For example, in a comparison between Al Gore and George W. Bush, respondents would be coded Yea if they rate Gore more favorably than Bush, Nay if they rate Bush more favorably than Gore, and missing if they rate them equally or do not rate both stimuli. This produces a cutting plane that divides Gore from Bush. This form of binary choice data can be analyzed with any of the methods discussed in this chapter, but we prefer OC because of its nonparametric properties.

Of course, this method can also be used to convert rank orders into a series of binary choices.† Rank order data includes individuals' rankings of a set of alternatives from "most preferred/important" to "least preferred/important." For example, the Major Party Activist Studies (Abramowitz et al., 2001) have asked party caucus attenders and convention delegates to rank the candidates seeking their party's presidential nomination. Party activists may like *all* of the candidates, but what is of interest is how voters or activists decide between several acceptable alternatives (see, e.g., Brady (1990)). Jacoby (2006, 2013) makes the same point in his work on value hierarchies and political behavior: rankings of values (e.g., "is freedom more or less important than economic security to you?") are far better measures of personal value *structures* or *hierarchies* than ratings of values (e.g., "how important is freedom to you?"). Value hierarchies are influential when individuals must decide between mutually desirable goals, as is the case with most public policy debates.

As an example of this procedure, we convert and analyze feeling thermometers data from the 2008 American National Election Study (ANES). The 2008 ANES asked respondents to rate their favorability towards nine political stimuli on a 0-100 scale: Sen. John McCain (`mccain`), Pres. George W. Bush (`bush`), Sen. Barack Obama (`obama`), Sen. Joe Biden (`biden`), Gov. Sarah Palin (`palin`), Sen. Hillary Clinton (`hclinton`), former Pres. Bill Clinton (`bclinton`), the Democratic Party (`demparty`), and the Republican Party (`repparty`). The data is stored in the matrix `candidatetherms2008` shown below. Missing values are coded as `NA`.

```
data("candidatetherms2008", package="asmcjr")
print(candidatetherms2008[1:5,1:5])
```

```
     mccain bush obama biden palin
[1,]     60   85    25    15    90
[2,]     70   60     0    50    70
[3,]     60   85    30    50    70
[4,]     70   50    55    NA    75
[5,]     70   70    30    50    85
```

Below we program a function (`binary.comparisons()`) that creates a series of binary comparisons for each pair of stimuli from a rectangular matrix of

†Coombs' original unfolding model dealt with the analysis of rank order preference data, and this model was later applied to the analysis of binary choice data (e.g., the NOMINATE model). We would actually be working backwards in this instance (i.e., converting rank order data to binary choice format), but note that the same information is contained in a complete set of pairwise binary comparisons between the alternatives and the OC algorithm has desirable properties lacking alternative unfolding methods (i.e., that is nonparametric and minimizes the number of classification errors).

preferential choice data. Missing values in the original matrix should be coded as `NA`, as they are in the 2008 ANES data. The `binary.comparisons()` function uses the `combn()` base function in `R` to generate all of the stimuli pairs (1 vs. 2, 1 vs. 3, 2 vs.3, etc.). Values of 1 are inserted if the first stimuli in the pair is ranked higher, values of 6 are inserted if the second stimuli is ranked higher, and values of 9 are inserted if both ratings are equal or if one or both of the ratings are missing.[‡] Column names are created by combining the stimuli names with an underscore (e.g., `mccain_obama`) and the resulting object of binary comparisons is returned.

Below we run the `binary.comparisons()` function on the 2008 data and store the new matrix in the object `T`. We print the first few rows and columns of `T` to compare them with the corresponding raw data in `candidatetherms2008`. Note, for example, that the first respondent rated McCain lower than Bush (60 vs. 85), so the value for `mccain_bush` in the first row is 6, meaning the second stimuli was preferred. The fifth respondent rated McCain and Bush equally (70), so on the fifth row the value for `mccain_bush` is 9, or missing. The same is true for the fourth respondent's `mccain_biden` value: it is also 9 because she did not rate Biden.

Because tied ratings (a frequent occurrence with feeling thermometers) are treated as missing, the proportion of missing data is fairly high in the `T` matrix. 27.6% of the comparisons are missing with an average of 10 missing comparisons per respondent. However, Poole (2000, Appendix A4)[§] reports the results of Monte Carlo tests that show OC performs very well even at high rates of missing data. Only once the proportion of missing entries reaches about 70% does the performance of the OC algorithm begin to deteriorate in the recovery of the true legislator ideal points. This is another reason why we prefer to use OC when analyzing public opinion data.

```
X <- binary.comparisons(candidatetherms2008)
print(X[1:5,1:4])
```

```
     mccain_bush mccain_obama mccain_biden mccain_palin
[1,]           6            1            1            6
[2,]           1            1            1            9
[3,]           6            1            1            6
[4,]           1            1            9            6
[5,]           9            1            1            6
```

The `T` matrix is then converted to the `rollcall` object `ANES08` below.

[‡]Our choice of 1, 6 and 9 values is based on the default coding system in the `rollcall()` function used to convert matrices to `rollcall` objects that can be analyzed by the `oc()`, `wnominate()` or `ideal()` functions. Of course, alternate values can be chosen for Yea, Nay and Missing votes.

[§]Available for download at: `https://legacy.voteview.com/pdf/paapp2.pdf`.

```
ANES08 <- rollcall(data=X, yea=1, nay=6, missing=9, notInLegis=0,
    legis.names=NULL, vote.names=colnames(T), legis.data=NULL,
    vote.data=NULL, desc="2008 American National Election Study")
```

The binary feeling thermometer data (`ANES08`) can now be analyzed by the `oc()` function. Respondent #3 (who rates the Republican stimuli most highly) is set as the conservative constraint on both dimensions. As with the first example, we run `oc()` separately in one and two dimensions to compare their fit statistics.

```
result1 <- oc(ANES08, dims=1, minvotes=20, lop=0.005,
    polarity=3, verbose=FALSE)
result2 <- oc(ANES08, dims=2, minvotes=20, lop=0.005,
    polarity=c(3,3), verbose=FALSE)
```

The OC fit statistics (the correct classification rate and APRE) are quite high in both dimensions given the "noise" usually associated with public opinion data. These results support the argument—first articulated by Weisberg and Rusk (1970)—that candidate evaluations are structured in low-dimensional space. The first dimension is most important, correctly classifying 87.7% of respondents' binary stimuli preferences with a moderately high APRE value of 0.651. The two-dimensional model, however, offers a definite improvement with a correct classification rate of 92.3% and an APRE value of 0.783.

```
fits <- cbind(result1$fits, result2$fits)
colnames(fits) <- c("1 Dim", "2 Dim")
rownames(fits) <- c("% Correct", "APRE")
fits
```

```
              1 Dim     2 Dim
% Correct 0.8768592 0.9231842
APRE      0.6513286 0.7824972
```

By inspecting the angles of the cutting lines of the stimuli comparisons (`theta4`, computed the same way as in Section 5.3.5.3), it is clear that the first dimension divides conservative and Republican stimuli from liberal and Democratic stimuli, while the second dimension is more important in modeling intra-party divisions (for example, the cutting line between Hillary Clinton and Barack Obama has an angle of about -6°). We also store the first and second dimension respondent coordinates (`oc1` and `oc2`), the normal vectors (`norm.vec.1` and `norm.vec.2`), and the cut point (`xpoint`,`ypoint`) below.

```
angles <- makeCutlineAngles(result2)
angles$comparison <- colnames(X)
print(angles[18, c("comparison", "angle")])
```

```
      comparison     angle
18 obama_hclinton -5.619652
```

In this example, we want to estimate John McCain's location by plotting only the cutting lines for the comparisons between him and the eight other stimuli. These are stored in the first eight pairwise comparisons. The function `plot.cutlines()` does this using the same method as described above. A preference for McCain is denoted as a Yea vote in each comparison, and so we need to determine whether the Yea alternative is above or below the cutting line on each vote. As before, the function identifies whether the Yea vote is above or below the line by considering whether `kerrors12` is greater than `kerrors34`, in which case Yea is above the cutting line; if `kerrors34` is greater than `kerrors12`, then Yea is below the cutting line. We also include arrows that indicate the direction of Yea (McCain) for each cutting line.

Figure 5.14 plots the eight cutting lines that bound the McCain polytope: a triangular region labeled with the letter "M". This polytope serves not only as the estimate of McCain's location, but is also the optimal polytope for a respondent ranking John McCain more favorably than all other stimuli (that is, the one that would minimize her classification errors).

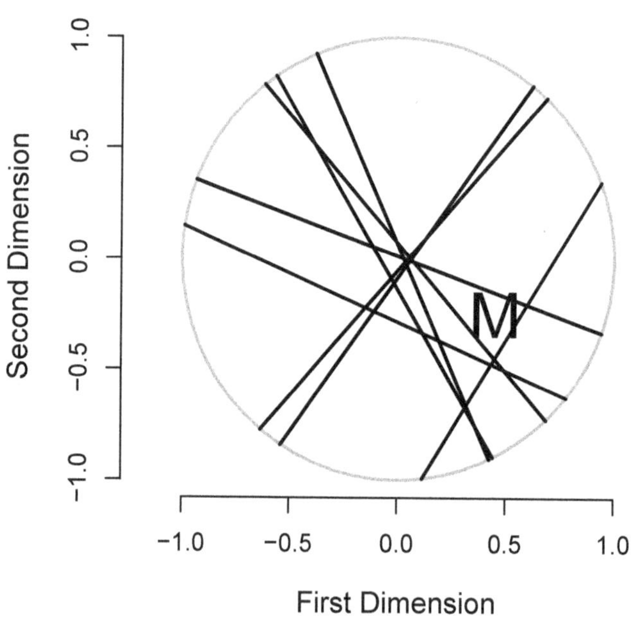

FIGURE 5.14: Optimal Classification Estimated Location of John McCain from Feeling Thermometer Rankings

Based on their stimuli rankings, OC also estimates coordinates for the respondents in the same space. Figure 5.15 plots the locations of McCain voters using the values from the variable **presvote2008**, in which a McCain vote is coded as 0 and an Obama voted coded as 1. Figure 5.15 shows that McCain is located near the center of the distribution of his supporters on the first, primary dimension. This result stands in contrast to past scaling work which has shown a tendency to push candidates to the edges of the latent space (see Poole and Rosenthal, 1984; Bakker and Poole, 2013). McCain is located closer to the exterior of his supporters on the second dimension, but this is not terribly meaningful as this dimension mostly represents intra-party divisions and the vote choice is between candidates separated on the first dimension. Hence, a respondent who rated Sarah Palin more favorably than John McCain would nonetheless be likely to vote for John McCain over Barack Obama.

```
data("presvote2008", package="asmcjr")
pts <- data.frame(
    oc1 = result2$legislators[,"coord1D"],
    oc2 = result2$legislators[,"coord2D"],
    vote = presvote2008[,1])
pts <- pts[which(pts$vote == 0), ]
```

5.5 Conclusion: Comparing Methods for the Analysis of Legislative Roll Call Data

In this chapter, we have demonstrated how to use two methods—NOMINATE and Optimal Classification—for the analysis of legislative (binary) roll call data and documented the theory underlying each procedure. We have thus far only tangentially addressed the more practical concern of which method is likely to be most appropriate for the analysis of different types of roll call data sets. In this section, we review the main differences between the two estimation procedures with a specific focus on how these differences manifest themselves when dealing with different legislative contexts. Our aim is to provide the reader some usable criteria to arbitrate between these methods given the nature of their data.

First, the parametric method NOMINATE treat errors differently than the nonparametric OC procedure. Namely, NOMINATE is more dependent on voting errors in the sense that they base their estimates on specific assumptions about the errors: that the errors are iid (independent and identically distributed), normally distributed, and less likely to occur as the distance from the cutting plane (or difficulty parameter) of the roll call vote increases. Conversely, OC treats all errors equally, and depends on the errors only in the

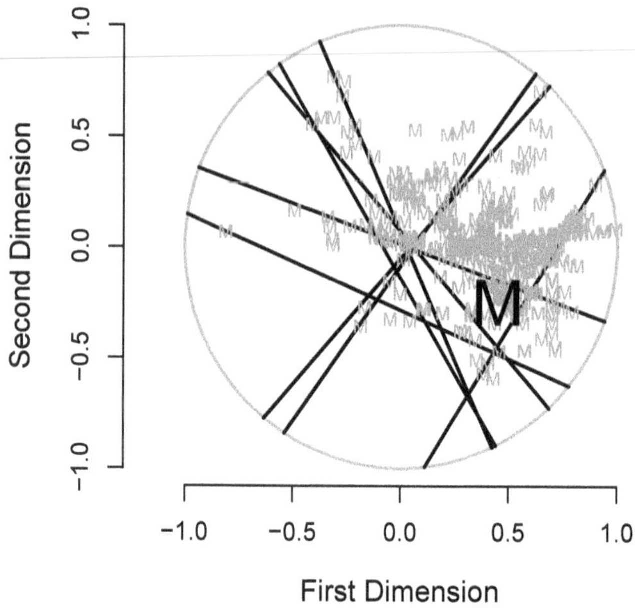

FIGURE 5.15: Optimal Classification Estimated Locations of John McCain and McCain Voters from Feeling Thermometer Rankings

sense that it aims to minimize the total number of errors in its classification of legislator choices. Hence, OC is preferable in contexts with low error rates and in situations where the distribution of the error is unknown.

For instance, Rosenthal and Voeten (2004) selected OC for their analysis of the French Fourth Republican based on unstable characteristics of the National Assembly during this period: variation in party cohesion, frequent party switching, proxy voting and strategic voting. These features make the parametric reliance on legislators' voting errors to recover their ideal points troublesome. Moreover, because roll call voting in the French Fourth Republic was strongly ideological, the spatial model, ironically, fits the data *too* well. When the error rate is small, parametric methods push ideal points to the edges of the space in order to maximize the log likelihood.

Below, we run W-NOMINATE (as an example of a parametric method) for the French Fourth Republic data and compare the estimates with those from the OC analysis (in Figure 5.11). The first and second dimension W-NOMINATE and OC coordinates are stored in the objects `wnom1`, `wnom2`, `oc1` and `oc2` (recall that the OC coordinates are rotated 45°). To rotate the **wnominate** estimates to the target matrix of `oc` coordinates to compare the two sets of results (as do Rosenthal and Voeten (2004)), we return to the `procrustes()` function in the **MCMCpack** package discussed in Chapter 3. We first remove missing data from the W-NOMINATE and OC coordinates and store them in the objects `wnom.comp` and `oc.comp`. Recall that in the `procrustes()` function, the `X` argument denotes the matrix to be rotated, the `Xstar` argument denotes the target matrix and the `translation` and `dilation` arguments denote whether the transformed matrix should be translated or dilated (by default, `FALSE`).

```
result.oc <- oc(rc, dims=2, minvotes=20, lop=0.025,
                polarity=c(2,2), verbose=FALSE)
result.wnom <- wnominate(rc, ubeta=15, uweights=0.5,
                         dims=2, minvotes=20, lop=0.025,
                         trials=3, polarity=c(2,2), verbose=FALSE)


library(MCMCpack)
oc.comp <- na.omit(cbind(oc1, oc2))
wnom.comp <- na.omit(cbind(wnom1, wnom2))
proc <- procrustes(X=wnom.comp, Xstar=oc.comp, translation=FALSE,
    dilation=FALSE)
```

Next, we pull together the results from both the rotated W-NOMINATE and the OC models. We attach party labels to the coordinates and then remove any missing data. With this data, we can plot both of the point configurations together as in Figure 5.16.

```
wnom1.new <- proc$X.new[,1]
wnom2.new <- proc$X.new[,2]
party <- rc$legis.data$PAR[!is.na(wnom1)]
plot.dat <- data.frame(
    coord1D = c(wnom1.new, oc.comp[,1]),
    coord2D = c(wnom2.new, oc.comp[,2]),
    party = rep(party, 2),
    model = factor(rep(c(1,2), each=length(wnom1.new)),
        labels=c("W-NOMINATE", "OC"))
)
plot.dat$party <- factor(plot.dat$party,
    levels=c(1,2,5,7), labels=c(
        "Communist", "Socialist", "Christian Dem", "Poujadist"))
plot.dat <- na.omit(plot.dat)
```

The rotated W-NOMINATE ideal points for the French Fourth Republic are then plotted in Figure 5.16 along with the OC configuration. Rosenthal and Voeten (2004, pp. 626-627) note several implausible features of the W-NOMINATE results. First, several parties (e.g., the Communists, Socialists, and Poujadists) are pushed to the edges of the latent space.[¶] Rosenthal and Voeten (2004) conclude that this is more a product of pushing deputies with few voting errors as far away from the cutting lines as possible than ideological extremity of the parties. As a result, W-NOMINATE leaves large regions of the space empty. Finally, the substantive meaning of the dimensions (left-right and pro-/anti-regime) are less clear in the W-NOMINATE estimates. Hence, in legislative contexts like the French Fourth Republic, nonparametric methods seem preferable to parametric options.

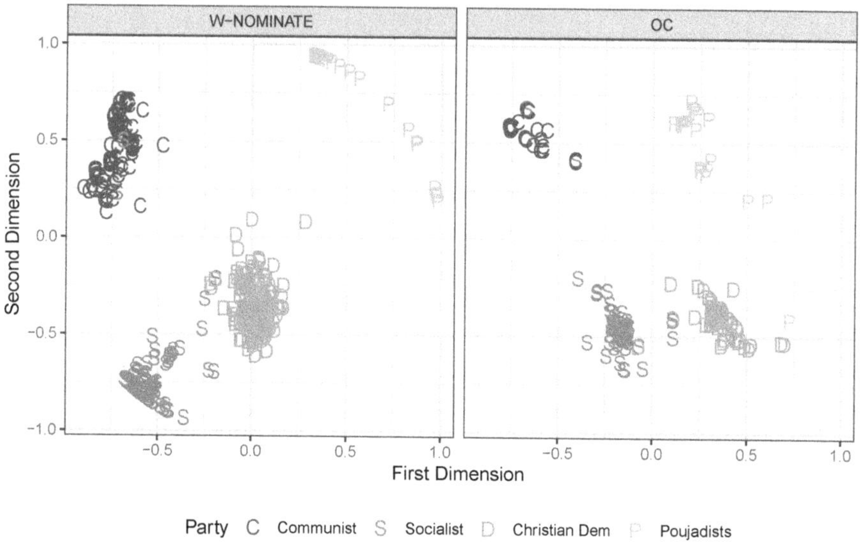

FIGURE 5.16: W-NOMINATE Ideal Point Estimates of Deputies of the French Fourth Republic (with Procrustes Rotation)

5.5.1 Identification of the Model Parameters

An important difference between the ideal point estimation methods discussed in this chapter concerns the means of identifying the parameters. Since ideal point estimation methods model distances between the points, the model is not *identified* so long as the parameters themselves (i.e., the ideal point and

¶As discussed earlier in the chapter, this is known as "rimming" and can occur when legislators are constrained to lie within the unit hypersphere (between the points -1 and 1 in one dimension, a circle with radius one in two dimensions, and a sphere with radius one in three dimensions) as in W-NOMINATE and OC.

roll call locations) remain unidentified (Rivers, 2003; Bakker and Poole, 2013). For instance, imagine producing a map of the United States from a set of inter-city distances. We can rotate (i.e., spin or flip) the two-dimensional map to produce an infinite number of configurations, and each configuration will be an equally valid reproduction of the inter-point distances (e.g., Chicago will be 983 miles or units from Boston in each).

NOMINATE achieves identification by constraining ideal points to lie within the unit hypersphere (that is, within 1 unit of the origin in s-dimensional space) and sets a direction of the configuration by constraining a right-wing legislator on each dimension to have a positive score on that dimension. Because of the unit hypersphere constraint, the log likelihood of the ideal point estimates will not necessarily be maximized for all legislators. As discussed, some legislators compile roll call voting records that fit the spatial model too well and commit too *few* voting "errors." Such a legislator will be pushed too far away from the remaining legislators to maximize the log likelihood.

This creates the "sag" problem: exaggerating the distances between ideologically extreme and interior legislators. The "sag" problem is not present in D-NOMINATE or DW-NOMINATE because many legislators serve in multiple legislatures so that they make enough voting errors to prevent them from being pushed to the edge of the space. Legislators serving in one legislature can be near-perfect voters and be at the edge. But there is unlikely to be a large gap between the ideologically extreme and interior legislators. W-NOMINATE addresses the "sag" problem by requiring the most extreme left and right-wing legislators to be within 0.1 units of the *second-* most extreme left and right-wing legislators. Practically, this is achieved by estimating the legislator coordinates, setting the most extreme legislator(s) at $-1/1$ and omitting them from a subsequent estimation, and then using those results (Poole, 2005, pp. 155-159).

Identification of the legislator ideal points and policy alternatives is most problematic in OC given its nonparametric characteristics: this is the price to pay for avoiding strict parametric assumptions about the data. Indeed, this has long been the Achilles' heel of nonmetric scaling methods (Jacoby, 1991). In one dimension, the OC results are identified only to a rank order. In two or more dimensions, the situation is improved. With multiple dimensions, OC is able to locate legislator ideal points and policy alternatives within polytopes formed by the intersection of cutting planes. This is sufficient to recover metric-level information about the parameters (Peress, 2012). Moreover, with a reasonable number of roll call votes (and thus cutting planes), the polytopes become small enough that the identification problem is not a serious one.

5.5.2 Comparing Ideal Point Estimates for the 111th US Senate

In selecting a method of roll call analysis, it is important to carefully consider the theory underlying their construction. The methodological differences between these methods can matter substantively. However, in many situations, theoretical considerations may be "full of sound and fury" with little practical significance. Specifically, in stable legislative bodies that conduct regular roll call votes with reasonable rates of voting errors, choice of scaling method appears to have little effect on the results (Poole, 2005; Hix, Noury and Roland, 2006).

As an example, we compare results from the `wnominate()` and `oc()` functions on roll call data from the 111th US Senate. Based on the presence of a dominant ideological dimension to the data and to facilitate comparison of the results, we estimate one-dimensional models for each method. The roll call matrix from the 111th Senate is stored in the `rollcall` object `hr111`, and the procedures are run below. The legislator ideal points are stored in the objects `wnom1` and `oc1`.

```
wnom.result <- wnominate(hr111, ubeta=15, uweights=0.5, dims=2,
    minvotes=20, lop=0.025, trials=1, polarity=c(2,2),
    verbose=FALSE)
oc.result <- oc(hr111, dims=2, minvotes=20, lop=0.025,
    polarity=c(2,2), verbose=FALSE)
weight <-  wnom.result$weights[2]/wnom.result$weights[1]
plot.dat <- data.frame(
    wnom = c(wnom.result$legislators$coord1D,
             wnom.result$legislators$coord2D*weight),
    oc = c(oc.result$legislators[,"coord1D"],
           oc.result$legislators[,"coord2D"]),
    dimension = factor(rep(c(1,2), each =
        nrow(oc.result$legislators)),
        labels=c("First Dimension", "Second Dimension")),
    party = party)
plot.dat$party <- factor(plot.dat$party, levels=c(100, 200),
    labels=c("Democrats", "Republicans"))
plot.dat <- na.omit(plot.dat)
```

The results are then plotted in Figure 5.17. The first dimensions are quite highly correlated for both parties (D=0.91, R=0.94). The second dimensions are a bit less highly correlated with each other (D=0.79, R=0.57).

```
ggplot(plot.dat, aes(x=wnom, y=oc)) +
      geom_point(pch=1) +
    facet_grid(party ~ dimension) +
    theme_bw() +
    theme(aspect.ratio=1) +
    xlab("W-NOMINATE Score") +
    ylab("OC Score")
```

```
: Democrats
: First Dimension
[1] 0.8980147
```

```
---------------------------------------------
: Republicans
: First Dimension
[1] 0.9354241
---------------------------------------------
: Democrats
: Second Dimension
[1] 0.7851153
---------------------------------------------
: Republicans
: Second Dimension
[1] 0.5740925
```

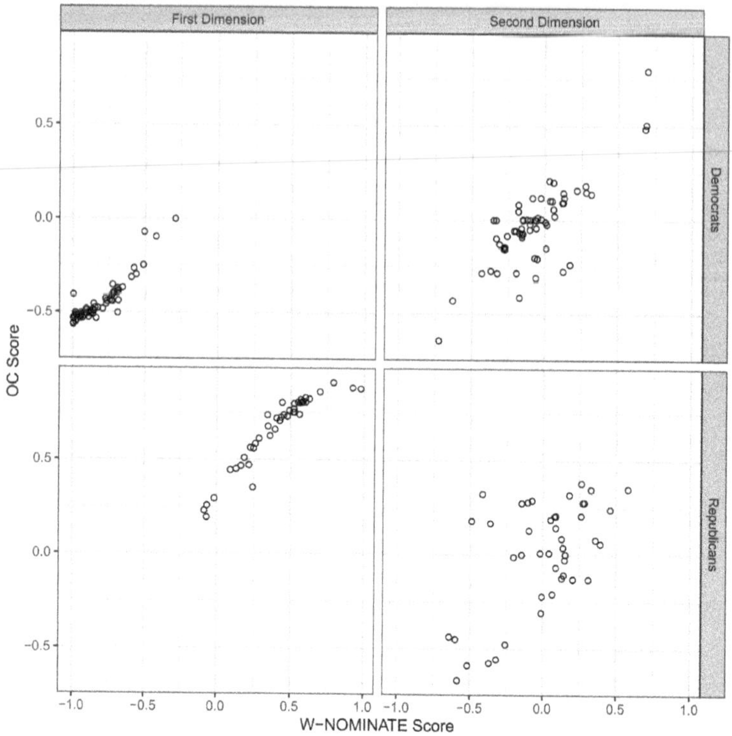

FIGURE 5.17: Ideal Points from W-NOMINATE and OC for Members of the 111th US Senate

5.6 Exercises

1. Use the `readKH()` function to read in the 104th US Senate roll data (`sen104kh.ord`) as the `rollcall` object `sen104`.

   (a) Provide the code used to create the `rollcall` object `sen104`.

   (b) How many legislators served and roll call votes are included in `sen104`?

   (c) Examine the first five rows of the matrix `sen104$legis.data`. Which Senator (with corresponding row number) would make a suitable conservative constraint on the first dimension? The second dimension?

2. Use the `rollcall()` function to read in roll call data from the Fourth European Parliament (`rcv_ep4.Rda`) as the `rollcall` object `ep4`. The first five columns are legislator data (including legislator name in the second column) and the sixth through final columns are roll call votes. Yea votes are coded as 1, Nay votes as 2, Abstentions and Present but not Voting as 3 and 4, and Absences and Not in Legislature as 0.

   (a) Provide the code used to create the `rollcall` object `ep4`.

   (b) Examine the matrix `ep4$legis.data`. What is the row number corresponding to the first member of the European Party Group "E" (for the European People's Party, a right-wing party group encompassing many members of national Christian Democratic parties)?

3. Use the `wnominate()` function to run W-NOMINATE on the 104th US Senate roll data (stored in the `rollcall` object `sen104kh` from Exercise 1) in two dimensions.

   (a) Neatly format and print the `summary(result)` object.

   (b) What are the correct classification rates, APRE, and GMP fit statistics in one and two dimensions?

   (c) Plot the estimated two-dimensional configuration of legislator coordinates from W-NOMINATE. Label Democratic Senators as "D", Republican Senators as "R", and President Clinton as "C".

4. Re-run W-NOMINATE on the `sen104kh` data using 250 trials of the parametric bootstrap to generate uncertainty estimates for the legislator coordinates.

   (a) Which legislator has the largest bootstrapped standard error on the first dimension? The second dimension?

   (b) Plot the estimated two-dimensional configuration of legislator coordinates with cross-hairs corresponding to their 95% confidence intervals on each dimension.

5. Use the W-NOMINATE results for the `sen104kh` data to plot the 104th Senate's vote on President Clinton's 1996 welfare reform bill (The Personal Responsibility and Work Opportunity Reconciliation Act of 1996). The Senate's 78-21 vote to pass the bill took place on August 1, 1996 and is stored in Vote #875.

   (a) Neatly format and print the W-NOMINATE estimates for the vote stored in `result$rollcalls`.

(b) Do a side-by-side plot of the vote with all legislators included in the left plot and only the errors (legislators who were incorrectly classified) in the right plot. Color Yea votes red and Nay votes blue and draw the cutting line dividing predicted Yeas from predicted Nays. Finally, print the number of Yea and Nay votes, the PRE statistic for the vote, and the number of errors in both plots.

6. Use the `oc()` function to run Optimal Classification on the 1993-1994 session of the California State Senate roll call data (stored in the file `CASEN.19931994.Rda`) separately in one and two dimensions.

   (a) Convert `CASEN.19931994.Rda` to a `rollcall` object using the `rollcall()` function. The first three columns of `CASEN.19931994` are legislator data (Name, Party, and Party Code [Democrat = 100, Republican = 200, and Independent = 328]), and the fourth through final columns are roll call data (Yea votes are coded as 1, Nay votes as 6, Abstentions and Missing as 9, and Not in Legislature as 0).

   (b) What are the fit statistics from Optimal Classification analysis of the 1993-1994 session of the California State Senate in one and two dimensions?

   (c) Plot the two-dimensional Optimal Classification coordinates of state Senators in the 1993-1994 session of the California State Senate, labeling the legislators by party.

6

Bayesian Scaling Models

In this chapter, we revisit some of the methods previously presented in this book through a Bayesian framework. The Bayesian approach has several advantages. The most important of these is that latent variables can be treated as variables that are entirely missing. In Bayesian modeling, missing data are treated as parameters to be estimated in the model in exactly the same way that any other parameters are treated. That is, after deriving the posterior distribution for these quantities, we can sample directly from these distributions. Thus, estimates of the latent trait have posterior distributions which yield estimates of uncertainty that are derived directly as a function of the model, rather than being computed *post hoc* with something like the bootstrap.

There are several other advantages to the Bayesian inferential framework as well. Because we can sample directly from the posterior, it is easy to make inferences about functions of parameters (e.g., predicted probabilities). This is obviously possible in frequentist models, too, but relies either on the delta method or on the boostrap to generate draws from the sampling distribution of the statistics. Another practical advantage of the Bayesian framework in general, and the software that we use to estimate these models (BUGS, JAGS or Stan) is that new models can be estimated with relative ease. All of these advantages will be visible as we move through the chapter below.

Before going into the specific models, let's put a bit of context around the Bayesian modeling approach. First, we can think of the problem in the generalized linear model framework:

$$E(y_{ij}) = g^{-1}(\beta, \xi), \tag{6.1}$$

where y_{ij} is the ith response on the jth indicator variable, β are a set of coefficients that relate the latent value ξ_i to the indicator variables and $g^{-1}(\cdot)$ is the inverse of the link function mapping the unbounded linear predictor into the response space of the dependent variable.

One of the main challenges in measurement models of all types is identification, that is, there often exist an infinite number of solutions that fit the data equally well. There are several strategies for achieving model identification.

1. Deterministically setting a value in β to an arbitrary constant (usually 1).

2. Putting directional priors on $\boldsymbol{\beta}$ such that β_j's posterior has a lower bound of zero for all or some j.

3. Fixing the mean and variance of ξ (usually zero and one, respectively).

In the following sections, we revisit some of the models we presented in previous chapters, highlighting the advantages of estimating these models from a Bayesian perspective.

6.1 Bayesian Aldrich-McKelvey Scaling

A Bayesian adaptation of A-M scaling was developed in Hare et al. (2015). This method retains the properties of A-M scaling described above but enables the handling of missing data, an alternative means of obtaining uncertainty for the stimuli estimates, and simultaneous estimation of stimuli locations and the individual distortion parameters.*

We demonstrate how to perform Bayesian A-M (BAM) scaling using Martyn Plummer's `rjags` package in `R` (Plummer, 2013). `rjags` facilitates Bayesian modeling within `R` by calling the separate JAGS (Just Another Gibbs Sampler) program (Plummer, 2003*b*). JAGS is compatible across platforms (Windows, Mac, etc.), and can be downloaded at `http://mcmc-jags.sourceforge.net/`. For this example, we again analyze respondents' ideological placements of French political parties in the 2009 European Election Studies (EES) so that we can compare results obtained from the classical maximum likelihood (ML) and Bayesian procedures.

With the `rjags` package, we can work entirely within `R` with the exception of writing the `.bug` model code in a separate text editor. We will access this file (`BAM_JAGScode.bug`) and assign values to the required parameters later with the `jags.model()` function in `rjags`. Note that `zhat` contains the estimated locations of the stimuli (with a normalization constraint, see Jackman (2009, chap. 9)) and `a` and `b` are the intercept and weight terms for the individuals (which map their placements onto the "true" stimuli positions) with diffuse uniform priors. Respondent self-placements are not included at this stage, but we will use them later to estimate respondent ideal points using samples of α_i and β_i. Finally, the model includes both stimuli and respondent-specific error terms (`tauj` and `taui`, respectively), which accounts for heteroskedasticity in respondents' placements of the stimuli (see Geweke (1993)).† You can get the

*Substantive applications of Bayesian Aldrich-McKelvey scaling can be found in Zakharova and Warwick (2014), Carroll and Kubo (2018), Struthers, Hare and Bakker (2020), and Butters and Hare (2020).

†The JAGS code can also be used to estimate this model in WinBUGS and OpenBUGS.

code with

```
system.file("templates/BAM_JAGScode.bug", package="asmcjr")
```

which will produce:

```
model{
for(i in 1:N){  ##loop through respondents
for(j in 1:q){  ##loop through stimuli
z[i,j] ~ dnorm(mu[i,j],tau[i,j])
mu[i,j] <-a[i] + b[i]*zhat[j]
tau[i,j] <-taui[i] * tauj[j]  ##respondent and stimuli
}}                           ##precision terms
for(i in 1:N){
a[i] ~ dunif(-100,100)  ##uniform priors on alpha
b[i] ~ dunif(-100,100)  ##uniform priors on beta
taui[i] ~ dgamma(ga,gb)  ##priors on variance (respondents)
}
ga ~ dgamma(.1,.1)  ##hyperpriors for taui
gb ~ dgamma(.1,.1)  ##hyperpriors for taui
for(j in 1:q){
tauj[j] ~ dgamma(.1,.1)  ##priors on variance (stimuli)
zhatstar[j] ~ dnorm(0,1)  ## priors on zhat (norm. constraint)
zhat[j] <-(zhatstar[j]-mean(zhatstar[]))/sd(zhatstar[])
}}
```

Next, we load the data and calculate the classical (ML) A-M result. We will compare these results to those obtained from the Bayesian approach and use them as initial values for the JAGS model. We recode missing values in the data as `NA` for `rjags`. We also subtract 5 from each row so that negative scores denote left-wing placements and positive scores denote right-wing placements. Self-placements (in the first column) are stored in the object `self` and the party placements are re-stored in the object `franceEES2009`.

```
data(franceEES2009)
MLE_result.france <- aldmck(franceEES2009, polarity=2, respondent=1,
     missing=c(77,88,89), verbose=FALSE)
bamdata <- bamPrep(franceEES2009,
    missing=c(77,88,89), self=1)
```

We include a cutoff parameter to set the minimum number of responses that an individual must provide to be included in the scaling. There is no such option in the `aldmck()` function because it eliminates rows (individuals) with any missing data. Bayesian A-M scaling will include all individuals in its estimation (even those who have provided no responses), but we think it is prudent to omit respondents with high levels of missing data. The following commands retain only respondents who have placed at least five of the eight parties:

```
#Aldrich-Mckelvey scaling, with cutoff of 5 or more
bamdata <- bamPrep(franceEES2009,
    missing=c(77,88,89), self=1, nmin=5)
```

Before using the `jags.model()` function in `rjags` to compile the model, within `R` we assign values to `N` (the number of rows or individuals), `q` (the number of columns or stimuli) and `z` (the data). All three of these parameters are referenced in the JAGS model file `BAM_JAGScode.bug`. We also write a function to generate initial values for `zhatstar` by adding random noise (drawn from a normal distribution with mean 0 and standard deviation 1) to the stimuli estimates from ML A-M scaling. We assign two Markov chains and a burn-in period—the number of iterations to be discarded from the beginning of the Markov chains—of 10,000 iterations. Finally, we store 5,000 samples from the posterior distributions of `zhat`, `a` and `b` into `mcmc` objects of the same name.

```
bam.france <- BAM(bamdata, polarity=2, n.adapt=2500, n.sample=5000,
      zhat=TRUE, ab=TRUE, resp.idealpts=TRUE)
```

Convergence diagnostics can then be run on the chains for the sampled parameters (see Section 6.4.4.1).

The command `summary()` can be used to display the summary statistics of an `MCMC` object (`bayes_result`). Below we compile the mean, standard deviation and lower and upper 95% credible intervals (the range within which a specified proportion of the posterior distribution falls) into the matrix `bayes.out` for the `zhat` values. The results have a high degree of face validity: the Extreme Left and National Front parties are the most extreme parties, with the left-right ordering of the stimuli the same as that produced by ML A-M scaling.

```
bam.france$zhat.ci

         stimulus idealpt    sd  lower  upper
1    Extreme Left  -1.209 0.011 -1.230 -1.187
2       Communist  -0.885 0.011 -0.906 -0.864
3       Socialist  -0.169 0.014 -0.195 -0.142
4          Greens  -0.070 0.014 -0.098 -0.042
5    UDF (Bayrou)   0.309 0.013  0.283  0.334
6  UMP (Sarkozy)    1.109 0.014  1.082  1.136
7 National Front    1.677 0.013  1.652  1.702
8      Left Party  -0.763 0.015 -0.793 -0.733
```

Respondent ideal points can be calculated by applying the transformation parameters α and β to their self-placements. Rearranging Equation 2.1 to solve for respondents' "true" locations ($z_{(self)}$) yields:

$$\frac{z_{i(self)} - \alpha_i}{\beta_i} \tag{6.2}$$

where $z_{i(self)}$ is respondent i's self-placement on the issue scale. Accordingly, we calculate ideal points for each sample m $(1,...,t)$ of α_i and β_i for each respondent i $(1,...,n)$. This forms a distribution with t values for each ideal point stored in the matrix **idealpt**. This is returned in the element **resp.samples** by the BAM function if the option **resp.idealpts=TRUE** is specified. We can also calculate summary statistics for the respondent ideal points. We recommend using the median rather than the mean value for the point estimate since the median is robust to long tails, which occur because as $\beta_i \to 0$ the ideal point goes to $\pm\infty$. These summaries are returned in the **resp.summary** by BAM if the option **resp.idealpts=TRUE** is specified.

The use of the Bayesian A-M model also allows us to assess the probability that the stimuli locations lie within certain intervals by examining their posterior distributions. One practical application is that we can determine the probability that two stimuli are in reverse order. For example, the estimates place the Socialist Party (−0.180) to the left of the Greens (−0.076). What is the probability that the Socialist Party is actually to the right of the Greens? To address this question, we run 100,000 simulations in which one random draw from the Socialist Party's posterior density is compared to one random draw from the Greens' posterior density. We then calculate the percentage of simulations where the Socialist Party draw is larger (to the right) of the Greens draw and store it in the object **p.wrong.order**. We find that in no trial is the draw from the Socialist Party's posterior density larger than the draw from the Greens' posterior density, meaning that the probability of incorrect ordering is negligible. If, on the other hand, the Socialist Party draw was larger than the Greens draw in 5,000 of the 100,000 simulations, we would conclude that the probability of incorrect ordering is 5%. Figure 6.1 displays the posterior densities of the two stimuli.

```
diffStims(bam.france$zhat, stims=c(3,4))

             Comparison Probability
1 Pr(Greens > Socialist)       1.000

soc_samples <- do.call("c", bam.france$zhat[,3])
greens_samples <- do.call("c", bam.france$zhat[,4])
df <- data.frame(x = c(soc_samples, greens_samples),
      stimuli = factor(rep(c(1,2), each=length(soc_samples)),
                labels=c("Socialists", "Greens")))
ggplot(df, aes(x=x, group=stimuli, color=stimuli)) +
  geom_line(stat="density") + xlim(range(df$x)) +
  scale_color_manual(values=gray.palette(2)) +
  xlab("Left-Right") + theme_bw() +
  theme(legend.text=element_text(size=12),
```

```
legend.position="bottom", aspect.ratio=1) +
guides(shape = guide_legend(override.aes = list(size = 4)))
```

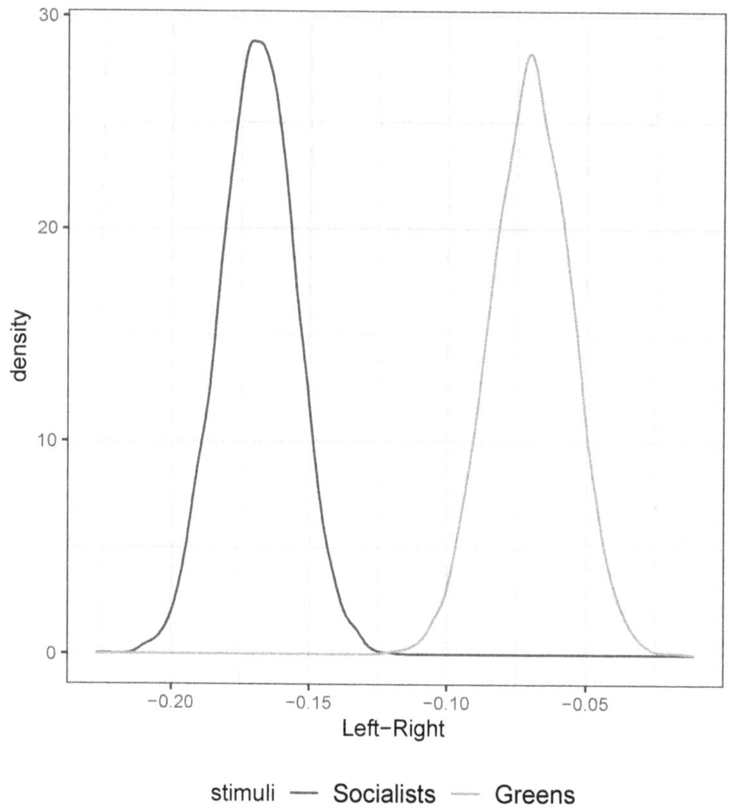

FIGURE 6.1: Posterior Densities of the Socialist Party and the Greens

In Figure 6.2, we compare the estimated stimuli locations obtained from ML and Bayesian A-M scaling. The point estimates are nearly identical, with a Pearson correlation of 0.999. There are some minor differences; namely, the National Front party is placed further to the right and the UMP party further to the left by ML estimation. With this in mind, all parties fall very close to the OLS regression line between the set of estimates.

```
#Comparison of estimated stimuli locations
df2 <- data.frame(MLE = MLE_result.france$stimuli,
```

```
    BAM = bam.france$zhat.ci$idealpt,
    stimname = names(MLE_result.france$stimuli))
ggplot(df2, aes(x=MLE, y=BAM)) +
  geom_point(pch=16, cex=2)  +
  theme_bw() +
  geom_text(aes(label=stimname),
     nudge_x=c(-0.204, 0.180, -0.144, 0.120,
               -0.204, 0.228, -0.228, -0.156)) +
  geom_smooth(method="lm", se=FALSE, fullrange=TRUE,
              color="black", lwd=.5) + xlim(-.8,1)
```

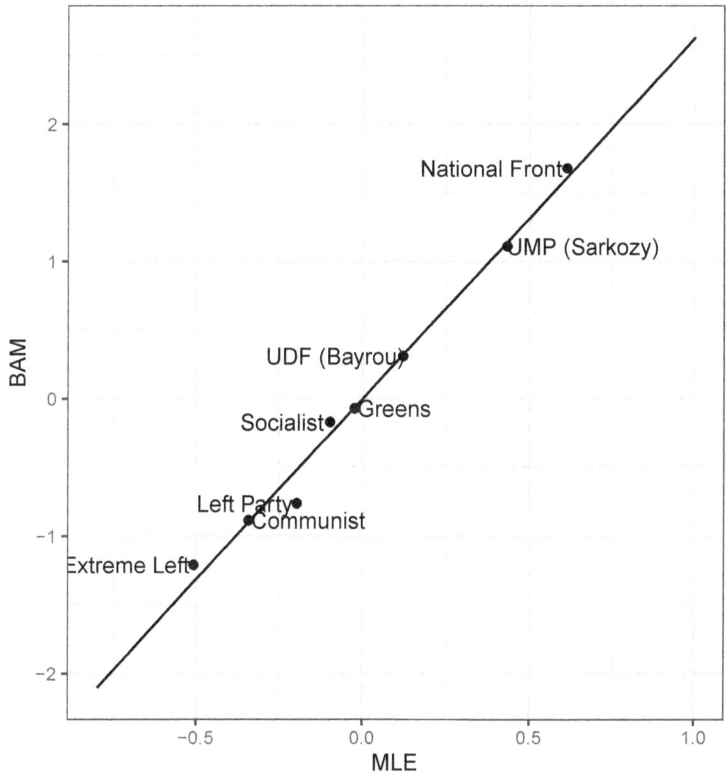

FIGURE 6.2: Comparison of Estimated Left-Right Positions of French Political Parties (2009 European Election Study) with ML and Bayesian Aldrich-McKelvey Scaling

6.1.1 Comparing Aldrich-McKelvey Standard Errors

We next compare the standard errors produced by Bayesian A-M scaling and the bootsrapped standard errors from ML A-M scaling. We normalize the point estimates (arranged on rows) from each method across trials with the function `apply(samples, 1, scale)` and `apply(boot.aldmck$t, 1, scale)`. We then calculate the standard error for each set of stimuli estimates by taking the standard deviation of the standardized distributions of `samples.scale` and `boot.scale`.

```
zhat.samples <- do.call("rbind", bam.france$zhat)
samples.france <- zhat.samples[,
    match(names(MLE_result.france$stimuli), colnames(zhat.samples))]
samples.scale.france <- apply(zhat.samples, 1, scale)
bayes.se.france <- apply(samples.scale.france, 1, sd)
boot.scale.france <- t(apply(boot.france$b$t, 1, scale))
# Extracting standard errors
# Since we're using normal theory confidence intervals,
# we first scale the bootstrapped stimuli and then
# get the square root of the variances.
boot.se.france <- sqrt(diag(var(boot.scale.france)))
df3 <- data.frame(stimname = names(MLE_result.france$stimuli),
                  BAM = bayes.se.france, Boot = boot.se.france
      )
```

Figure 6.3 plots the Bayesian and ML (bootstrapped) A-M standard errors with a cardinal, 45° line. All points fall below the line, meaning that they all have higher bootstrapped standard errors. The most important point made by Figure 6.3 is that the Bayesian approach produces smaller standard errors than the bootstrap method. The bootstrap standard errors tend to be from 20% to 50% bigger, depending on the party. This is probably due to the fact that the Bayesian approach has a larger sample size for each stimulus than the MLE approach which must use listwise deletion because of missing responses.

```
#Bootstrapped standard errors plot syntax
ggplot(df3, aes(x=Boot, y=BAM)) +
  geom_point() +
  theme_bw() +
  xlim(.01, .023) +
  ylim(.01,.023) +
  geom_abline(slope=1, intercept=0) +
  geom_text(aes(label=stimname),
      nudge_x=c(.0015,-.00125, -0.001,-.001, .0012,.001,0,.0012),
      nudge_y=c(.0003, -.0003, 0,0,0,-.0003,-.0003,0))
```

Finally, below we show Aldrich and McKelvey's (1977, p. 116) method of calculating the standard errors for the stimuli point estimates using Equation 6.3. We then compare the standard errors produced from this method

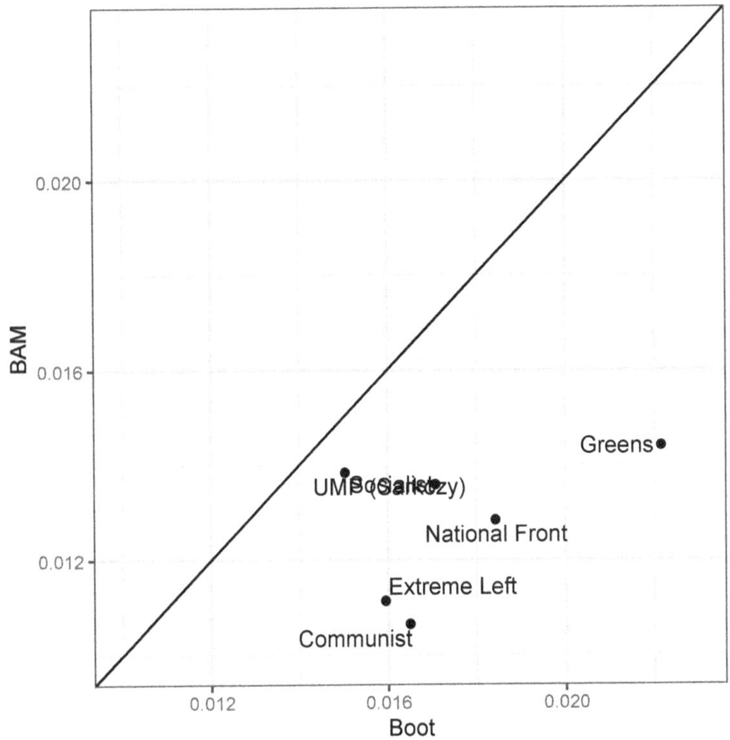

FIGURE 6.3: Comparison of Estimated Standard Errors of Left-Right Positions of French Political Parties (2009 European Election Study) with ML and Bayesian Aldrich-McKelvey Scaling

with the ML bootstrapped and Bayesian standard errors. Aldrich and McKelvey (1977, p. 117) note that their approach of calculating uncertainty bounds is almost certainly biased, and indeed these standard errors are an order of magnitude larger than the ML bootstrapped and Bayesian standard errors. This comparison highlights the pitfalls of analyzing residuals.

$$\hat{\sigma}_j = \sqrt{\frac{\sum(\hat{\alpha}_i + \hat{\beta}_i \hat{z}_j - z_{ij})^2}{N}} \tag{6.3}$$

```
df3$ASE <- aldmckSE(result.france, franceEES2009)
#Standard error comparisons
print(df3)
```

```
       stimname        BAM       Boot       ASE
1  Extreme Left 0.01113152 0.01594624  6.106526
2     Communist 0.01064539 0.01649372 11.768966
```

```
3       Socialist 0.01358508 0.01707336 19.400465
4          Greens 0.01439929 0.02215227 14.865599
5    UDF (Bayrou) 0.01291426 0.02412545 20.333114
6  UMP (Sarkozy) 0.01384110 0.01503619 24.653032
7 National Front 0.01283669 0.01842635 15.931198
8      Left Party 0.01513860 0.03088582 19.816097
```

For the bootstrap standard errors the key is the matrix:

$$A = \left[\sum_{i=1}^{n} X_i (X'_i X_i)^{-1} X'_i \right] \tag{6.4}$$

The stimuli configuration is the second eigenvector from the q by q matrix $A - nI$. The A matrix will not change much from bootstrap trial to bootstrap trial because it is summed over a sample of n respondents with replacement. The 100 bootstrap trials will produce roughly the same A matrix each time. The stimuli coordinates are an eigenvector of $A - nI$. The eigenvector is a direction through the point cloud defined by $A - nI$ so even though the point cloud changes slightly from trial to trial, the direction does not change much. This is why the standard errors are so small.

In the Bayesian result, the credible intervals for the stimuli are small because the stimuli coordinates are multiplied by the respondents' α_i and β_i values and placed in the mean of a normal distribution for the respondents' perceptions. Hence, the variation of the respondents' perceptions is where it belongs—in the noisy estimates of the α_i and β_i. This is why the Bayesian standard errors are almost identical to the bootstrap standard errors.

Let $\hat{y}_{ij} = \hat{\alpha}_i + \hat{\beta}_i \hat{z}_j$. Computing standard errors for the stimuli from the $\hat{y}_{ij}$ produces much larger values because these are computed over the n individuals and the computation is aggregating all the noise in the α_i and β_i values. Going back to the A matrix in Equation 6.4, note that it does not have the α_i and β_i in it. A is simply constructed directly from the raw responses. Because A is a sum, there will be a substantial smoothing of the individual level noise. This is why both Aldrich and McKelvey (1977) and Palfrey and Poole (1987) find that the stimuli configuration is exceptionally stable under all sorts of error conditions. The α_i and β_i values are estimated via a simple OLS regression. Each respondent's α_i and β_i values will be imprecisely estimated because typically q is a number less than 10. Using these to compute $\hat{y}_{ij}$ values and then computing standard errors from them will almost certainly overestimate the true standard errors of the stimuli.

6.2 Bayesian Multidimensional Scaling

There has been a recent surge of interest in applying Bayesian methods to metric MDS analysis of proximities data (Oh and Raftery, 2001; Okada and Shigemasu, 2010; Bakker and Poole, 2013). In the Bayesian framework, the observed distances (δ_{jm}) are assumed to be drawn from some distribution. Bakker and Poole (2013) use the log-normal distribution (because distances are inherently positive and variances should increase with the size of the distances) with mean $ln(d_{jm})$ (the estimated distances) and variance σ^2. That is

$$ln(\delta_{jm}) \sim \mathcal{N}(ln(d_{jm}), \sigma^2) \tag{6.5}$$

and

$$d_{jm} = \sqrt{\sum_{k=1}^{s}(z_{jk} - z_{mk})^2} \tag{6.6}$$

Where j and m index the stimuli and z_{jk} and z_{mk} are the coordinates of stimuli j and m on the kth dimension.

The coordinates (z_{jk}, z_{mk}) are treated as having simple normal prior distributions:

$$\xi(z_{jk}) = \frac{1}{(2\pi\kappa^2)^{\frac{1}{2}}} e^{-\frac{z_{jk}^2}{2\kappa^2}} \tag{6.7}$$

and the variance term σ^2 is given a uniform prior distribution where, empirically, b is no greater than 2:

$$\xi(\sigma^2) = \frac{1}{c}, 0 < c < b \tag{6.8}$$

And after dropping unnecessary constants, the log of the posterior is

$$-\frac{q(q-1)/2}{2}\ln(\sigma^2) - \frac{1}{2\sigma^2}\sum_{j=1}^{q-1}\sum_{m=j+1}^{q}\left(\ln\left(d_{jm}^*\right) - \ln\left(d_{jm}\right)\right)^2 \tag{6.9}$$
$$-\frac{1}{2\kappa^2}\left(\sum_{j=1}^{q}\sum_{k=1}^{s} Z_{jk}^2\right) - \ln(c)$$

Markov chain Monte Carlo (MCMC) methods are then used to "explore" the posterior distribution and calculate summary statistics.

Because non-informative priors are used, Bayesian MDS is closely related to maximum likelihood (ML) MDS methods of Ramsay (1977) and Takane

(1981). However, the Bayesian framework offers two important attractive properties. First, with the use of posterior distributions, we can develop more direct uncertainty measures for the point estimates. Second, the means of the coordinates' posterior distributions are preferable to the modes (which are used in ML MDS methods), particularly in the case of asymmetric or multi-modal distributions.

6.2.1 Example 1: Nations Similarities Data

Below we demonstrate Bayesian MDS on the `nations` data set. As in Section 6.1, the BUGS code below (`nations_JAGS_model.bug`) can be estimated in R using the `rjags` package (Plummer, 2013) and the `jags.model()` function. We fix three coordinates (the USSR at the origin and the USA's second dimension coordinate at zero) to achieve identification (see Bakker and Poole (2013, Appendix A1)). We also constrain four coordinates to be positive (Congo's second dimension and Israel's first dimension coordinates) or negative (Japan's second dimension and China's first dimension coordinates) to correct for sign flips. You can get the code with:

```
system.file("templates/BMDS_JAGS_model.bug", package="asmcjr")
```

which will produce the following code:

```
model{
z[10,1] <-z[10,2] <-0.000 #fix USSR at origin
z[11,2] <-0.000 #fix USA second dimension coordinate at 0
for (i in 1:(N-1)){ #loop through stimuli
for (j in (i+1):N){
dstar[i,j] ~ dlnorm(mu[i,j],tau)
mu[i,j] <-log(sqrt((z[i,1]-z[j,1])*(z[i,1]-z[j,1])+
(z[i,2]-z[j,2])*(z[i,2]-z[j,2])))
}}
tau ~ dunif(0,10) #prior on precision (1/variance)
#
for(l in 1:2){ #priors on coordinates
z[i,l] ~ dnorm(0.0,0.01)
}
for(l in 1:2){
for(k in 3:6) {
z[k,l] ~ dnorm(0.0,0.01)
}}
z[7,2] ~ dnorm(0.0,0.01)
z[8,1] ~ dnorm(0.0,0.01)
z[9,2] ~ dnorm(0.0,0.01)
z[11,1] ~ dnorm(0.0,0.01)
```

```
for(l in 1:2) {
z[12,l] ~ dnorm(0.0,0.01)
}
z[2,2] ~ dnorm(0.0,0.01)I(0,) #set polarity of first/second
z[7,1] ~ dnorm(0.0,0.01)I(0,) #dimension coordinates for four
z[8,2] ~ dnorm(0.0,0.01)I(,0) #nations (Congo,Israel,Japan,
z[9,1] ~ dnorm(0.0,0.01)I(,0) #and China) to fix the sign flips
}
```

There is a function in the `asmcjr` package called `BMDS` which is a function that calls the above file using whatever polarity and fixed constraints you wish to include. The polarity constraints (the last four lines before the end of the printed file above) are specified with `posStims` and `negStims` each should be a vector of two integer values giving the row numbers of observations on the first and second dimensions that are constrained to be positive and negative, respectively. Fixed constraints (e.g., the top two lines of the model definition) can be accomplished by providing a matrix to the argument `z`. This should be a matrix with the same number of rows as the data and two columns. All unconstrained entries should be missing (`NA`) and all non-missing entries should take on the desired constrained value. In our example above, we would specify

```
z <- matrix(NA, nrow=nrow(nations), ncol=2)
z[10, ] <- 0
z[11,2] <- 0
```

We could then use this in the call to the function:

```
nations.bmds <- BMDS(nations, posStims=c(7,2), negStims=c(9,8), z=z,
       fname="nations_jags.txt", n.sample=10000, n.adapt=10000)
```

a nice byproduct of this approach is that the routine saves in `fname` the model file called by jags, so you could use it directly if so desired.

Figure 6.4 plots the nations at the posterior means of their first and second dimension coordinates (stored in the object `jags.df`). By constraining some of the nations to lie in certain regions of the space, Bayesian MDS achieves a configuration where the substantive dimensions (pro-West/pro-Communist and economic development) are closely aligned with the recovered dimensions.

```
jags.df <- data.frame(
  x = nations.bmds$zhat.ci$idealpt[1:12],
  y = nations.bmds$zhat.ci$idealpt[13:24],
  country = rownames(nations)
  )
ggplot(jags.df, aes(x=x, y=y, group=country)) +
  geom_point() +
  geom_text(aes(label=country),
```

```
            nudge_y=c(rep(-.15,11), .175))+
    theme_bw() +
    xlab("First Dimension") +
    ylab("Second Dimension") +
    theme(aspect.ratio=1) +
    xlim(-3,6) +
    ylim(-3,6)
```

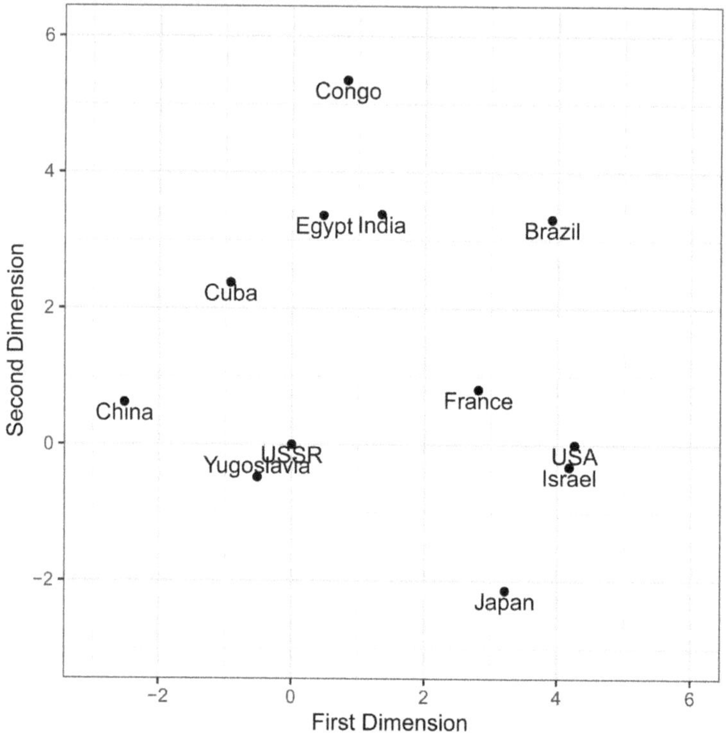

FIGURE 6.4: Bayesian Multidimensional Scaling of the Nations Similarities Data

We can also plot the densities of the coordinates' posterior distributions. Figure 6.5 shows the posterior densities of Congo's and Cuba's first dimension coordinates. Since this dimension represents pro-West/pro-Communist alignment, we expect that not only will Cuba's posterior mean be to the left of the Congo's, but that Cuba's distribution should be less dispersed than Congo's (since there is presumably less uncertainty about Cuba's Communist affiliation in 1968). Indeed, this is precisely what Figure 6.5 shows. This example highlights how the direct estimation of uncertainty is a valuable component of the Bayesian MDS approach.

```
samples <- do.call(rbind, nations.bmds$zhat)
dens.df <- data.frame(
  dim1 = c(samples[,2:3]),
  country = rep(c("Congo", "Cuba"), each=nrow(samples)))
ggplot(dens.df, aes(x=dim1, group=country)) +
  geom_line(aes(color=country, lty=country), stat="density") +
  scale_color_manual(values=gray.palette(2)) +
  theme_bw() +
  xlab("First Dimension") +
  theme(aspect.ratio=1, legend.position="bottom") +
  guides(shape = guide_legend(override.aes = list(size = 4)))
```

Recall that what is actually being modeled here are distances between the points, not the locations of the points themselves. In order to achieve identification, we fixed the USSR at the origin. The effect of this is to "transmit" the uncertainty in the point representing the USSR to all the other points in the configuration (cf., Lewis and Poole, 2004). There is no "cure" for this problem. It is inherent in MDS methods.

6.3 Bayesian Multidimensional Unfolding

The least squares unfolding procedures discussed above are quite effective, but the Bayesian multidimensional unfolding model developed by Bakker and Poole (2013) offers two distinct advantages. First, the Bayesian multidimensional unfolding model uses the log-normal distribution to model the estimated distances. Bakker and Poole (2013) argue that the log-normal distribution is preferable because distances are inherently positive and because intuitively, smaller observed distances should have smaller variances (see also Aitchison and Brown, 1957; Johnson, Kotz and Balakrishnan, 1994, chap. 14; Limpert, Stahel and Abbt, 2001). Second, Bayesian (MCMC-based) estimation "illuminates" the posterior distributions of the parameters, which produces better estimates of uncertainty for the parameters. The Bayesian framework also effectively deals with multimodal distributions, whereas maximum likelihood or least squares optimization methods will settle on one of the modes and report it as the point estimate. If we know that a respondent or stimuli's ideal point is multimodal, we can use the posterior mean rather than a single mode as the point estimate and express the uncertainty in that point estimate that is revealed by the existence of the multimodal distribution.

Recall that x_{ik} represents the ith $(i = 1,...,n)$ individual coordinate on the kth $(k = 1,...,s)$ dimension and z_{jk} represents the jth $(j = 1,...,q)$ stimuli co-

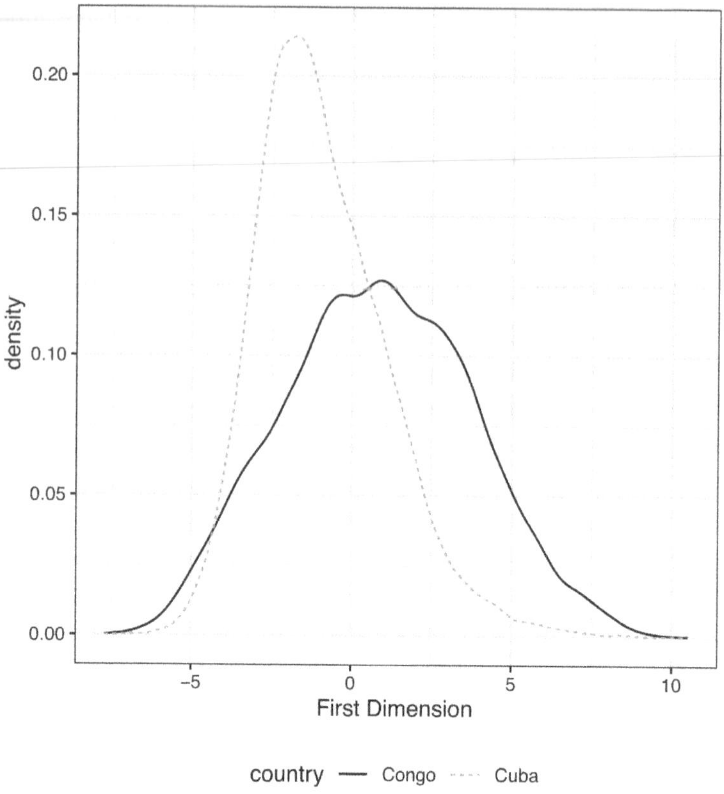

FIGURE 6.5: Posterior Densities of the First Dimension Coordinates of Congo and Cuba from Bayesian Multidimensional Scaling

ordinate on the kth dimension, and that d_{ij} represents the Euclidian distance between individual i and stimulus j in the s-dimensional space:

$$d_{ij} = \sqrt{\sum_{k=1}^{s} (x_{ik} - z_{jk})^2} \tag{6.10}$$

The observed distances are assumed to be drawn from the log-normal distribution:

$$\ln(d_{ij}^*) \sim N(\ln(d_{ij}), \sigma^2) \tag{6.11}$$

which produces the likelihood function:

$$L^*(x_{ik}, z_{jk}|D^*) = \frac{1}{(2\pi\sigma^2)^{\frac{nq}{2}}} \left(\prod_{i=1}^{n} \prod_{j=1}^{q} \frac{1}{d_{ij}^*} \right) e^{\frac{-1}{2\sigma^2} \sum_{i=1}^{n} \sum_{j=1}^{q} \left(\ln(d_{ij}^*) - \ln\left(\sqrt{\sum_{k=1}^{s} (x_{ik} - z_{jk})^2} \right) \right)^2} \tag{6.12}$$

Bakker and Poole (2013) use simple normal prior distributions for the individual and stimuli coordinates:

$$\xi(x_{ik}) = \frac{1}{(2\pi\zeta^2)^{\frac{1}{2}}} e^{\frac{-x_{ik}^2}{2\zeta^2}} \tag{6.13}$$

$$\xi(z_{jk}) = \frac{1}{(2\pi\kappa^2)^{\frac{1}{2}}} e^{\frac{-z_{jk}^2}{2\kappa^2}} \tag{6.14}$$

and a uniform prior for the variance term:

$$\xi(\sigma^2) = \frac{1}{c}, 0 < c < b \tag{6.15}$$

where b, empirically, is no greater than 2.

Hence, the posterior distribution is:

$$\xi(x_{ik}, z_{jk}|D^*) \propto \prod_{i=1}^{n} \prod_{j=1}^{q} \{ f_{ij}(x_{ik}, z_{jk}|d_{ij}^*) \} \, \xi(x_{11}) \ldots \xi(x_{ns}) \xi(z_{11}) \ldots \xi(z_{qs}) \xi(\sigma^2) \tag{6.16}$$

6.3.1 Example 2: 1968 American National Election Study Feeling Thermometers Data

The Bakker and Poole (2013) Bayesian unfolding procedure is executed using the following three steps:

1. Run SMACOF metric unfolding to get starting coordinates for the L-BFGS (Limited-Memory Broyden-Fletcher-Goldfarb-Shanno) optimization procedure.

2. Run L-BFGS to find the posterior modes of the parameters and use this result as the target configuration for the slice sampler.

3. Run the slice sampler to estimate the posterior densities of the parameters. At each iteration of the sampler, rotate the results to the target L-BFGS configuration obtained in Step 2.

We show how to perform Bayesian multidimensional unfolding using feeling thermometers data from the 1968 American National Election Study (ANES).

The 1968 ANES asked 1673 respondents to use a 100-point scale to rate how warmly they felt about 12 political figures: George Wallace, Hubert Humphrey, Richard Nixon, Eugene McCarthy, Ronald Reagan, Nelson Rockefeller, Lyndon Johnson, George Romney, Robert Kennedy, Edmund Muskie, Spiro Agnew, and Curtis LeMay. These data are interesting because they encompass a diverse set of figures from three political parties (with George Wallace and Curtis LeMay representing the American Independent Party) and are drawn from a tumultuous time in American politics.

The dataframe `ANES1968` includes the feeling thermometers data and whether respondents' reported voting in the 1968 elections (`vote.turnout`, 1 = voted, 0 = did not vote) and their reported presidential vote choice (`presidential.vote`, 1 = Humphrey, 2 = Nixon, 3 = Wallace). We store the feeling thermometers in the object `anes.input`. The feeling thermometers range from 0 ("very cold or unfavorable") to 100 ("very warm or favorable"), but values of 98, 99, and 100 were recoded in the original ANES file as 97 values (to allow these values to represent refused/missing responses). Hence, the values in `anes.input` range between 0 and 97. Missing values are coded as `NA`. We print a portion of `anes.input` below:

```
data("ANES1968", package="asmcjr")
anes.input <- ANES1968[,1:12]
anes.input <- as.matrix(anes.input)
vote.turnout <- ANES1968[,13]
print(anes.input[1:5,1:6])
```

```
     Wallace Humphrey Nixon McCarthy Reagan Rockefeller
[1,]      15       40    70       60     60          60
[2,]      15       60    40       85     30          97
[3,]      NA       NA    NA       NA     NA          NA
[4,]       0        0    50        0     60          60
[5,]       0       30    60       50     50          50
```

We delete respondents (rows) who provided less than five feeling thermometer ratings. This leaves 1392 respondents to analyze. We also transform the ratings to distances with the standard method (subtracting them from 100 and dividing by 50).

We next program the parameters that are required for the subsequent estimation procedures. `nrowX` and `ncolX` are the number of individuals and stimuli, respectively, in the data. The burn-in period is specified with the `nburn` parameter. `nslice` is the number of iterations of the Markov chains to be executed and retained following the burn-in period. `NS` is the number of dimensions to be estimated. `N` and `NDIM` are the total number of model parameters to be estimated by the Bakker and Poole (2013) Bayesian unfolding procedure. The difference between the two is that `N` includes $\frac{NS(NS+1)}{2}$ fixed constraints on the number of individual and stimuli coordinates on each dimension ($NS(nrowX + ncolX)$), while `NDIM` only subtracts $NS - 1$ as it fixes a single parameter on each dimension beyond the first to set the rotation. `UNFOLD` is set to 1 if Bayesian unfolding is being performed and 0 if Bayesian multidimensional scaling (discussed in Chapter 3) is being performed.

The `C` code requires that the rows (individuals) have fewer than 5 missing entries and that at least 25% of the column (stimuli) values be non-missing. Within the algorithm, missing values set to −999. The `bayesunfold()` function will perform all of the necessary prepatory steps and feed the required data to the `C` routines.

```
result <- bayesunfold(anes.input, dims=2,
    burnin=1000, nsamp=2000)
```

Our experience with the stability of the Bayesian unfolding algorithm is somewhat inconsistent. It seems to work well on Windows and in Linux, but is inconsistent on macOS. Our experience suggests that installing the package from source (preferably with the GNU version of the gcc and g++ compilers) has a higher probability of working. Further, opening and running R directly in the terminal, rather than through the R-GUI, radian or the like produced the best results.[‡]

The algorithm first runs SMACOF metric unfolding to get good starting values for the parameters. The estimated configuration of candidates and voters is shown in Figure 6.6. Note, we are only plotting the presidential candidates to keep the display as uncluttered as possible. The candidates are labeled with their names and the voters are labeled "H" if they voted for Humphrey, "N" for Nixon, and "W" for Wallace.

The SMACOF result seems reasonable: The coalitions of Humphrey, Nixon and Wallace voters are fairly distinct from one another, and the candidates themselves are mostly clustered into the three party groups exterior to the distribution of the voters. The first dimension corresponds, more or less, to the liberal-conservative continuum and the second dimension appears to capture things like social issues and a partisan-based divide between Democrats and Republicans. Wallace, a staunch and visible segregationist during his tenure as Democratic governor of Alabama, found his strongest support among southern and working-class white Democrats and former Democrats. Hence, it is not surprising that he and his supporters would be proximately located on a dimension that taps into partisan sentiment.

There are some anomalies here. First, Nixon appears to be surrounded by his voters, but close to very few of them. Wallace seems to be pushed to the edge of his electoral coalitions contrary to the predictions of the classic spatial voting model. Humphrey seems to be toward the South-eastern edge of his electoral coalition, too. Configurations in which candidates flank their

[‡]Further still, on macOS, the routine often produces a degenerate solution the first time it is run. An indication of this is two columns of zeros on the right-hand side of the L-BFGS output. Our experience is that running the function again will resolve the problem.

supporters has been a recurring phenomenon in studies that use least squares metric unfolding to estimate candidate and voter locations from feeling thermometers data (e.g., Poole and Rosenthal, 1984).

```
row.dat <- as.data.frame(result$smacof.result$conf.row)
col.dat <- as.data.frame(result$smacof.result$conf.col)
col.dat$candidate <- factor(1:nrow(col.dat), labels=rownames(col.dat))
col.dat <- col.dat %>% filter(candidate %in%
  c("Humphrey", "Wallace", "Nixon"))
row.dat$vote <- NA
row.dat$vote  <- factor(ANES1968[,14], levels=1:3,
    labels=c("Humphrey", "Nixon", "Wallace"))[result$retained.obs]
ggplot(na.omit(row.dat), aes(x=D1, y=D2)) +
    geom_point(aes(shape=vote, colour=vote), size=3) +
    scale_shape_manual(values=c("H", "N", "W"),
        labels=c("Humphrey", "Nixon", "Wallace"),
        name="Vote") +
    scale_colour_manual(values=c("gray50", "gray65", "gray80"),
        labels=c("Humphrey", "Nixon", "Wallace"),
        name = "Vote") +
    geom_text(data = col.dat, aes(label=candidate),
        size=5, color="black") +
    theme_bw() +
    theme(aspect.ratio=1, legend.position="bottom") +
    xlab("First Dimension") +
    ylab("Second Dimension")
```

Because of the scale of the optimization problem, the Bakker and Poole (2013) procedure uses the limited-memory BFGS optimizer with starting values from SMACOF. L-BFGS uses the likelihood function based on the log-normal distribution in Equation 6.12 as its loss function. The function `bayesunfold()` pulls these results together and does a Procrustes rotation, allowing both dilation and translation, of each posterior draw using the L-BFGS result as its target. There are two elements in the result object - `rotated` and `unrotated` (the raw draws from the MCMC algorithm). Within each element are several different results. First, there are `stim.samples` and `indiv.samples` - the chain values for both individuals and stimuli. There are also `simuli` and `individuals` which are the posterior mean configurations for their respective groups.

The L-BFGS result is plotted in Figure 6.7. Because the log-normal distribution is a better model of the data, the configuration in Figure 6.7 is more palatable from the standpoint of classical spatial voting theory. The candidates are not pushed so far to the edges of the ideological space and are located closer to the center of their voters.

```
ind.dat <- as.data.frame(result$lbfgs.result$individuals)
names(ind.dat) <- c("D1", "D2")
ind.dat$vote <- NA
```

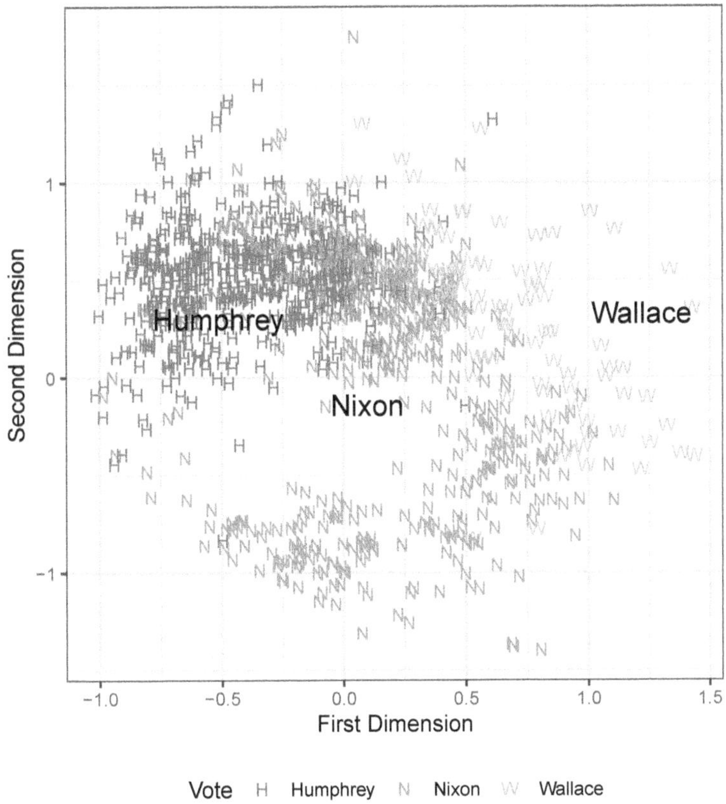

FIGURE 6.6: SMACOF Metric Unfolding of ANES 1968 Feeling Thermometers Data

```
ind.dat$vote <- factor(ANES1968[,14], levels=1:3,
    labels=c("Humphrey", "Nixon", "Wallace"))[result$retained.obs]
stim.dat <- as.data.frame(result$lbfgs.result$stimuli)
names(stim.dat) <- c("D1", "D2")
stim.dat$candidate <- factor(1:nrow(stim.dat),
    labels=rownames(result$unrotated$stimuli$mean))
ind.dat <- na.omit(ind.dat)
stim.dat <- stim.dat %>% filter(candidate %in%
  c("Humphrey", "Nixon", "Wallace"))
ggplot(ind.dat, aes(x=D1, y=D2)) +
    geom_point(aes(shape=vote, colour=vote), size=2) +
    scale_shape_manual(values=c("H", "N", "W"),
        labels=c("Humphrey", "Nixon", "Wallace"),
        name="Vote") +
```

```
scale_color_manual(values=c("gray50", "gray65", "gray80"),
    labels=c("Humphrey", "Nixon", "Wallace"),
    name="Vote") +
geom_text(data = stim.dat, aes(label=candidate),
    size=6, color="black") +
theme_bw() +
theme(aspect.ratio=1, legend.position="bottom") +
xlab("First Dimension") +
ylab("Second Dimension")
```

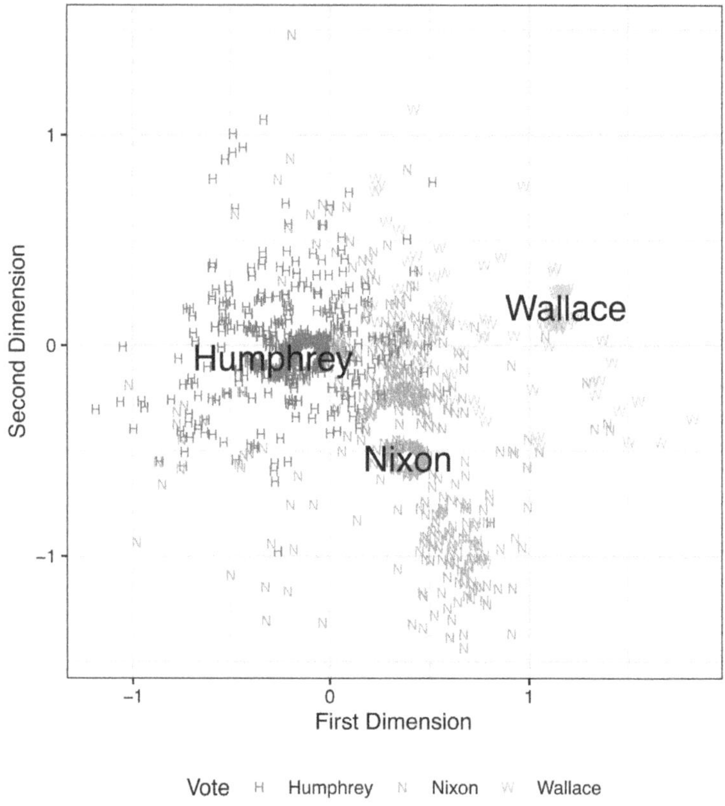

FIGURE 6.7: L-BFGS Optimization of the ANES 1968 Feeling Thermometers Data

The L-BFGS solution serves as the target configuration for the Bayesian unfolding procedure itself, which is executed using slice sampling (Neal, 2003).

Figure 6.8 shows the stimuli and individuals' mean first and second dimension coordinates from Bayesian unfolding using the slice sampler. This configuration is much more satisfying from the perspective of classical spatial voting theory. Nixon, Humphrey and Wallace are each located near the center of their voters.

```
ind.dat <- as.data.frame(result$rotated$individuals$mean)
names(ind.dat) <- c("D1", "D2")
ind.dat$vote <- NA
ind.dat$vote <- factor(ANES1968[,14], levels=1:3,
    labels=c("Humphrey", "Nixon", "Wallace"))[result$retained.obs]
stim.dat <- as.data.frame(result$rotated$stimuli$mean)
names(stim.dat) <- c("D1", "D2")
stim.dat$candidate <- factor(1:nrow(stim.dat),
    labels=colnames(anes.input))
stim.dat <- stim.dat %>% filter(candidate %in%
  c("Humphrey", "Wallace", "Nixon"))
ind.dat <- na.omit(ind.dat)
ggplot(ind.dat, aes(x=D1, y=D2)) +
    geom_point(aes(shape=vote, colour=vote), size=2) +
    scale_shape_manual(values=c("H", "N", "W"),
        labels=c("Humphrey", "Nixon", "Wallace"),
        name="Vote") +
    scale_color_manual(values=c("gray50", "gray65", "gray80"),
        labels=c("Humphrey", "Nixon", "Wallace"),
        name="Vote") +
    geom_text(data = stim.dat, aes(label=candidate),
        size=6, color="black") +
    theme_bw() +
    theme(aspect.ratio=1, legend.position="bottom") +
    xlab("First Dimension") +
    ylab("Second Dimension")
```

One of the clear advantages offered by a Bayesian approach to unfolding is the estimation of uncertainty about the parameter locations. In Figure 6.9 we plot the contours of the posterior distribution of the three presidential candidates in the 1968 thermometers data and their 90% credible intervals. We see that the credible intervals for the stimuli are slightly larger along the second dimension than the first. It is clear from the plot below that the posterior overlap is zero for any pair of candidates. So, we can say that all three candidates reliably occupy different spaces in the two dimensions of electoral competition.

```
## Difference between Main Candidates
stim.mat <- as.matrix(result$rotated$stim.samples)
d <- stim.mat[,3] -  #Nixon
     stim.mat[,2] #Humphrey
md1 <- mean(d > 0)
```

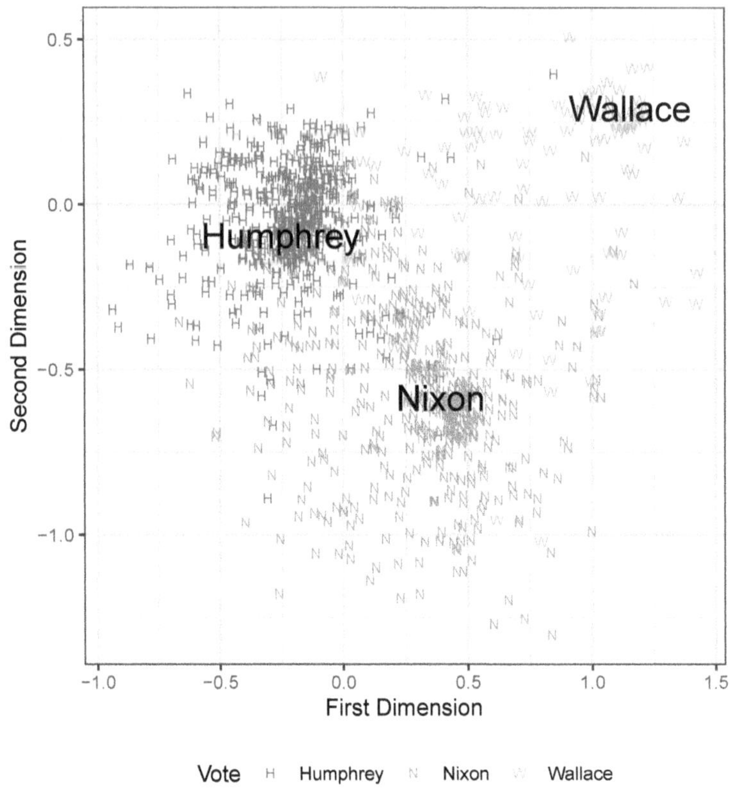

FIGURE 6.8: Bayesian Unfolding of the ANES 1968 Feeling Thermometers Data

```
d <- stim.mat[,1] -  #Wallace
     stim.mat[,2] #Humphrey
md2 <- mean(d > 0)
d <- stim.mat[,1] -  #Wallace
     stim.mat[,3] #Nixon
md3 <- mean(d > 0)
c(md1, md2, md3)

h.df <- as.data.frame(
  as.matrix(result$rotated$stim.samples)[,c(2,14)])
n.df <- as.data.frame(
  as.matrix(result$rotated$stim.samples)[,c(3,15)])
w.df <- as.data.frame(
```

```
  as.matrix(result$rotated$stim.samples)[,c(1,13)])
names(h.df) <- names(w.df) <- names(n.df) <- c("D1", "D2")
h.df$candidate = factor(1, levels=1:3,
                     labels=c("Humphrey", "Nixon", "Wallace"))
n.df$candidate = factor(2, levels=1:3,
                     labels=c("Humphrey", "Nixon", "Wallace"))
w.df$candidate = factor(3, levels=1:3,
                     labels=c("Humphrey", "Nixon", "Wallace"))
plot.df <- rbind(h.df, n.df, w.df)
z.df <- data.frame(
    x=result$rotated$stimuli$mean[,1],
    xl=result$rotated$stimuli$lower[,1],
    xu=result$rotated$stimuli$upper[,1],
    y=result$rotated$stimuli$mean[,2],
    yl=result$rotated$stimuli$lower[,2],
    yu=result$rotated$stimuli$upper[,2])
z.df$candidate <- factor(1:12, labels=colnames(anes.input))
z.df$short <- substr(as.character(z.df$candidate), 1, 3)
z.df <- z.df %>% filter(candidate %in%
  c("Humphrey", "Nixon", "Wallace"))
ggplot(plot.df, aes(group=candidate)) +
  geom_density_2d(aes(x = D1, y = D2), bins=4, color="gray75") +
  geom_segment(data=z.df, aes(x=xl, xend=xu, y=y, yend=y)) +
  geom_segment(data=z.df, aes(x=x, xend=x, y=yl, yend=yu)) +
  geom_text(data=z.df, aes(x=c(-.08, .4, 1), y=c(-0.1, -0.52, 0.3)),
          label=c("Humphrey", "Nixon", "Wallace")) +
  theme_bw() +
  theme(aspect.ratio=1) +
  labs(x="First Dimension",
       y="Second Dimension")
```

Uncertainty will necessarily be greater for the respondents than the stimuli. This is simply because there is more information about the stimuli than the voters. In this case, individuals rate at most 12 candidates, while there are hundreds of ratings from which to estimate the stimuli locations. We illustrate this in Figure 6.10 below. These plots show the size of the 90% credible intervals for the coordinates on the first (column 1) and second (column 2) dimensions. The distributions of the credible intervals for the respondent coordinates are plotted as smoothed histograms in the second row of the display, while the density of the widths of the 90% credible intervals for the 12 stimuli are plotted in the first row. What Figure 6.10 makes clear is that the uncertainty around the point estimates of the stimuli is smaller than for nearly all of the respondents on both dimensions.

It would be impractical to plot the credible intervals for all 1,392 respondents, but we do isolate two respondents with different profiles to show how those different profiles manifest in posterior uncertainty. As seen below, Respondent #8 in our original data provides an unconventional set of feeling

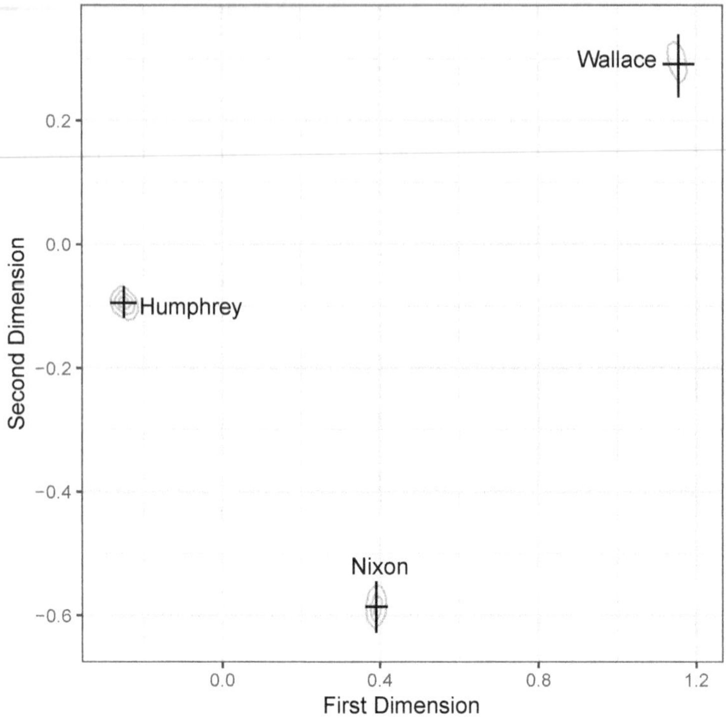

FIGURE 6.9: Stimuli Means and Credible Intervals from Bayesian Unfolding of the ANES 1968 Feeling Thermometers Data

thermometer ratings. He rates all of the candidates at 50 with the exceptions of Robert F. Kennedy (his most-preferred figure at 90), George Wallace (80) and President Lyndon Johnson (60). Because nine of the stimuli are rated equally, we have little information to glean about his political preferences.

```
anes.input <- anes.input[which(result$retained.obs), ]
print(anes.input[6,])
```

```
    Wallace    Humphrey       Nixon    McCarthy      Reagan
         80          50          50          50          50
Rockefeller     Johnson      Romney     Kennedy      Muskie
         50          60          50          90          50
      Agnew       LeMay
         50          50
```

Because of respondent #8's profile, we would expect the respondent's ideal point estimate to have a wide posterior density. We contrast Respondent 8 with respondent 730. This person uses nearly the whole range of the thermometer scores, giving Wallace, Reagan, Rockefeller and LeMay all ratings of

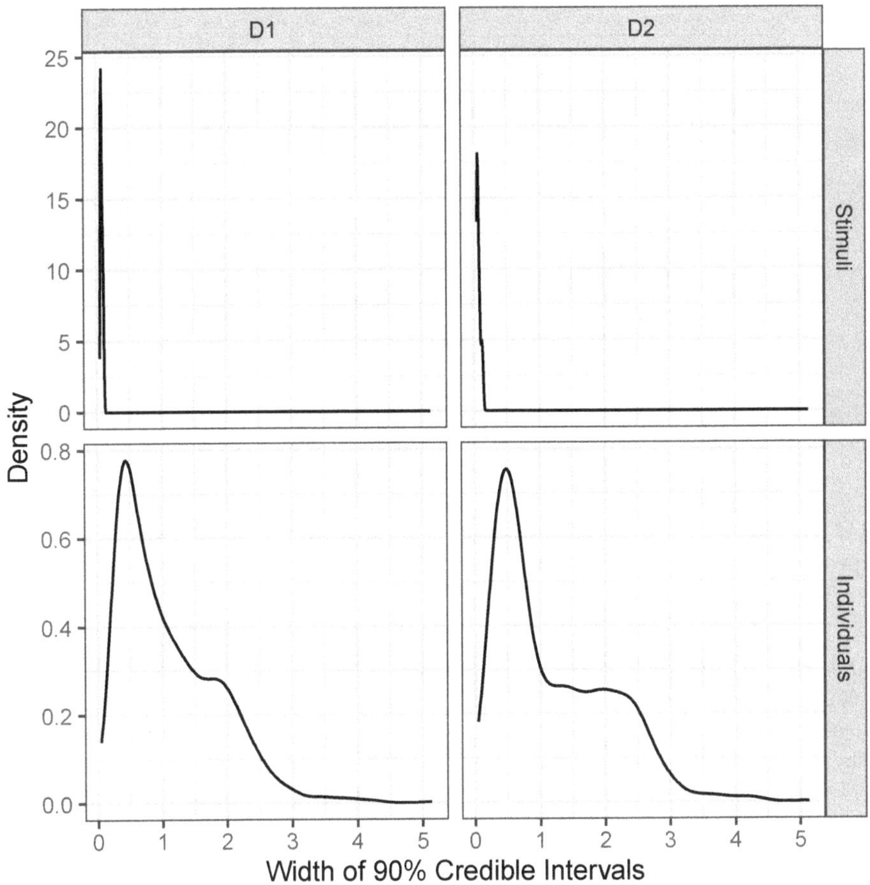

FIGURE 6.10: Uncertainty about Point Coordinates from Bayesian Unfolding of the ANES 1968 Feeling Thermometers Data

zero. On the other hand, Kennedy, Muskie, McCarthy, Agnew, Romney and Johnson - all receive scores of 97. Nixon, was rated 80 on the thermometer scale. This leads to an individual with the smallest credible interval on the first dimension. This is precisely what Figure 6.11 shows. The contours indicate a wide posterior distribution for respondent #8 and a narrow one for respondent #730.

The Bakker and Poole (2013) Bayesian unfolding model addresses the problems we have noted with least squares or maximum likelihood unfolding methods. Methodologically, it utilizes the log-normal distribution to model distances. The log-normal distribution is widely used in the natural sciences to

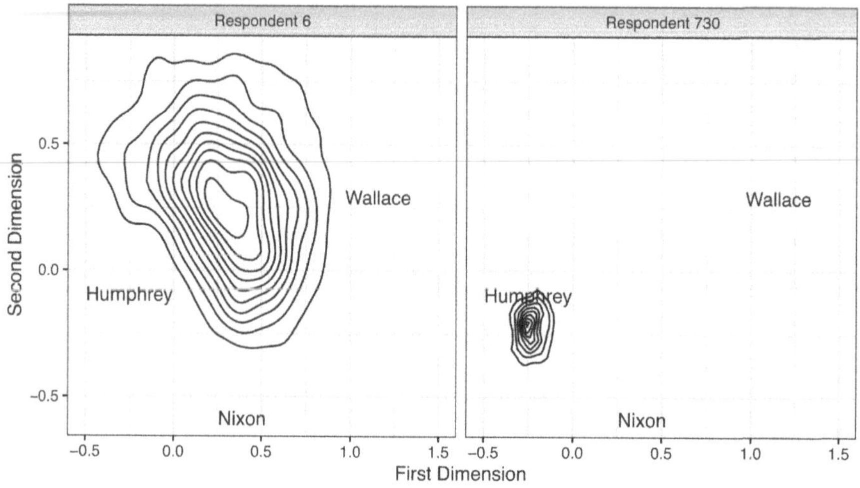

FIGURE 6.11: Posterior Density of Respondent Ideal Points from Bayesian Unfolding of the ANES 1968 Feeling Thermometers Data

model objects and processes that are inherently positive: for example, the concentration of elements in the Earth's crust and latent periods of infectious diseases (Limpert, Stahel and Abbt, 2001). It is also a more realistic model of the error process because the size of an observed distance should be proportional to the size of the *variance* of that distance. Unlike least squares or maximum likelihood optimization techniques (mode-finders), the Bayesian approach "illuminates" the entire distribution and allows us to summarize parameter locations with mean values and construct credible intervals around those point estimates. These features of the Bayesian unfolding model mean that it more plausibly locates stimuli closer to the center of the space, long the Achilles' heel of least squares and maximum likelihood unfolding techniques.

6.4 Parametric Methods - Bayesian Item Response Theory

The Item Response Theory (IRT) model was developed in psychometrics for the measurement of skills-based tests and test subjects. In the IRT model, subjects possess a latent level of *ability* that is measured through a series of test questions (items). For each item, subjects are predicted to answer correctly if they have a level of ability above some threshold, and incorrectly otherwise. This threshold is known as the item *difficulty* parameter. The individual ability and item difficulty variables comprise the one-parameter IRT model (also known as the Rasch model). The two-parameter IRT model adds a *discrimination* parameter for each item that measures how well each item discriminates subjects based on their latent ability. For instance, a completely unambiguous question would have a discrimination parameter of ∞ (that is, a step function), while an unclear question would have a discrimination parameter near 0 (a nearly flat slope).

The original IRT model can be extended to the ideal point estimation of legislators if one accepts that we can substitute political ideology for ability (both being latent individual attributes), legislators for subjects, and roll calls for test items (Ladha, 1991). For example, while a student who correctly answers a battery of calculus questions would be predicted to correctly answer an arithmetic question, a legislator who votes in favor of a series of very left-wing proposals would be predicted to support a moderately left-wing proposal over a right-wing status quo. In this application, the difficulty parameter represents the cutting plane (i.e., how left- or right-wing a legislator needs to be in order to be classified as a Yea/Nay vote), and the discrimination parameter indicates how well a roll call vote separates legislators based on ideology. However, while the focus in testing applications of the IRT model is on the estimated values of the item parameters (to determine how well test items are constructed), political scientists' quantity of interest is usually the individual parameters: the legislator ideal points (Clinton, Jackman and Rivers, 2004, p. 356).

Scalogram analysis or Guttman scaling (Guttman, 1944, 1950) is intimately related to the IRT model, as both are a form of *cumulative scaling*. Guttman scaling is a set of items (questions, problems, etc.) that are ranked in order of difficulty so that those who answer correctly (agree) on a more difficult (or extreme) item will also answer correctly (agree with) to all less difficult (less extreme) items that preceded it. The difference is that Guttman scaling uses a deterministic step function to model responses. For each item, the probability of a correct response is 0 for individuals with scores below a cutting point and 1 for individuals whose scores are higher than the cutting point. In contrast, the IRT model uses a smoothed function in which the probability of a correct response increases monotonically over the latent scores such that the probability of a correct response is only 0 or 1 asymptotically.

Both unfolding analysis and cumulative scaling deal with an individual's responses to a set of stimuli. But cumulative scaling is very different from the unfolding model. In terms of utility theory, unfolding analysis assumes a single-peaked (usually symmetric) utility function. That is, utility (the degree of preference) declines with distance from the individual's ideal point. In contrast, cumulative scaling (the classical IRT model) is based on an Item Characteristic Curve (ICC, also referred to as an item response function or a trace line) that monotonically increases or decreases over the relevant dimension or space. Above some threshold the individual is predicted to respond Yea/correct, and below the threshold the individual is predicted to respond

Nay/incorrect. The counterpart to an ideal point is the position on the scale where the individual's responses switch from Yea/correct to Nay/incorrect.

Interestingly, these two very different models are observationally equivalent in the context of parliamentary voting (Weisberg, 1968; Ladha, 1991; Poole, 2005). In the unfolding model, there are two outcomes for every parliamentary motion—one corresponding to Yea and one corresponding to Nay. Legislators vote for the option closest to their ideal points. In one dimension this forms a perfect scalogram (Weisberg, 1968). Hence, Guttman scaling methods and their IRT descendants can be used to analyze legislative roll call data.

The IRT model has been adapted by political scientists (Clinton, Jackman and Rivers (2004), Martin and Quinn (2002)) to perform unfolding analysis using the random utility model with quadratic deterministic utility. In this usage of the IRT model, the recovered dimensions are *policy*, not *ability* dimensions. The individuals' ideal points and the midpoints of the Yea and Nay alternatives are recovered (see Equation 6.20 below). In this context, the IRT model is isomorphic with the Quadratic-Normal (QN) unfolding method developed by Poole (2001, 2005).

Since IRT models involve the joint estimation of the ability, difficulty, and (in the two-parameter model) discrimination parameters, Bayesian methods are attractive because they allow for the simultaneous estimation of the parameters. More specifically, the Gibbs sampler ("the workhorse of the MCMC world" (Robert and Casella, 2010, p. 199)) offers an efficient means of analyzing high-dimensional posterior densities, like those regularly produced from legislative roll call data. The Gibbs sampler breaks down the high-dimensional posterior density into a series of more tractable components and samples from the *conditional* densities for each component, improving the approximation to the posterior density at each iteration (Jackman, 2009, pp. 214-221).

Our focus in this section is on the extension of the two-parameter (Bayesian) IRT model to the analysis of legislative roll call data developed by Clinton, Jackman and Rivers (2004). In the Clinton-Jackman-Rivers (CJR) model, x_i is the legislator ideology score, α_j is the roll call difficulty parameter and β_j is the roll call discrimination parameter. As with the NOMINATE model, let i index legislators, $(i = 1,..,n)$, j index roll calls $(j = 1,...,q)$ and k index dimensions $(k = 1,...,s)$.

The CJR model, like the NOMINATE model, is parametric. However, unlike NOMINATE, the CJR model uses the quadratic functional form to model legislators' utility functions. Hence, in the unidimensional case, legislator i's utility for the Yea outcome on the jth roll call (compare with Equation 5.7) is:

$$U_{ijy} = u_{ijy} + \varepsilon_{ijy} = \sum_{k=1}^{s} -(x_{ij} - O_{jky})^2 + \varepsilon_{ijy} \tag{6.17}$$

As with NOMINATE, the errors in the IRT model are assumed to follow a normal distribution. Assuming quadratic utility and normally distributed errors, the probability that legislator i votes Yea on the jth roll call (compare with Equation 5.9) is:

$$
\begin{aligned}
P_{ijy} &= P(U_{ijy} > U_{ijn}) \quad (6.18)\\
&= P(\varepsilon_{ijn} - \varepsilon_{ijy} < u_{ijy} - u_{ijn})\\
&= P(\varepsilon_{ijn} - \varepsilon_{ijy} < \|x_i - O_{jn}\|^2 - \|x_i - O_{jy}\|^2)\\
&= P(\varepsilon_{ijn} - \varepsilon_{ijy} < 2(O_{jy} - O_{jn})'x_i + O'_{jn}O_{jn} - O'_{jy}O_{jy})\\
&= \Phi(\beta_j' x_i - \alpha_j)
\end{aligned}
\quad (6.19)
$$

Where, by construction, $\alpha_j = O'_{jn}O_{jn} - O'_{jn}O_{jy} = (O_{jn} - O_{jy})'(O_{jn} + O_{jy})$ and $\beta_j = 2(O_{jy} - O_{jn})$ so that the midpoint of the jth roll call is:

$$M_j = \frac{(O_{jn} + O_{jy})}{2} = \frac{(O_{jn} - O_{jy})'(O_{jn} + O_{jy})}{2(O_{jy} - O_{jn})} = \frac{\alpha_j}{\beta j} \quad (6.20)$$

Since M_j is equal to $\frac{\alpha_j}{\beta j}$, when $x_i = \frac{\alpha_j}{\beta j}$ then P_{ijy} and P_{ijn} is 0.5, since legislator i is equally distant and hence indifferent between the two outcomes. The midpoint of roll call j is not simply the difficulty parameter $alpha_j$ because in this context, we are estimating a latent *ideological* dimension under the assumption of single-peaked (quadratic) utility functions rather than an ability or dominance-based dimension. What is being recovered, then, are the legislator ideal points and the midpoints of the roll calls.

This yields the likelihood function (compare with Equation 5.10):

$$L(\mathbf{B}, \alpha, \mathbf{X}|\mathbf{Y}) = \prod_{i=1}^{n}\prod_{j=1}^{q}(P_{ijy})^{y_{ij}}(1 - P_{ijy})^{1-y_{ij}} \quad (6.21)$$

Where y_{ij} are 0 or 1, B is a q by s matrix of β_j values, α is a q length vector of α_j values, and X is a n by s matrix of x_i values (Jackman, 2001; Clinton, Jackman and Rivers, 2004).[§] Priors (usually vague and normal) are assigned for the unknown parameters B, α, and X. This yields the posterior distribution:

$$p(\mathbf{B}, \alpha, \mathbf{X}|\mathbf{Y}) \propto p(\mathbf{B}, \alpha, \mathbf{X}) \times L(\mathbf{B}, \alpha, \mathbf{X}|\mathbf{Y}) \quad (6.22)$$

[§]Note that because a single item difficulty parameter (α_j) is estimated in the multidimensional case, this is a *compensatory* multidimensional IRT model (Bolt and Lall, 2003). α_j can be interpreted as providing the location where the slope of the ICC (item characteristic curve, which models the probability of a given response across values along the latent dimension) is at its steepest; that is, where the item is at its most discriminating (Reckase, 1985, 2010). The meaning of the discrimination parameters (β_{jk}) and legislator ideal points (x_{ik}) is unchanged from the unidimensional model. In addition, the β_{jk} values can be used to assess how strongly items load onto each of the latent dimensions in much the same way as the factor loadings are used in factor analysis (Jackman, 2001, pp. 229-230).

6.4.1 The MCMCpack and pscl Packages in R

The `MCMCpack` package (Martin, Quinn and Park, 2011) allows for Bayesian (MCMC-based) estimation of a suite of models common in the social sciences, from the classical linear regression model to more complex models involving hierarchical regression, limited dependent variables, and the measurement of latent variables (e.g., factor analysis and IRT). The sampling of the posterior density is implemented via the `C++` programming language, allowing for very efficient estimation of the model. In this section we demonstrate the package's `MCMCirt1d()` and `MCMCirtKd()` functions for the Bayesian estimation of unidimensional and multidimensional IRT models.

The `pscl` package (Jackman, 2012) includes the aforementioned `readKH()` and `rollcall()` functions to create `rollcall` objects in `R`. It also includes the versatile `ideal()` function that performs Bayesian IRT in one or more dimensions using the Clinton, Jackman and Rivers (2004) model discussed above.

6.4.2 Example 3: The 2000 Term of the US Supreme Court (Unidimensional IRT)

We first demonstrate how to perform Bayesian unidimensional IRT with the `MCMCirt1d()` function using data from the 2000 Term of the US Supreme Court. The data set (`SupremeCourt`) is included in the `MCMCpack` package, and contains the nine Justices' votes on 43 non-unanimously decided cases. We transpose the matrix so that the Justices are on the rows and the votes are on the columns and store it in the object `SupCourt2000`. The `MCMCirt1d()` and `MCMCirtKd()` functions require that the values in the roll call matrix be either 0, 1, or *NA*, as they are in the `SupremeCourt` matrix.

```
library(MCMCpack)
data(SupremeCourt)
SupCourt2000 <- t(SupremeCourt)
print(SupCourt2000[1:6,1:6])
```

```
           1 2 3 4 5 6
Rehnquist  0 0 0 0 1 0
Stevens    1 1 1 0 1 1
O'Connor   1 0 0 0 0 0
Scalia     0 0 0 0 0 0
Kennedy    1 0 0 0 1 0
Souter     1 1 1 0 0 0
```

The first argument required by the `MCMCirt1d()` function is the name of the roll call matrix (`SupCourt2000`). The polarity of the result can be set by constraining the sign of one of the Justice's ideal points with the argument `theta.constraints` (we constrain Justice Scalia to have a positive score). We can also achieve identification here by fixing $s+1$ points (e.g., `theta.constraints=list(Scalia=1,Ginsburg=-1)`, but with unidimensional IRT, we prefer to identify the result via a normalization constraint (see Section 6.4.2.2). This can be done via post-processing which we discuss below.

We set the sampler to run for 55,000 iterations with the argument `mcmc=55000`,discarding the first 5,000 draws with `burnin=5000` and thinning the samples by 10 with `thin=10`. This will produce 5,000 samples of the joint posterior distribution. Starting values for the parameters can be set with the arguments `theta.start`, `alpha.start`, and `beta.start`. If set to `NA` (the default value), starts are automatically generated. Specifically, the values for `theta.starts` are generated from an eigenvalue-eigenvector decomposition of the agreement score matrix between legislators.¶ The values for `alpha.starts` and `beta.starts` are generated through a series of probit regressions using the values from `theta.start`.

Next, the `MCMCirt1d()` function requires values for the prior means and prior precisions (the inverse variances) of the parameters. The prior means of the ideal points (x_i), item difficulty parameters (α_j) and item discrimination parameters (β_j) are set with the `t0`, `a0`, and `b0` arguments, respectively. The prior precisions are assigned with the `T0`, `A0`, and `B0` arguments, respectively. By defaults, all prior means are set to 0, the prior precision of the ideal points is set to 1 (variance of 1) and the prior precision of the item difficulty and discrimination parameters is set to 0.25 (variance of 4).

Finally, the seed for the random number generator is set with the `seed` argument (the `NA` default means that the Mersenne Twister (Matsumoto and Nishimura, 1998) is used with the seed value of 12345). If the `verbose` argument is set to a value greater than 0, it will print the status of the sampler at each nth iteration. The draws from the marginal posterior densities of the item and ability (ideal point) parameters are stored in the result if `store.item` and `store.ability` are set to `TRUE`. If there are many subject or item parameters being estimated and computer memory is limited, the applicable argument may need to be set to `FALSE`. Parameters that are constrained to a constant value (e.g., if a legislator's ideal point is constrained to be -1), these items can be dropped from the result with the argument `drop.constant.items=TRUE`.

```
posterior1d.unid <- MCMCirt1d(SupCourt2000,
    theta.constraints=list(Scalia="+"),
    mcmc=55000, burnin=5000, thin=10,
    theta.start=NA, alpha.start=NA, beta.start=NA,
    t0=0, T0=1, a0=0, A0=0.25, b0=0, B0=0.25,
    seed=NA, verbose=0, store.item=TRUE,
    store.ability=TRUE, drop.constant.items=TRUE)
```

¶As discussed in Chapter 3, this is an efficient way to generate high-quality initial values for the legislator parameters.

We identify the model through normalization, which first requires us to standardize the posterior distribution for the ideal points. The ideal points for the justices (or more generally for the latent variable values) are those that start with **theta**. We can pull those from the model with a simple regular expression.[‖] We can scale the rows of the matrix with the `apply` function. We store these scaled points in the **justicesZ** object. We also calculate the standard deviations for each of the iterations of the Markov chain for subsequent use.

```
justices <- posterior1d.unid[,
  grep("^theta", colnames(posterior1d.unid))]
justicesZ <- apply(justices, 1, scale)
justices_sd <- apply(justices, 1, sd)
justices_mean <- apply(justices, 1, mean)
```

Next, we recognize that **MCMCpack** uses the regression parameterization, so the quantities that get returned come from the following model:

$$F^{-1}(Pr(Y_{ij}=1)) = \alpha_j + \beta_j x_i \tag{6.23}$$

If we standardize the x_i variable, then we need to make an adjustment to the other model parameters. That is, if we have $x_i^* = \frac{(x_i - \bar{x})}{s_x}$, then we want to know what α_j^* and β_j^* are in the equation below:

$$F^{-1}(Pr(Y_{ij}=1)) = \alpha_j^* + \beta_j^* x_i^* \tag{6.24}$$

Considering a single item j, if we are only standardizing the x_i variable, then we know that β^* is βs_x. Since the standardized and unstandardized models produce the same predictions, we can equate the right-hand sides of both equations and then solve for α^*.

$$\begin{aligned}
\alpha + \beta x_i &= \alpha^* + \beta^* \frac{(x - \bar{x})}{s_x} && (6.25)\\
&= \alpha^* + \frac{\beta^* x}{s_x} - \frac{\beta^* \bar{x}}{s_x} && (6.26)\\
&= \alpha^* + \frac{\beta x s_x}{s_x} - \frac{\beta \bar{x} s_x}{s_x} && (6.27)\\
&= \alpha^* + \beta x - \beta \bar{x} && (6.28)\\
-\alpha^* &= \beta x - \beta \bar{x} - \beta x - \alpha && (6.29)\\
\alpha^* &= \beta \bar{x} + \alpha && (6.30)
\end{aligned}$$

So, we can use these to adjust the other model parameters given that we have scaled the latent variable.

[‖] These are also the first ideal points in the matrix.

```
alphas <- posterior1d.unid[,
  grep("^alpha", colnames(posterior1d.unid))]
betas <- posterior1d.unid[,
  grep("^beta", colnames(posterior1d.unid))]
disc.parameters <- apply(betas, 2, function(x)x*justices_sd)
diff.parameters <- apply(betas, 2, function(x)x*justices_mean) +
  alphas
```

What we have just done above is to force the variance of the ideal points to be one for each iteration of the MCMC sampler. That identifies the model and makes the draws comparable across iterations. This is important because the standard deviations of the ideal point estimates varied from just below 1 to over 2.2, indicating that changes in ideal point locations from one draw to another were a function not only of true movement, but also of changes in the scale of the distribution from iteration to iteration.

We calculate the point estimates of the subject and item parameters as the means of their respective marginal posterior distributions.

```
idealpt.means <- colMeans(justices)
diff.means <- colMeans(diff.parameters)
disc.means <- colMeans(disc.parameters)
```

In Figure 6.12 we plot the sampled posterior distributions of the first four Supreme Court Justices in the roll call matrix `SupCourt2000`: Justices John Paul Stevens, Sandra Day O'Connor, Antonin Scalia and Chief Justice William Rehnquist. We limit the number of Justices to four for presentation purposes; we could also label the posterior means with the first three characters of their last names with the command `substring(rownames(SupCourt2000)[i],1,3)`. Based on their voting decisions in the 2000 Term of the Supreme Court, Justice Stevens is the most liberal member of the Court, Justice Scalia is the most conservative, and Justice O'Connor occupies the median ideological position. This result accords with widespread belief (and some empirical evidence) that Justice O'Connor constituted the "swing" vote on the Court during this period (Martin, Quinn and Epstein, 2005).

```
sc.df <- data.frame(
    ideal.pt = c(t(justicesZ)),
    justice = factor(rep(1:9, each=nrow(justices)),
                     labels=colnames(SupremeCourt))
)
sc.df <- sc.df[which(sc.df$justice %in%
  c("Stevens", "O'Connor", "Rehnquist", "Scalia")), ]
dens <- by(sc.df$ideal.pt, list(sc.df$justice), density)
x <- c(unlist(sapply(dens, function(z)z$x)))
y <- c(unlist(sapply(dens, function(z)z$y)))
dens.df <- data.frame(x=x, y=y, justice = factor(rep(1:4, each=512),
    labels=colnames(SupremeCourt)[1:4])
)
```

```
ggplot(dens.df) +
  geom_line(aes(x=x,y=y, linetype=justice), size=.75) +
  theme_bw() +
  theme(aspect.ratio=1, legend.position="bottom") +
  xlim(-2,3) +
  labs(x="Ideology (Liberal - Conservative)",
       y = "Posterior Density", linetype="Justice") +
  guides(linetype=guide_legend(nrow=2))
```

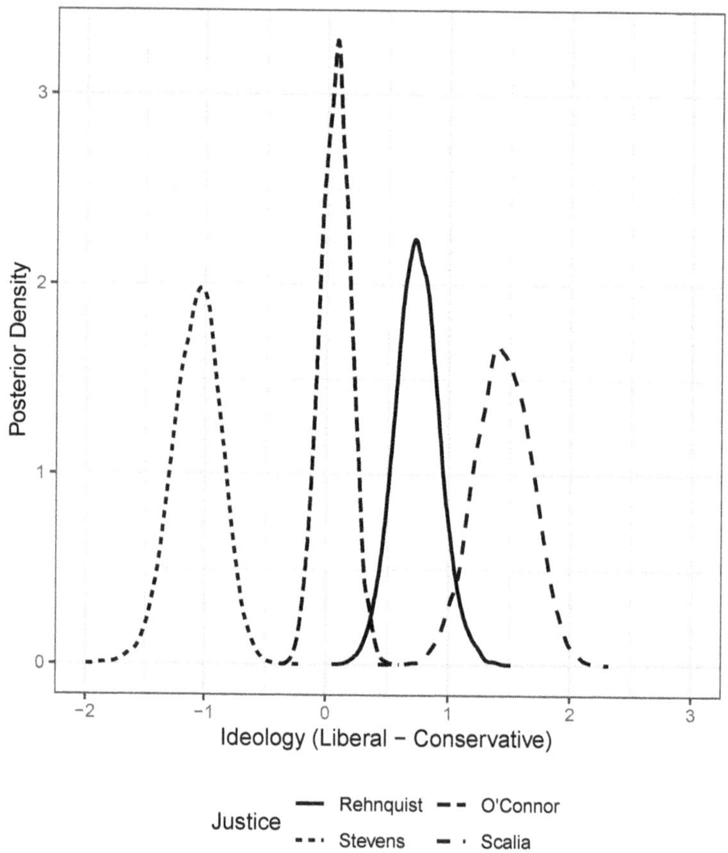

FIGURE 6.12: Posterior Densities of Supreme Court Justices

From Figure 6.12, it is clear that the posterior distributions of the ideologically centrist Justices (like Justice O'Connor) are less widely dispersed than those on the ideological periphery (Justices Stevens and Scalia). It is more challenging to estimate the ideal points of extreme legislators because there are fewer votes that divide legislators on the edge of the space. Hence, it is difficult to determine just *how* extreme they are. Poole and Rosenthal (1997, Appendix) and Jackman (2000*a*, p. 324) have also noted this point.

We illustrate this pattern in Figure 6.13 where we plot the location of the 43 midpoints on the latent ideological dimension with a vertical line superimposed on the each justice's posterior distribution. As can be seen, most votes divide Justices near the center of the space (31 of the 43 cases were decided by a 5-4 or 6-3 vote). Of course, this problem is most acute when there is a relatively sparse number of roll call votes.

```
cl.df <- lapply(1:4, function(x)data.frame(cutline = diff.means/disc.
means))
for(i in 1:4){
  cl.df[[i]]$justice <- factor(i, levels=1:9,
                               labels=levels(sc.df$justice))
}
cl.df <- do.call(rbind, cl.df)
cl.df <- cl.df %>% filter(cutline > -4)
ggplot(sc.df) +
  geom_vline(data=cl.df, aes(xintercept=cutline), size=.035,
             color="black") +
  geom_density(aes(x=ideal.pt), fill="gray50", linetype=0, alpha=.5) +
  facet_wrap(~justice, scales="free_y") +
  theme_bw() +
  theme(panel.grid.major=element_blank(),
        panel.grid.minor=element_blank()) +
  labs(x="Ideology (Liberal - Conservative)",
       y = "Posterior Density") +
  theme(aspect.ratio=1)
```

6.4.2.1 Item Characteristic Curves

Above, we discussed Item Characteristic Curves (ICCs). These are the functions that relate the latent trait to the probability of a positive response on each outcome variable. We can also produce these curves from the output of the model. We use the `diff.means` and `disc.means` objects to construct the ICCs We do so over the range of values in the ideal point estimates. These can all be seen in Figure 6.14. Note that in the figure some items have a positive slope and some a negative one. This indicates that a vote in favor means different things depending on the particular case. It indicates a liberal ruling in some cases and a conservative one in others. Also note that there are differences in slope - some have a quite steep slope (i.e., they are good at discriminating on the latent dimension), and others are nearly flat.

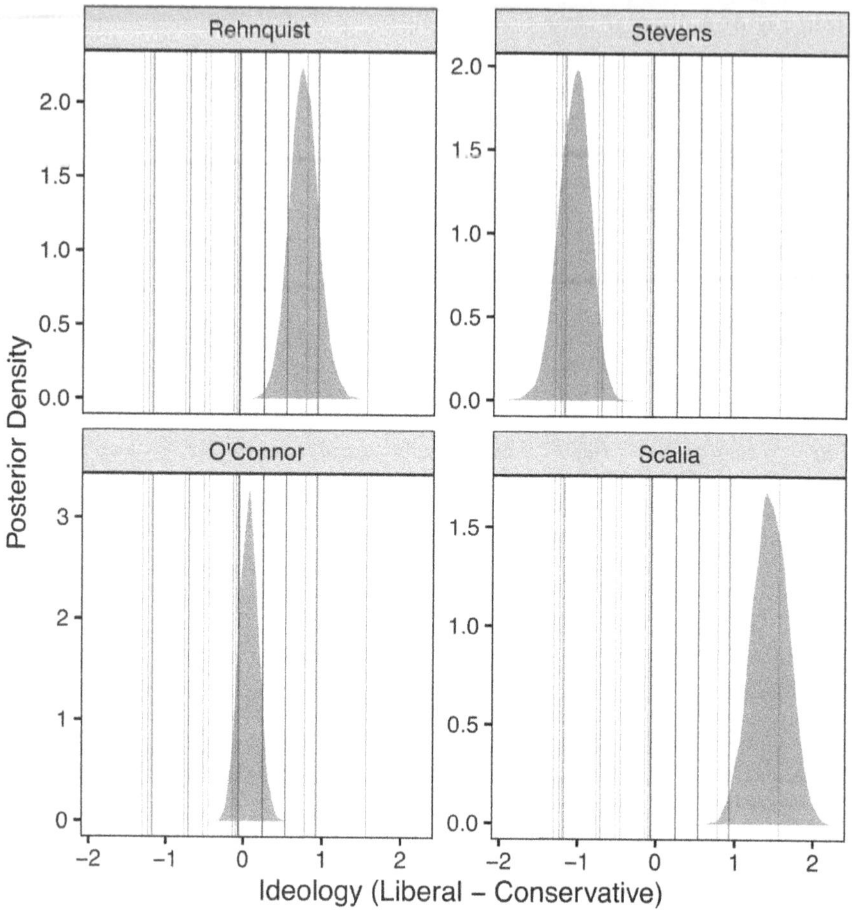

FIGURE 6.13: Item Difficulty Parameters (Cut Lines) of US Supreme Court 2000 Term Votes

```
rg <- range(c(justicesZ))
s <- seq(rg[1], rg[2], length=100)
probs <- pnorm(cbind(1, s) %*% rbind(diff.means, disc.means))
icc.df <- data.frame(
  probs = c(probs),
  s = rep(s, 43),
  indicator = as.factor(rep(1:43, each=100))
)
ggplot(icc.df, aes(x=s, y=probs, group=indicator)) +
  geom_line() +
  theme_bw() + theme(aspect.ratio=1) +
  labs(x="Ideology (Liberal - Conservative)",
       y = "Posterior Probability of Vote")
```

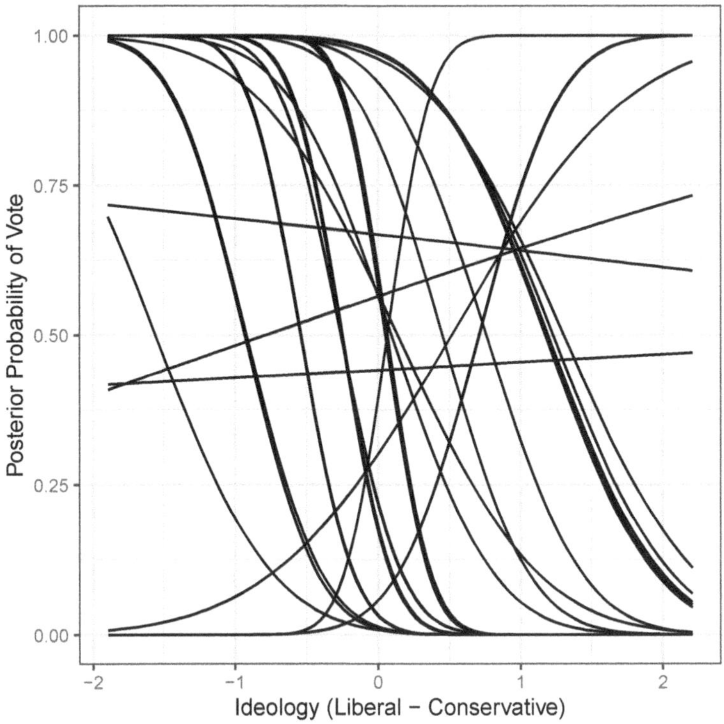

FIGURE 6.14: Item Characteristic Curves of Supreme Court Items

6.4.2.2 Unidimensional IRT using the ideal() Function in the pscl Package

We next demonstrate how to perform unidimensional Bayesian IRT with the `ideal()` function in the `pscl` package (Jackman, 2012) using the same data from the 2000 term of the US Supreme Court. The `ideal()` function first requires that the data be formatted as a `rollcall` object, which we do below by creating the object `SC.rc`.

The `ideal` function requires several arguments, the first of which is the `rollcall` object to be analyzed. `dropList` is a list of legislators and roll call votes to be dropped from the analysis. The default value, shown below, is to drop unanimous roll call votes from the analysis. We also add the parameter `legisMin=20` that drops legislators who have cast less than 20 scalable roll call votes. `d` represents the number of dimensions to be estimated. The total number of iterations to be run is set with `maxiter`, with the first `burnin` iterations discarded and the remaining number of iterations thinned by the value of `thin`. The default values, used below, produce 95 samples from the conditional posterior densities of the parameters.

Yea/Nay votes can be imputed for missing votes if `impute` is set to `TRUE`, although we recommend against doing so. The normalization constraint (i.e., requiring the ideal points to have a mean of 0 and a standard deviation of 1) is sufficient for *local* identification of the one-dimensional IRT model, and is implemented via post-processing by setting the `normalization` argument to `TRUE`. Normal vague priors (mean of 0 and variance of 25) are assigned by setting `priors` to `NULL`. Alternatively, `priors` can be a list of matrices specifying the means (`xp`) and precisions (`xpv`) of the legislator ideal points and the means (`bp`) and precisions (`bpv`) of the roll call vote (item) parameters, with the discrimination parameters first and the difficulty parameter last.

Starting values for the parameters are assigned with the `startvals` argument. The default value of `"eigen"` means that the starting values are calculated from a double-centered agreement score matrix (see Chapter 3). `startvals` can also be a list of ideal point and roll call parameter locations. `store.item` indicates whether the roll call parameters (the difficulty and discrimination parameters) should be stored in the result. With large roll call matrices, the number of item parameters can balloon and be a strain on memory resources. `file` specifies the file directory where the MCMC output should be written, if desired. Finally, the progress of the Gibbs sampler will be printed to the screen if `verbose` is set to `TRUE`.

```
library(pscl)
SC.rc <- rollcall(SupCourt2000, legis.names=rownames(SupCourt2000))
ideal.SupCourt <- ideal(SC.rc,
    dropList=list(codes="notinLegis",lop=0,legisMin=20),
    d=1, maxiter=10000, thin=100, burnin=500,
    impute=FALSE, normalize=TRUE, priors = NULL, startvals="eigen",
    store.item=TRUE, file=NULL, verbose=FALSE)
```

The `ideal()` function stores eight objects in the list `ideal.SupCourt`:

n Number of legislators.

m Number of roll call votes.

d Number of dimensions estimated.

codes Codes for roll call votes.

x An array of ideal point samples with the chains on the rows and the legislators on the columns. The results for each dimension can be accessed with `$x[,,1]`, `$x[,,2]`, etc.

beta An array of roll call vote samples with the chains on the rows and the roll call votes on the columns. The results for each dimension can be accessed with `$beta[,,1]`, `$beta[,,2]`, etc.

xbar The mean ideal point from the sampled values for each legislator.

betabar The mean difficulty and discrimination parameter from the sampled values for each roll call vote.

call The arguments passed to the `ideal()` function.

In one dimension, we can identify the result either through the above-discussed normalization constraint or by fixing two legislator ideal points via post-processing. That is, we can run the model without constraints, obtain an unidentified result, and then identify the result by restricting two legislators to fixed locations. Below we re-run the `ideal()` function on the Supreme Court data in one dimension, but this time we omit the normalization constraint so that the obtained result is unidentified. We then use the `postProcess()` function in the `pscl` package to fix the locations of two ideal points and achieve identification of the unidimensional result. The first argument in the `postProcess()` function is the `ideal` object itself and the second argument is a list of constraints. Below we fix Justice Stevens' ideal point to -1 and Justice Scalia's ideal point to 1, which identifies the `ideal` result (now stored in the object `identified.result`).

For the unidimensional IRT model, we recommend use of the normalization constraint because it does not force as rigid of a configuration of the ideal points. As we discuss in Section 6.4.5, however, the normalization constraint is insufficient for identification of multidimensional IRT results. In the multidimensional case, constraints on the legislator and/or roll call parameters are required.

```
unidentified.result <- ideal(SC.rc, d=1, maxiter=10000,
    thin=100, burnin=500, normalize=FALSE)
identified.result <- postProcess(unidentified.result,
    constraints=list(Stevens=-1,Scalia=1))
```

The `pscl` package includes a useful `plot.ideal()` function that can be used to plot the estimated legislator ideal points and surrounding credible intervals from an `ideal` result. We do so below, setting the credible interval at 0.95 (i.e., the 95% ranges of the posterior densities). Not surprisingly, the results are essentially identical to those from the `MCMCirt1d()` function earlier, with the two configurations correlated at $r = 0.998$.

```
plot(ideal.SupCourt, conf.int=0.95)
```

6.4.3 Running Multiple Markov Chains in MCMCpack and pscl

The use of multiple chains with dispersed starting values and a sufficient number of iterations produces a more comprehensive search of the posterior densities of the parameters. Aside from preferring a more fulsome exploration of the posterior distribution for its own sake, there are several more concrete

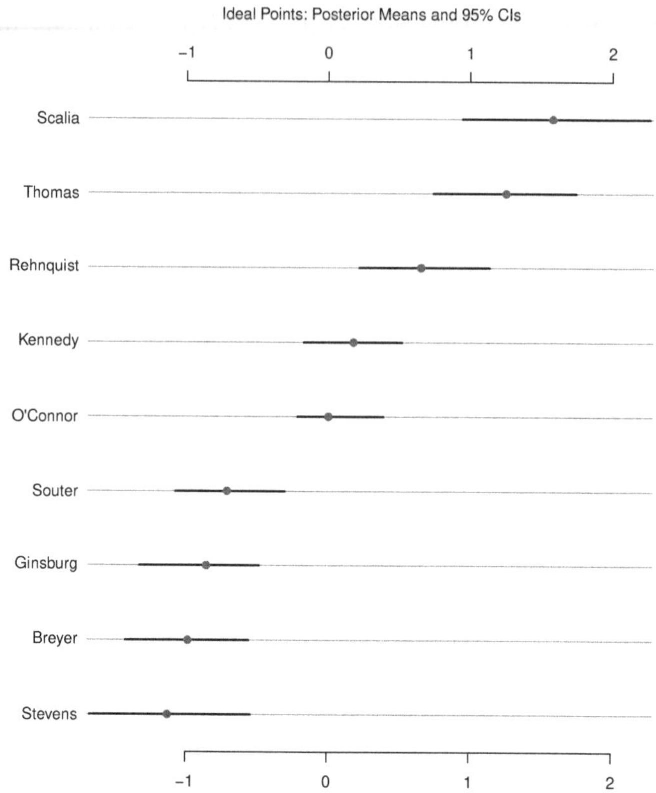

FIGURE 6.15: Point Estimates and 95% Credible Intervals for Supreme Court Justices

advantages. Multiple chains can reveal identification problems that cannot be diagnosed with a single chain. This often results in parallel chains (both with otherwise good properties) that straddle zero. Further, one of the most common diagnostic procedures, the Potential Scale Reduction Factor (PSRF) (Gelman and Rubin, 1992) requires two chains and is technically only valid for chains started in different parts of the multivariate posterior distribution.

The `MCMCirt1d()` and `MCMCirtKd()` functions in the `MCMCpack` package and the `ideal()` function in the `pscl` package do not include an option to run multiple Markov chains. However, we can execute the functions multiple times with separate initializations. This could include different starting values for the parameters or, in `MCMCpack`, a different seed for the (pseudo-) random number generator or an alternate random number generator itself.

First, we re-run the `MCMCirt1d()` function on the `SupCourt2000` data set and store the results in the MCMC object `posterior1d.unid.2`. The command itself is identical to that used in Example 3 with the exception of the starting values.

Below we change the seed for the Mersenne Twister random number generator from its default of 12345 to 23456. For the starting values of the Justice ideal points (`theta.start`), we store nine random draws from the uniform distribution bounded between −2 and 2 in the vector `theta`. Because Justice Scalia (the fourth Justice) is constrained to have a positive score, we take the absolute value of the fourth number in `theta`. We then use random draws from the standard normal distribution as the starting values for the difficulty and discrimination parameters (`alpha.start` and `beta.start`). As before, to identify the result, we normalize the values of each iteration of the chain and store the identified result in the object `posterior1d.2`.

```
theta <- runif(9, min=-2, max=2)
theta[4] <- abs(theta[4])

posterior1d.unid.2 <- MCMCirt1d(SupCourt2000,
    theta.constraints=list(Scalia="+"),
    mcmc=55000, burnin=5000, thin=10,
    theta.start=theta, alpha.start=rnorm(1), beta.start=rnorm(1),
    t0=0, T0=1, a0=0, A0=0.25, b0=0, B0=0.25,
    seed=23456, verbose=0, store.item=TRUE,
    store.ability=TRUE, drop.constant.items=TRUE)
posterior1d.2 <- t(apply(posterior1d.unid.2, 1, scale))
```

In order to merge the results from the two chains in an `mcmc.list` object, we must first format the identified results `posterior1d` and `posterior1d.2` as `mcmc` objects with the command `as.mcmc`. We also transfer the names of the estimated parameters, which are stored as the column names of the original (un-identified) results `posterior1d.unid` and `posterior1d.unid.2`. The results are then merged in the `mcmc.list` object `chains`. `chains[[1]]` stores the first result, and `chains[[2]]` stores the second result.

```
posterior1d <- as.mcmc(posterior1d)
colnames(posterior1d) <- colnames(posterior1d.unid)
posterior1d.2 <- as.mcmc(posterior1d.2)
colnames(posterior1d.2) <- colnames(posterior1d.unid.2)
chains <- mcmc.list(posterior1d,posterior1d.2)
```

MCMC convergence diagnostics (discussed in Section 6.4.4.1) can be run on the object `chains`. However, in order to calculate statistics about the posterior densities of the parameters based on both chains, we combine the two sets of simulations into a single matrix (`full.posterior1d`) using the `rbind` command. Then, as before, the marginal posterior densities of the legislator and roll call parameters are extracted into the objects `legislators` and `rollcalls`.

```
full.posterior1d <- rbind(chains[[1]],chains[[2]])
nlegislators <- nrow(SupCourt2000)
```

```
legislators <- full.posterior1d[,1:nlegislators]
rollcalls <- full.posterior1d[,
  (nlegislators+1):ncol(full.posterior1d)]
```

To run multiple chains with the `ideal()` function, we can use different starting values. Below, we use default starts from the double-centered agreement score matrix in `ideal.SupCourt1` and random starts in `ideal.SupCourt2`.

```
ideal.SupCourt1 <- ideal(SC.rc, d=1, normalize=TRUE,
    store.item=TRUE, startvals="eigen")
ideal.SupCourt2 <- ideal(SC.rc, d=1, normalize=TRUE,
    store.item=TRUE, startvals="random")
```

Then, we combine both sets of chains for the legislator and roll call parameters in the MCMC objects `ideal.chains.legislators` and `ideal.chains.rollcalls` below. This facilitates monitoring the values of both chains across iterations (e.g., with the command `plot(ideal.chains.legislators)`) or the estimation of parameter summary statistics from both chains (e.g., with the command `summary(ideal.chains.legislators)`).

```
ideal.chains.legislators <- mcmc.list(
  as.mcmc(ideal.SupCourt1$x[,,1]),
  as.mcmc(ideal.SupCourt2$x[,,1]))
ideal.chains.rollcalls <- mcmc.list(
  as.mcmc(ideal.SupCourt1$beta[,,1]),
  as.mcmc(ideal.SupCourt2$beta[,,1]))
```

6.4.4 Example 4: The Confirmation Vote of Robert Bork to the US Supreme Court (Unidimensional IRT)

As an example of how to plot voting patterns on individual roll call votes in the Bayesian IRT framework, we use the 100th US Senate's vote on the confirmation of President Reagan's nominee Judge Robert H. Bork to the US Supreme Court. The confirmation hearings of Judge Bork were intensely ideological, pitting Bork's advocacy of judicial restraint and conservative views on civil rights and liberties issues against the liberal positions espoused most vocally by Senators Joe Biden (D-DE) and Ted Kennedy (D-MA). On October 23, 1987, his nomination was defeated in a mostly party-line $42-58$ vote (2 Democrats voted Yea and 6 Republicans voted Nay).

We first load the roll call data from the 100th Senate.

```
library(pscl)
hr <- readKH("sen100kh.ord",
       dtl=NULL,
       yea=c(1,2,3),
       nay=c(4,5,6),
       missing=c(7,8,9),
```

```
        notInLegis=0,
        desc="100th Senate Roll Call Data",
        debug=FALSE)
```

```
Attempting to read file in Keith Poole/Howard Rosenthal (KH) format.
Attempting to create roll call object
100th Senate Roll Call Data
102 legislators and 799 roll calls
Frequency counts for vote types:
rollCallMatrix
    0     1     2     3     4     5     6     8     9
  800 53203    30   416   134    29 21116     6  5764
```

Because the MCMCirt1d() and MCMCirtKd() functions require that the values in the roll call matrix be either 0, 1, or *NA*, we create a new matrix dat composed of the roll calls stored in the **rollcall** object **hr$votes**. We replace missing values (7, 8, 9, and 0) in the dat matrix with NA. We replace the values for Yea votes (1, 2, and 3) with 1, and the values for Nay votes (4, 5, and 6) with 0.

Finally, spaces and punctuation marks need to be absent from the legislator and roll call identifiers (the row and column names of dat) in order to assign them as ability and item constraints. For example, KENNEDY (D MA) needs to be formatted as KENNEDYDMA in the calls to the MCMCirt1d() and MCMCirtKd() functions. We do so with the use of the gsub() commands below. The first argument in the gsub() command identifies the string or punctuation to be replaced in a character string (e.g., "\\s" denotes spaces). The second argument identifies the string that should replace the specified text or punctuation (e.g., "" in the first example means that spaces should be replaced with nothing; that is, deleted). The final argument in the gsub() command specifies the object on which to make the replacements. Below we delete all spaces, parentheses, and hyphens from the row and column names of the matrix dat.

```
dat <- hr$votes
dat[dat==7 | dat==8 | dat==9 | dat==0] <- NA
dat[dat==1 | dat==2 | dat==3] <- 1
dat[dat==4 | dat==5 | dat==6] <- 0
colnames(dat) <- gsub("\\s","",colnames(dat))
rownames(dat) <- gsub("\\s","",rownames(dat))
rownames(dat) <- gsub("\\(","",rownames(dat))
rownames(dat) <- gsub("\\)","",rownames(dat))
rownames(dat) <- gsub("\\-","",rownames(dat))
```

The call to the MCMCirt1d() function below is the same as for the first example, except that sign constraints are placed on Senators Jesse Helms (R-NC) and Ted Kennedy (D-MA) (placing Kennedy on the left of the space and Helms on the right). We achieve identification by post-processing the result, normalizing the samples (with mean 0 and standard deviation 1) from the posterior densities of the legislator ideal points at each iteration. The identified result is stored in the matrix `posterior1d`, to which we also assign the column names from the original result.

```
posterior1d.unid <- MCMCirt1d(dat,
    theta.constraints=list(HELMSRNC="+",KENNEDYDMA="-"),
    mcmc=55000, burnin=5000, thin=10,
    theta.start=NA, alpha.start=NA, beta.start=NA,
    t0=0, T0=1, a0=0, A0=0.25, b0=0, B0=0.25,
    seed=NA, verbose=0, store.item=TRUE,
    store.ability=TRUE, drop.constant.items=TRUE)
posterior1d <- t(apply(posterior1d.unid, 1, scale))
colnames(posterior1d) <- colnames(posterior1d.unid)
```

The columns of the matrix `posterior1d` are arranged such that the ideal points are followed by the roll call parameters. Below we store the first n columns of `posterior1d` in the matrix `legislators`, and the $n+1$ through final columns of `posterior1d` in the matrix `rollcalls`.

```
nlegislators <- nrow(dat)
legislators <- posterior1d[,1:nlegislators]
rollcalls <- posterior1d[,(nlegislators+1):ncol(posterior1d)]
```

The Bork confirmation vote is Roll Call #348. The draws from the posterior densities of the difficulty and discrimination item parameters for this roll call vote are stored as the columns named `alpha.Vote348` and `beta.Vote348` in the matrix `rollcalls`. We assemble those column names with the `paste` function as below. The difficulty and discrimination parameters are calculated as the mean of the sampled values from their posterior densities.

```
nrollcall <- 348
diff.mean <- mean(rollcalls[,paste("alpha.Vote",nrollcall,sep="")])
disc.mean <- mean(rollcalls[,paste("beta.Vote",nrollcall,sep="")])
```

The legislator ideal point estimates (`idealpt.mean`) are also calculated as the mean of each legislator's simulated posterior density. The legislator-specific variables—`party` and `state`—are drawn from the original `rollcall` object `hr`. Legislator vote choices are stored in the object `vote`.

```
idealpt.mean <- colMeans(legislators)
party <- hr$legis.data$partyCode
state <- hr$legis.data$icpsrState
vote <- as.integer(dat[,nrollcall])
```

The identity and number of spatial voting errors are calculated in the same manner as with W-NOMINATE. However, instead of a `polarity` variable, we identify errors by comparing the legislator ideal points with the roll call difficulty parameter for Yea and Nay voters. On this roll call vote, legislators with ideal points less (more liberal) than the value of the difficulty parameter are predicted Nay votes, and legislators with ideal points greater (more conservative) than the difficulty parameter are predicted Yea votes. Hence, `errors12` (the sum of `errors1` and `errors2`) contains the correct number of voting errors. The PRE statistic of the vote is calculated as below.

```
kpyea <- sum(vote==1, na.rm=T)
kpnay <- sum(vote==0, na.rm=T)
errors1 <- !is.na(vote) & vote==1 &
  (idealpt.mean < (diff.mean/disc.mean))
errors2 <- !is.na(vote) & vote==0 &
  (idealpt.mean > (diff.mean/disc.mean))
errors3 <- !is.na(vote) & vote==1 &
  (idealpt.mean > (diff.mean/disc.mean))
errors4 <- !is.na(vote) & vote==0 &
  (idealpt.mean < (diff.mean/disc.mean))
errors12 <- sum(errors1==1) + sum(errors2==1)
errors34 <- sum(errors3==1) + sum(errors4==1)
nerrors <- min(errors12, errors34)
PRE <- (min(kpyea,kpnay) - nerrors)/min(kpyea,kpnay)
```

Figure 6.16 shows the results from the unidimensional IRT model for the Bork confirmation vote. Legislators are plotted by their ideal point along the x-axis and their vote choice along the y-axis: Yea votes at 1.0 position and Nay votes at the 0.0 position. The midpoint of the vote ($\frac{\alpha_j}{\beta_j}$) is represented as the vertical dotted line in the plot.

The 14 voting errors are composed of the Nay votes who are to the right (i.e., more conservative) of the difficulty parameter and the Yea votes who are to the left (i.e., more liberal) of the difficulty parameter. Given the low number of errors and high value of the PRE statistic (0.67), a single ideological dimension performs well in classifying legislator vote choices. This is also reflected in the high value of the discrimination parameter (2.83), which indicates that this vote did a good job of discriminating legislators on the basis of their ideological position. We also plot the ICC in Figure 6.16, which illustrates the substantive meaning of the discrimination parameter. In the two-parameter IRT model, the ICC models the probability of a Yea vote as a function of the values of the legislator ideal points and the difficulty and discrimination parameters of the roll call vote.

Since this is a nonlinear model, we use the probit link function to generate the ICC. That is, the probability that the ith legislator votes Yea is: $p_{ij} = \Phi(\beta_j x_i - \alpha_j)$, where Φ is the cumulative distributive function of the standard normal distribution. We generate a sequence of `x` values spanning the range $(-3,3)$ to represent potential legislator ideal points. p_{ij} is calculated with the commands:

```
x <- seq(-3, 3, by=0.01)
y <- pnorm((disc.mean * x) - diff.mean)
```

The value of the discrimination parameter (β_j) influences the slope of the ICC: higher values produce a steeper (i.e., better *discriminating*) ICC. Note that, with a symmetric utility function applied to the binary choice problem in the IRT framework, the ICC predicts a 0.5 probability of a Yea vote at the midpoint $\frac{\alpha_j}{\beta_j}$ (see Equation 6.20). Because the midpoint is equal to the ratio of α_j to β_j, a roll call with the same β_j but a larger α_j will have the same ICC but it will be shifted to the right.

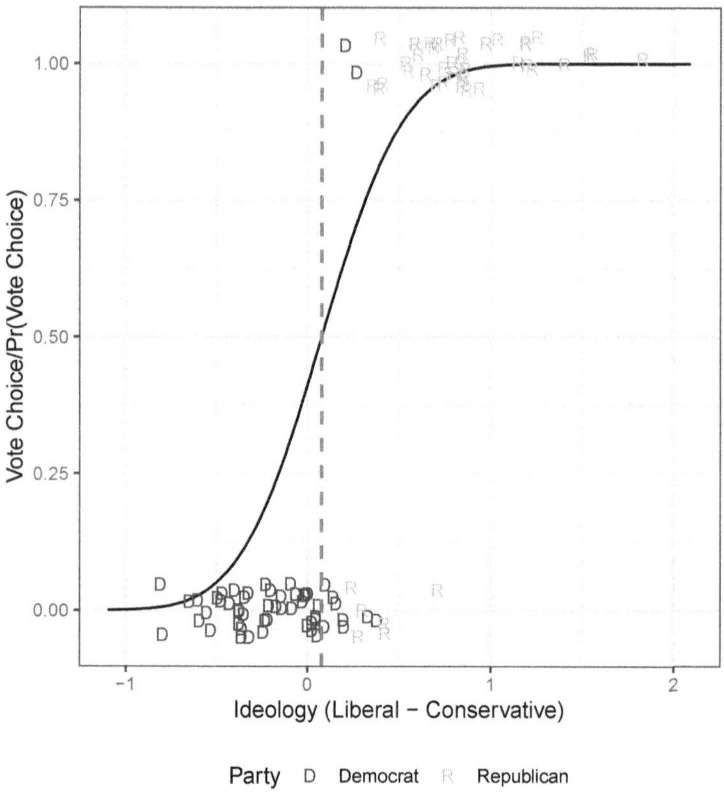

FIGURE 6.16: Unidimensional IRT Analysis of the 100th Senate Vote on the Confirmation of Robert Bork, All Legislators

We can also sort all roll call votes in the 100th Senate by the value of their discrimination parameter. The following commands sort the discrimination parameters and prints the first and last four. For example, Roll Call #73 (a $68-29$ vote to invoke cloture on an amendment to The Homeless Act) has a discrimination parameter of 0.09, indicating very little (unidimensional) ideological structure to this vote. Roll Call #671 (a $35-58$ vote on defense funding, including President Reagan's Strategic Defense Initiative (SDI) program) has a discrimination parameter of 3.49, indicating that Yea and Nay voters were strongly sorted along the liberal-conservative spectrum. The PRE values from W-NOMINATE for these votes are 0.17 and 0.94, respectively, also indicating a substantial difference in the ideological structure underlying these votes.

```
disc.parameters <- rollcalls[,seq(2,ncol(rollcalls), by=2)]
print(sort(colMeans(abs(disc.parameters)))[1:4])
```

```
beta.Vote345 beta.Vote614  beta.Vote73 beta.Vote744
  0.09065590   0.09166908   0.09208390   0.09215196
```

```
print(sort(colMeans(abs(disc.parameters)))[
  (ncol(disc.parameters)-3):ncol(disc.parameters)])
```

```
beta.Vote671 beta.Vote106 beta.Vote108 beta.Vote245
    3.442122     3.619691     3.630886     3.697324
```

6.4.4.1 Bayesian IRT Convergence Diagnostics

We next discuss three methods to assess convergence of the Markov chains that are available in the `coda` package in `R` (Plummer et al., 2006). The `coda` package is automatically loaded as a dependency of the `MCMCpack` package. The first method is a straightforward graphic inspection of the sampled values from the chains by iteration (`traceplot`) and the cumulative density of the sampled values (`densplot`) for each parameter. The `plot.MCMC()` function in the `coda` package plots both trace and density plots for a `mcmc` object or each of the variables in a `mcmc.list` object.

Below we convert the sampled values for President Reagan to the `mcmc` object `reagan` and then produce Figure 6.17 with the `plot()` command (which automatically executes the `plot.MCMC()` function when used on an `mcmc` object). Figure 6.17 indicates that the Markov chain has converged on the posterior density of President Reagan's ideal point. The trace plot on the left shows that—following the burn-in period of 5,000 draws—the Markov chain stays within a limited range between 0.96 and 1.69. In the density plot, we are looking for unimodality in the distribution of the sampled parameter values as evidence of convergence. Multimodality may indicate an identification problem leading to non-convergence.

```
reagan <- as.mcmc(legislators[,"theta.REAGANRUSA"])
plot(reagan)
```

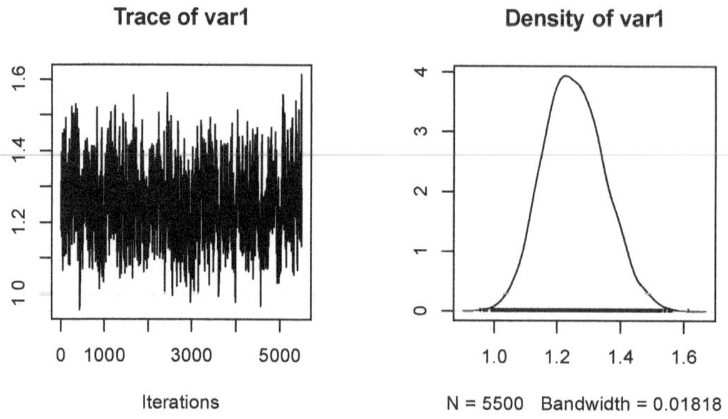

FIGURE 6.17: Trace and Density Plots for President Ronald Reagan

Next, the Geweke diagnostic (Geweke, 1992) conducts a difference of means test for the sampled values from two sections of the chain. The default setting in the `geweke.diag()` function is to compare the values of the Markov chain from the first 10% of iterations with those from the final 50% of iterations (these values are recommended by Geweke (1992)). If a difference of means test is statistically significant at some conventional level (we use 0.05), we can reject the null that the sampled values are drawn from the same distribution. That is, a significant result indicates that the chain has not settled over the course of the iterations.

Below we sort the parameters (legislators) by their absolute Geweke diagnostic values. We print only the first two legislators in order to conserve space, but the values for *all* parameters should be inspected. The Geweke diagnostic for Senator Inouye (D-HI) is 0.310458526079253, above the 1.96 value that would indicate significance at the $p \leq 0.05$ level. To investigate this finding, we use the Geweke-Brooks plot (implemented as the `geweke.plot` function in the `coda` package) that re-runs the Geweke diagnostic test as successive iterations of the chains are dropped. The three troublesome Z-values (greater than 1.96 or less than -1.96) for Senator Inouye in Figure 6.18 occur mostly within the first 10,000 iterations (that is, 15,000 iterations thinned by 10 for 1,500 samples) of the chain. After this point in the chain, the Geweke diagnostic values are all within the non-significant range, save for one. This result suggests that we should run more iterations with a longer burn-in period (perhaps 15,000 iterations rather than 5,000).

```
sort(abs(geweke.diag(legislators)$z), decreasing=T)[1:2]
```

```
theta.INOUYEDHI  theta.BIDENDDE
       3.201869        3.182309
```

```
inouye <- as.mcmc(legislators[,"theta.INOUYEDHI"])
geweke.ggplot(inouye) +
  theme_bw() +
  guides(colour=FALSE)
```

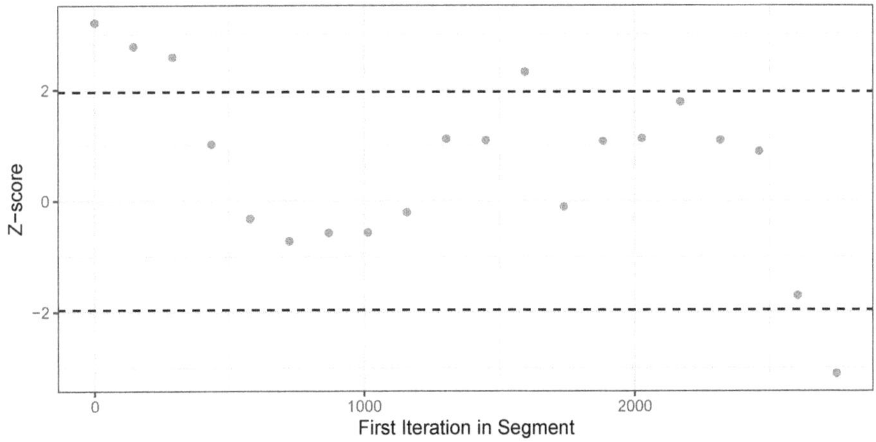

FIGURE 6.18: Geweke Diagnostic Plot for Senator Phil Gramm (R-TX)

Finally, the Gelman-Rubin diagnostic (from Gelman and Rubin (1992)) can be used to diagnose non-convergence when multiple Markov chains are run. As we discussed in Section 6.4.3, running multiple chains with dispersed starting values and a sufficient number of iterations is a best practice in Bayesian analysis that facilitates a more complete search of the posterior distributions of the parameters and makes convergence on the highest density regions of the posterior distributions more likely. For each parameter, the diagnostic calculates the potential scale reduction factor which measures the within-chain and between-chain variance. Values larger than 1 indicate greater variance and mean that it is less likely that the chains have converged to the stationary distribution. As a general rule, values above 1.1 or 1.2 indicate non-convergence, with higher values providing greater evidence of convergence failure (Gelman and Rubin, 1996).

The Gelman-Rubin diagnostic is implemented in the `gelman.diag()` function in the `coda` package. To demonstrate its use, we use data from the 2000 term of the US Supreme Court that was analyzed in Example 3. Recall that we ran two separate simulations which were stored in the `mcmc.list` object `chains`, as below. In general, two is the minimum number of chains that should be used and three or more chains is advisable; in this example, we stick with two for presentation purposes.

Next, we could standardize the chains to ensure that they are identified.

```
chains <- mcmc.list(posterior1d.unid, posterior1d.unid.2)
justices <- lapply(chains, function(x)x[,grep("^theta", colnames(x))])
justicesZ <- lapply(justices, function(x)apply(x, 1, scale))
rownames(justicesZ[[1]]) <- rownames(justicesZ[[2]]) <-
  grep("^theta", colnames(posterior1d.unid), value=TRUE)
justices_sd <- lapply(justices, function(x)apply(x, 1, sd))
justices_mean <- lapply(justices, function(x)apply(x, 1, mean))
alphas <- lapply(chains, function(x)x[,grep("^alpha", colnames(x))])
betas <- lapply(chains, function(x)x[,grep("^beta", colnames(x))])
disc.parameters <- lapply(1:2, function(i)apply(betas[[i]], 2,
  function(x)x*justices_sd[[i]]))
diff.parameters <- lapply(1:2, function(i)apply(betas[[i]], 2,
  function(x)x*justices_mean[[i]]) + alphas[[i]])
```

Finally, we could put the chains back together in an `mcmc.list` object.

```
chain1 <- cbind(t(justicesZ[[1]]), alphas[[1]], betas[[1]])
chain2 <- cbind(t(justicesZ[[2]]), alphas[[2]], betas[[2]])
chains <- mcmc.list(as.mcmc(chain1),as.mcmc(chain2))
```

The `mcmc.list` object that contains the results of multiple Markov chains (in this example, `chains`) is the first argument passed to the `gelman.diag()` function. The `confidence` argument specifies the probability that the value of the diagnostic lies below the upper confidence limit (by default, 0.95). `transform` can be switched to `TRUE` from its default value of `FALSE` to transform the sampled values from the posterior densities of the parameters so that they more closely approximate a normal distribution. Only the variance of the second half of the chain is used to assess convergence if `autoburnin` is set to `TRUE` (the default value). Finally, the `multivariate` argument specifies whether the multivariate version of the Gelman-Rubin diagnostic should be used. By default, it is set to `TRUE`; however, this version produces errors when the covariance matrix is not positive definite. Because it has been our experience that this problem arises with some frequency with this data set, we set the argument to `FALSE`.

The results of the Gelman-Rubin diagnostic are stored in the object `gr`. `gr` contains the point estimate and upper credible interval of the diagnostic for each parameter. We print only the first two variables for purposes of space, but all of the parameters should be examined. Below we show the parameter with the largest upper 95% credible interval for its Gelman-Rubin diagnostic value. This turns out to be the β term for the 24th vote, which has a point estimate of 1.035 and an upper 95% credible interval of 1.125. Since the latter value remains below 1.2, we are more confident of convergence of the chains.

```
gr <- gelman.diag(chains, confidence=0.95, transform = FALSE,
    autoburnin = TRUE, multivariate=FALSE)
print(gr$psrf[1:2,])
```

```
                 Point est. Upper C.I.
theta.Rehnquist   1.0056228   1.005704
theta.Stevens     0.9998853   1.002574
```

```
which.max(gr$psrf[,2])
```

```
beta.24
     57
```

```
gr$psrf[which.max(gr$psrf[,2]),]
```

```
Point est. Upper C.I.
  1.035217   1.125293
```

6.4.5 Example 5: The 89th US Senate (Multidimensional IRT)

In instances in which roll call voting taps into separate ideological dimensions, the `MCMCirtKd()` function in the `MCMCpack` package allows for the estimation of multidimensional Bayesian IRT models.

The call to the `MCMCirtKd()` function is the same as that to `MCMCirt1d`, with two exceptions. First, the number of dimensions to be estimated is set with the `dimensions` argument. Second, constraints are placed on the item parameters rather than the subject parameters with the `item.constraints` argument. As an identifying restriction for the two-dimensional model, we constrain one item to load only onto the first dimension (see Jackman, 2001, pp. 230-233). In practical terms, this means that this item's discrimination parameter is set to 0 on the other dimension. We also set the polarity of the space by constraining the signs of a discrimination parameter on each of the dimensions. A positive discrimination parameter means that legislators higher on the dimension will be more likely to vote Yea on the item.

In this example, we use roll call data from the 89th US Senate. During this period of congressional history (1965-1967), roll call voting was two-dimensional (Poole and Rosenthal, 2007, pp. 139-142). The first dimension represents the standard liberal-conservative divide and the second dimension picks up regional differences within the parties, particularly within the Democratic Party on racial issues. We use two landmark votes in the 89th Senate for the item constraints. The roll call vote (#151) to create Medicare (the Social Security Amendments of 1965) is constrained to have a negative item discrimination parameter on the first dimension and have a discrimination parameter of 0 on the second dimension. The roll call vote (#78) on the Voting Rights Act of 1965 is constrained to have a negative item discrimination parameter on the second dimension.

The remainder of the arguments in the MCMCirtKd() function are the same as those in the MCMCirt1d() function used in previous examples, except that we do not store the roll call item parameters here.

```
posterior2d <- MCMCirtKd(dat, dimensions=2,
    item.constraints=list(Vote151=list(2,"-"), Vote151=c(3,0),
    Vote78=list(3,"-")), mcmc=15000, burnin=5000, thin=10,
    theta.start=NA, alpha.start=NA, beta.start=NA,
    t0=0, T0=1, a0=0, A0=0.25, b0=0, B0=0.25,
    seed=NA, verbose=0, store.item=FALSE,
    store.ability=TRUE, drop.constant.items=TRUE)
```

We store the legislator attributes (party, code and state) and ideal points below. The first and second dimension ideal points are staggered as columns in the matrix posterior2d, so we assign every other column to idealpt1 and idealpt2 with the commands seq(1,ncol(posterior2d),by=2) and seq(2,ncol(posterior2d),by=2) as below:

```
party <- hr$legis.data$partyCode
code <- hr$legis.data$icpsrLegis
state <- hr$legis.data$icpsrState
idealpt1 <- colMeans(posterior2d[,seq(1, ncol(posterior2d), by=2)])
idealpt2 <- colMeans(posterior2d[,seq(2, ncol(posterior2d), by=2)])
```

We calculate the 90% HPD (highest posterior density) region (which is equivalent to the credible interval) of the legislator ideal points using the HPDinterval function below. The HPDinterval function calculates the upper and lower bounds of a region of the posterior density in which the "true" value of the parameter resides with specified probability (prob). This value is equivalent to the proportion of the area beneath the curve that resides between the upper and lower bounds. The ceiling(i/2) argument divides i by two and rounds up to the nearest integer. This is necessary because the number of columns in posterior2d is twice the size of the number of legislators (since there are two coordinates estimated for each legislator).

```
nlegislators <- nrow(dat)
idealpt1.10 <- idealpt1.90 <- idealpt2.10 <- idealpt2.90 <-
    rep(NA, nlegislators)
for (i in seq(1,ncol(posterior2d),2)){
idealpt1.10[ceiling(i/2)] <- HPDinterval(posterior2d[,i],
                                          prob=0.9)[1]
idealpt1.90[ceiling(i/2)] <- HPDinterval(posterior2d[,i],
                                          prob=0.9)[2]
idealpt2.10[ceiling(i/2)] <- HPDinterval(posterior2d[,i+1],
                                          prob=0.9)[1]
idealpt2.90[ceiling(i/2)] <- HPDinterval(posterior2d[,i+1],
                                          prob=0.9)[2]
}
```

The 90% HPD regions are plotted in Figure 6.19.

Three distinct clusters (hence the term "three-party system" (Poole and Rosenthal, 2007, pp. 54-55)) of legislators are present in Figure 6.19. The second dimension divides Southern Democrats from both Northern Democrats and Republicans. Clearly, a single dimension would be insufficient to capture voting dynamics in the 89th Senate.

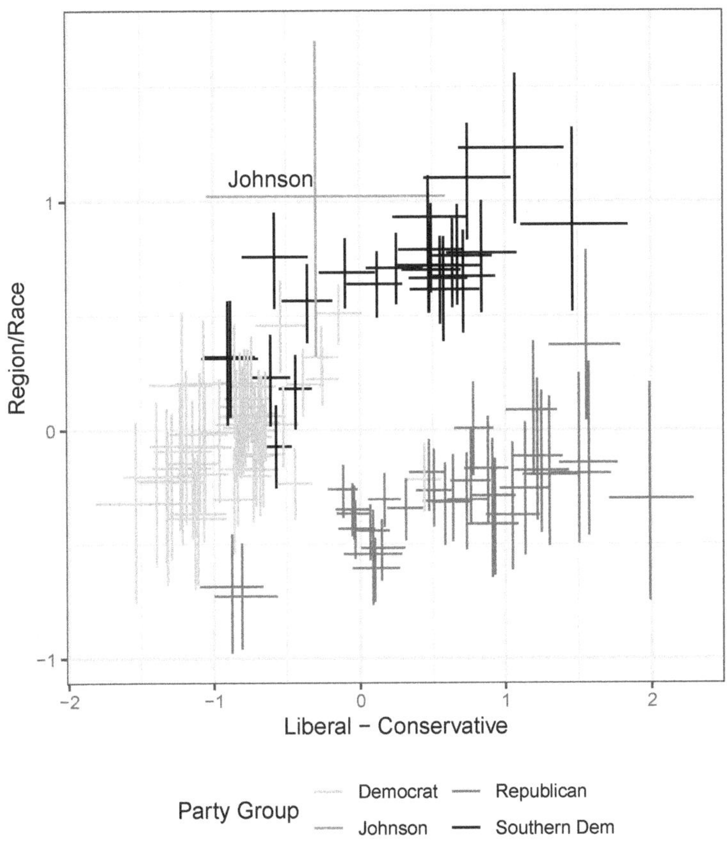

FIGURE 6.19: Multidimensional IRT Ideal Point Estimates of Members of the 89th US Senate with 90% HPD Regions

Note the sizable 90% HPD regions for Senator Olin D. Johnston (D-SC) in the top left corner of Figure 6.19. This is because Senator Johnston cast only nine votes in the 89th Senate before his death on April 18, 1965. This produces a considerable level of uncertainty about Senator Johnston's ideological position that is reflected in the posterior densities of his ideal point coordinates. If we want to delete legislators who have cast less than a specified number of votes, we can adapt the code used for the same purpose with Bayesian Aldrich-McKelvey scaling in this chapter. The commands below retain only legislators who have cast more than 20 valid (i.e., non-missing) votes in the roll call matrix `dat`.

```
cutoff <- 20
dat <- dat[rowSums(!is.na(dat)) >= 20,]
```

In addition, the `MCMCirt1d()` and `MCMCirtKd()` functions omit only unanimous roll call votes from the analysis. In order to drop additional lopsided votes (e.g., 99-1 votes), the `lop` function can be used to retain only votes in a roll call matrix `x` with less than some proportion `y` of voting legislators are in the minority. Below, votes are deleted from the `dat` matrix if they fail to meet the 2.5% threshold.

```
lop <- function(x,y){
    lop.votes <- rep(0,ncol(x))
    for (i in 1:ncol(x)){
        if ((length(x[,i][!is.na(x[,i]) & x[,i]==1]) /
            length(x[,i][!is.na(x[,i])])) < y) lop.votes[i] <- 1
        if ((length(x[,i][!is.na(x[,i]) & x[,i]==0]) /
            length(x[,i][!is.na(x[,i])])) < y) lop.votes[i] <- 1
        }
    return(lop.votes)
    }
dat <- dat[,lop(dat,0.025)==0]
```

The 89th Senate was noteworthy in that it included Senators Edward M. "Ted" Kennedy (D-MA) and Robert F. Kennedy (D-NY), brothers and liberal icons of twentieth century American politics.** IRT analysis indicates that Senator Robert Kennedy (with a mean first dimension score of −1.42) was more liberal than Senator Ted Kennedy (−1.20). This result has face validity because of the few votes on which the two Senators disagreed, Senator Robert Kennedy (RFK) tended to take the liberal side. This includes, for example, a vote (#161) to decrease funding for a rent supplement program administered by the Department of Housing and Urban Development (RFK voted Nay, his brother voted Yea) and a vote (#369) to deny foreign aid to Latin American countries with defense expenditures greater than 3.5% of their GDP (RFK voted Yea, his brother voted Nay).

**Senator Robert F. Kennedy was a freshman member of the 89th Senate, having been elected to represent New York in the 1964 election. He also served in the 90th Senate before his assassination on June 6, 1968.

The Bayesian framework allows us to directly assess the probability that this is the correct ideological ordering by comparing their posterior densities. Below we run 50,000 simulations in which draws from both Senators' first dimension posterior densities are compared, and the number of trials in which the draw from Senator Robert Kennedy's posterior density is less (more liberal) than Senator Ted Kennedy's draw.[††] Based on the simulations, we find that the probability that Senator Robert Kennedy was a more liberal legislator than Senator Ted Kennedy in the 89th Senate is 84.34%.

```
RFKennedy <- posterior2d[,"theta.KENNEDYDNY.1"]
TKennedy <- posterior2d[,"theta.KENNEDYDMA.1"]
nsims <- 50000
x1 <- RFKennedy[sample(1:length(RFKennedy), nsims, replace=T)]
x2 <- TKennedy[sample(1:length(TKennedy), nsims, replace=T)]
correct.order <- NULL
for (i in 1:nsims){
correct.order[i] <- x1[i] < x2[i]
}
p.correct.order <- mean(correct.order)

print(p.correct.order)

[1] 0.8434
```

6.4.5.1 Multidimensional IRT using the ideal() Function in the pscl Package

Multidimensional Bayesian IRT can also be performed with the `ideal()` function in the `pscl` package. To demonstrate its use, we again use the data set of roll call votes from the 89th US Senate, which is stored in the `rollcall` object `hr`. The Clinton, Jackman and Rivers (2004) (CJR) approach to identification of the multidimensional IRT model is to fix $s+1$ legislator ideal points (e.g., 3 ideal points in the two dimensions).

Below we run IDEAL on the 89th Senate data in two dimensions with no constraints, producing an unidentified result (`ideal.unidentified`). We then post-process `ideal.unidentified` using the `postProcess()` function, fixing three ideal points to achieve identification in the CJR framework. We use the W-NOMINATE coordinates of three legislators because the W-NOMINATE model is identified in one or more dimensions using only "soft" constraints (i.e., legislator coordinates are constrained to lie within the unit hypersphere and one legislator is constrained to have a positive score on each dimension in order to fix the rotation). We chose Senators distant from one another in the latent space: Senators Walter Mondale (D-MN), a solid liberal

[††]We conduct the same experiment to test the ordering produced by Bayesian Aldrich-McKelvey scaling in this chapter.

on the first dimension, Herman Talmadge (D-GA), an opponent of civil rights legislation with a high second dimension score, and Milward Simpson (R-WY), a staunch conservative and father of Senator Alan Simpson (R-WY). These three fixed points are the vertices of a large simplex (in the two-dimensional case, a triangle) that identifies the multidimensional IRT result.

```
ideal.unidentified <- ideal(hr, d=2, maxiter=55000,
    thin=50, burnin=5000, normalize=FALSE,
    store.item = TRUE)
ideal.identified <- postProcess(ideal.unidentified,
    constraints=list(MONDALE=c(-0.774,-0.188),
    TALMADGE=c(0.266,0.794), SIMPSON=c(0.971,-0.238)))
```

We then store the first and second dimension legislator coordinates (i.e., the means of the ideal point posterior densities) from the identified result in the objects `ideal1` and `ideal2`.

```
ideal1 <- ideal.identified$xbar[,1]
ideal2 <- ideal.identified$xbar[,2]
```

Figure 6.20 shows the estimated two-dimensional result from IDEAL in the left panel and W-NOMINATE in the right panel. The IDEAL and W-NOMINATE configurations of the 89th Senate are very similar as the two sets of inter-point distances correlate at 0.929. The differences are largely the consequences of the different ways that the two methods model the deterministic and stochastic components of legislator utility (IDEAL assumes quadratic utility and normally distributed errors, while W-NOMINATE assumes normal/Gaussian utility and uses the logistic distribution to model the error process).

We can also examine the first and second dimension discrimination parameters from IDEAL to determine how well each dimension taps into voting patterns on given roll call votes. Below we print the mean posterior value of the item parameters (stored in `$betabar`) of roll call votes of four pieces of legislation passed by the 89th Senate: the Voting Rights Act of 1965, the Housing and Urban Development Act of 1965, the Social Security Amendments of 1965 (which created Medicare and Medicaid), and the Immigration and Nationality Act of 1965.

Of interest to us are the `Discrimination D1` and `Discrimination D2` values for each roll call, as they tell us how well legislator positions on the first (`D1`) and second (`D2`) dimensions predict their vote choices on the roll call or, technically, the steepness of the ICC on each dimension. Note that the *absolute* value of the discrimination parameter communicates the substantive importance of the dimensions; the sign of the discrimination parameters only tells whether higher legislator ideal point values on the dimension correspond to an increased or decreased probability of voting Yea.

The two votes that correspond most closely to the classic economic liberal-conservative divide—the Housing and Urban Development Act of 1965 and the

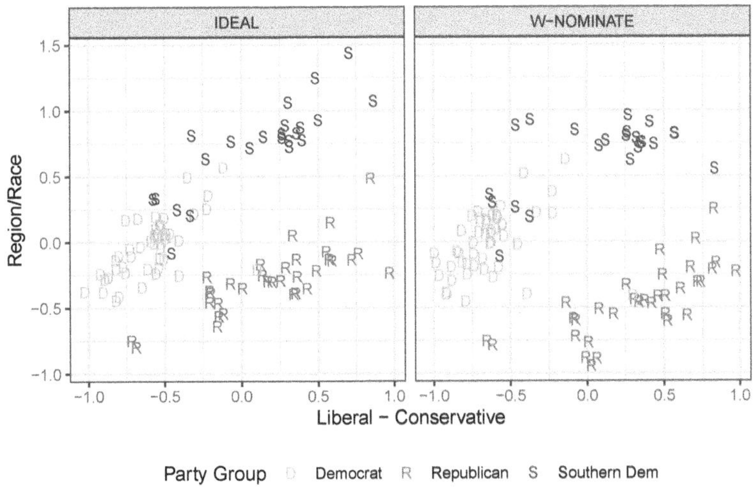

FIGURE 6.20: IDEAL (Multidimensional IRT) and W-NOMINATE Scaling of the 89th US Senate

Social Security Amendments of 1965 (two keystone "Great Society" programs)—have very high first dimension discrimination parameters and low second dimension discrimination parameters, as we would expect given that the second dimension represents regional divides mostly over racial issues. Indeed, two votes that tap into racial sentiment—the Voting Rights Act of 1965 and the Immigration and Nationality Act of 1965 (which abolished the use of race-based quotas)—have higher discrimination parameters on the second dimension than the first dimension (particularly on the Voting Rights Act vote).

```
 # Voting Rights Act of 1965
 print(ideal.identified$betabar["Vote 78",])

Discrimination D1 Discrimination D2        Difficulty
        -2.089022         -5.077468         -2.295737

 # Housing and Urban Development Act of 1965
 print(ideal.identified$betabar["Vote 162",])

Discrimination D1 Discrimination D2        Difficulty
      -5.87859655        0.84807481       -0.06055037

 # Social Security Amendments of 1965
 print(ideal.identified$betabar["Vote 174",])

Discrimination D1 Discrimination D2        Difficulty
       -6.4777012         0.7749098        -1.4744434
```

```
# Immigration and Nationality Act of 1965
print(ideal.identified$betabar["Vote 232",])
```

```
Discrimination D1 Discrimination D2        Difficulty
        -1.711669         -2.042896         -1.283641
```

6.4.6 Identification of the Model Parameters

In earlier chapters, we spoke about the problems of identification in scaling models. Scaling methods use different types of constraints on the solution to address the identification problem. Identification is trickiest in the Bayesian framework (Bakker and Poole, 2013). The Clinton, Jackman, and Rivers (CJR) (2004) Bayesian IRT model requires fixing the locations of $s+1$ legislator ideal points to achieve identification. This amounts to 2 constraints in the one-dimensional model, 6 constraints in the two-dimensional model, and so forth. This is known as the "Kennedy-Helms" restriction in a single dimension: setting a left-wing legislator (i.e., Sen. Ted Kennedy (D-MA)) at -1 and a right-wing legislator (Sen. Jesse Helms (R-NC)) at 1. However, pinning legislator locations means that the distances are no longer elastic, and thus uncertainty propagates from the fixed points to the locations of other legislators (Lewis and Poole, 2004; Bakker and Poole, 2013). For instance, there is some uncertainty around Kennedy's "true" location. We do not eliminate this uncertainty when we pin Kennedy at -1; rather, we transfer it to the point estimates of the other legislators, since what we really are modeling is the inter-point distances.

In the one-dimensional IRT model, we can also achieve *local* identification with a normalization constraint: require the ideal points to have mean 0 and standard deviation 1. The result is only defined up to a choice of rotation (that is, left- and right-wing legislators may be flipped on the scale), but this can be easily corrected. However, identification in the multidimensional IRT model is not achieved via the normalization constraint. When a two-dimensional model is estimated, three points (or, more to the point, three *distances*) are fixed in the CJR model.

However, in the context of *metric* MDS, Bakker and Poole (2013) have a general solution to identification in any number of dimensions. The Bakker-Poole solution is to "freeze" the posterior distributions of the legislators in certain quadrants. In two dimensions, this is achieved by setting one point at the origin and the first or second dimension coordinate of another point at 0. The Nelder-Mead hill-climbing method (Nelder and Mead, 1965) is used to analyze the log-posterior and estimate the target configuration. Finally, at the end of each iteration of the slice sampler (Neal, 2003), the sign of the draws from the legislators' posterior densities are compared to the target configuration and rotated if they don't match. Three coordinates must nonetheless be fixed in the two-dimensional case, but this method produces a *minimally* restricted, identified solution.

6.5 MCMC or α-NOMINATE

Each method discussed in this chapter assumes a different functional form to model legislators' utility functions. DW-NOMINATE uses the Gaussian (normal) function to model legislator utility, while Bayesian IRT (Section 6.4) uses the quadratic function. Optimal Classification (Section 5.4) does not make any assumptions about the utility function other than that it is single-peaked and symmetric. Figure 6.21 illustrates the Gaussian and quadratic utility functions for a legislator whose most preferred policy is at 0.

It is worthwhile to consider the practical meaning of these two functional forms. First, both functions are very similar in the immediate vicinity of the legislator's ideal point. The difference is mostly in the tails: the quadratic form posits an accelerating drop in utility as the policy moves away from the ideal point, while the normal form posits a *de*celerating decrease in utility as the policy moves outward. In the quadratic case, the legislator in Figure 6.21 is more sensitive to a change in policy between 2 and 3 than for a change from 1 to 2. For the Gaussian form, the reverse is true. On tax policy, for example, a legislator with a Gaussian utility function who most prefers a 40% top marginal income tax rate has already lost most utility by the time the policy has moved to a 70% rate so that she is mostly indifferent when faced with a vote between the 70% rate and an 80% rate. It is important to stress that the Gaussian legislator is not indifferent between the 40% rate and the 70% rate, but rather is largely indifferent between two extreme policy alternatives. This phenomenon has been described as *alienation from indifference* by Riker and Ordeshook (1973, pp. 324-330).

The relative merits of the Gaussian and quadratic functional forms to study legislative and voting behavior have been contested on theoretical grounds (see, for example, Poole and Rosenthal, 1983; Merrill and Grofman, 1999; Carroll et al., 2009; Clinton and Jackman, 2009). However, Carroll et al. (2013) have recently developed α-NOMINATE in order to empirically estimate the structure of legislator utility from choice data and determine which functional form best represents political actors' preferences over latent policy space.

α-NOMINATE uses a mixture model that nests the Gaussian and quadratic forms in the legislator utility function. That is, legislator utility is allowed to take on components of both the quadratic and Gaussian forms as a function of an added (α) parameter. In the quadratic utility model, the utility that legislator i derives from voting Yea on roll call j is:

$$U_{ijy} = -(X_i - O_{jy})^2 + \varepsilon_{ijy} \tag{6.31}$$

And recall from Equation 5.7 that in the Gaussian utility model, the utility that legislator i derives from voting Yea on roll call j is:

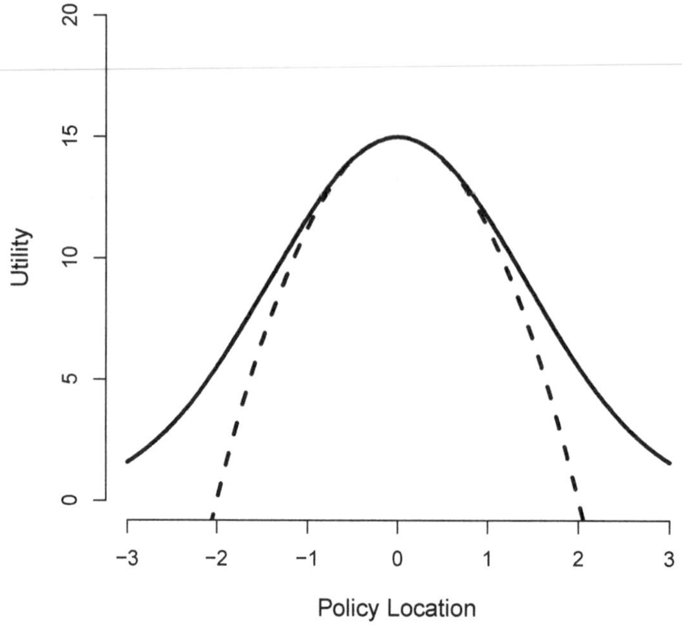

FIGURE 6.21: Gaussian (solid line) and Quadratic (dashed line) Utility Functions for Legislator with Ideal Point of 0

$$U_{ijy} = \beta e^{(-\frac{1}{2}w^2(X_i - O_{jy})^2)} + \varepsilon_{ijy} \tag{6.32}$$

The exponential expansion of Equation 6.32 is:

$$U_{ijy} = \beta \sum_{i=0}^{\infty} \frac{(-\frac{1}{2}w^2(X_i - O_{jy})^2)^i}{i!} + \varepsilon_{ijy} \tag{6.33}$$

The mixed utility model estimated by α-NOMINATE takes the exponential expansion of the Gaussian utility function and separates the constant and quadratic terms from the higher powered terms, and includes an α parameter that is allowed to vary between 0 and 1:

$$U_{ijy} = \beta \sum_{i=0}^{1} \frac{(-\frac{1}{2}w^2(X_i - O_{jy})^2)^i}{i!} + \alpha\beta \sum_{i=2}^{\infty} \frac{(-\frac{1}{2}w^2(X_i - O_{jy})^2)^i}{i!} + \varepsilon_{ijy} \tag{6.34}$$

When $\alpha = 1$, the utility function of the mixture model in Equation 6.34 is identical to the Gaussian model. When $\alpha = 0$, the utility function of the mixture model in Equation 6.34 is identical to the quadratic model plus a constant. The constant vanishes when the Yea and Nay utilities are subtracted from one another. When α takes on a value between 0 and 1, some combination of the quadratic and Gaussian functional forms is needed to model the structure of legislator utility, though α values close to 0 or 1 can be interpreted as evidence supporting the quadratic or Gaussian forms, respectively. α-NOMINATE bounds the values of α to lie within the unit interval with use of the prior $p(\alpha) \sim$ Uniform $(0,1)$.

Given the $n \times q$ matrix of observed roll call choices Y, the likelihood is proportional to the likelihood function given in Equation 5.10:

$$p(Y|\alpha,\beta,X,O) \propto \prod_{i=1}^{n}\prod_{j=1}^{q}\prod_{\tau=1}^{2} P_{ij\tau}^{C_{ij\tau}} \tag{6.35}$$

Bayesian inference for the familiar parameters (the ideal point, roll call, and β parameters) and the α parameter proceeds by simulating the posterior density given by:

$$p(\alpha,\beta,X,O|Y) \propto p(\alpha,\beta,X,O) \times p(Y|\alpha,\beta,X,O) \tag{6.36}$$

Diffuse priors are chosen for $p(\alpha,\beta,X,O)$. The parameter w is not estimated and is fixed at 0.5.

Table 6.1 is reproduced from Carroll et al. (2013). It shows the estimated means and standard deviations of α from roll call data in a variety of legislative and judicial settings: early and recent US Congresses, the US Supreme Court, the European Parliament, and the California Legislature. Carroll et al. (2013) find that the posterior mean of α is extremely close to 1 in all cases outside of the 5th US Senate and the 6th US House (which they attribute to the small number of legislators and roll call votes in early US Congresses). The other results strongly point towards the Gaussian functional form as the more appropriate model of legislator utility in spatial voting models. However, the ideal points recovered from each model—pure quadratic utility, pure Gaussian utility, and the mixture model—correlate highly. Thus, though the quadratic form appears to be an inferior model of utility, it appears to have little effect on the estimation of legislator ideal points.

The assumption of a particular utility function increases the precision of the estimates. For instance, in a single dimension, OC is limited to the recovery of rank-orders of the legislators and roll calls, while W-NOMINATE and Bayesian IRT are able to recover metric-level estimates. However, there may be cases (for instance, when dealing with public opinion data) in which we are more hesitant to make the leap of faith that individuals possess identically structured preference functions (e.g., Brady and Ansolabehere, 1989). Here, the imposition of a single utility function on the data may be too large a price to pay for greater precision. The analyst should seriously consider how realistic the assumption of a particular functional form of legislator utility is given the nature of individual behavior that produced the data.

Table 6.1: α-NOMINATE Estimates of α from Roll Call Data

Source	Posterior mean of α	σ_α
5th Senate, 1797-1799	0.215	0.166
6th House, 1799-1801	0.607	0.170
109th Senate, 2005-2007	0.996	0.004
US Supreme Court, 1953-2008	0.998	0.002
European Parliament, 1979-1984	0.986	0.001
European Parliament, 1994-1999	1.000	0.000
France First Legislature, 1946-1951	1.000	0.000
France Second Legislature, 1951-1956	1.000	0.000
France Third Legislature, 1956-1958	0.999	0.001
California State Assembly, 1993-1994	0.998	0.002
California State Assembly, 1997-1998	0.998	0.002
California State Assembly, 2001-2002	0.999	0.001
California State Assembly, 2005-2006	0.999	0.001
California State Senate, 1993-1994	1.000	0.000
California State Senate, 1997-1998	1.000	0.000
California State Senate, 2001-2002	1.000	0.000
California State Senate, 2005-2006	1.000	0.000

6.5.1 The anominate Package in R

The α-NOMINATE model can be estimated in `R` using the recently developed `anominate` package (Lo et al., 2013). The `anominate()` function uses slice sampling (Neal, 2003) to sample from the posterior densities of the parameters as described in Carroll et al. (2013).

Below we use the `anominate()` function to analyze the structure of legislators' utility functions in the 111th US Senate (this roll call data is included in the `anominate` package and can be loaded with the command `data(sen111)`, as below). The `anominate()` function requires 10 arguments. The first is the `rollcall` object to be analyzed (`sen111`). The number of dimensions to be estimated is set with `dims`. The `nsamp`, `thin` and `burnin` arguments set the total number of iterations, thinning interval and burn-in period for the slice sampler.[‡‡]

As with the `wnominate()` and `oc()` functions, `minvotes` establishes how many scalable votes required for a legislator to be included in the analysis, `lop` is the proportion of voting legislators in the minority at which a vote is excluded as lopsided, and `polarity` constrains a specified conservative or right-wing legislator to have a positive score on each dimension. The argument `random.starts` is a logical argument set to `TRUE` if the initial values for the legislator and bill parameters should be randomly drawn from a uniform

[‡‡] $\frac{nsamp}{thin}$ must be larger than `burnin`.

distribution between −1 and 1 and set to FALSE if the W-NOMINATE estimates should be used as the initial values. Finally, if verbose is set to TRUE the progress of the sampler at each 100th iteration is printed to the screen.

```
library(anominate)
data(sen111)
result <- anominate(sen111, dims=1, nsamp=2000, thin=1, burnin=1000,
    minvotes=20, lop=0.025, polarity=2, random.starts=FALSE,
    verbose=FALSE)
```

The anominate() function returns six objects that are stored in the list result:

alpha The sampled values of α.

beta The sampled values of *beta*.

legislators The sampled values of the legislator ideal points, with the dimensions stored in separate lists (e.g., legislators[[1]] stores the first dimension scores).

yea.locations The sampled values of the Yea locations for each vote, with the dimensions stored in separate lists.

nay.locations The sampled values of the Nay locations for each vote, with the dimensions stored in separate lists.

wnom.result The results from wnominate() analysis of the roll call data.

All of the objects except wnom.result are mcmc objects, which facilitates their analysis with functions in the coda package (Plummer et al., 2006); for example, to perform convergence diagnostics.

The anominate package includes functions to summarize and plot the sampled values of the main quantity of interest: the α parameter. Below, we see that α is very close to 1 for the 111th Senate: the mean value is 0.998 with a 95% credible interval between 0.994 and 1 (this information is also plotted in the title of Figure 6.22). This result points strongly towards the normal (Gaussian) function as the correct model of legislator utility in the 111th Senate.

```
summary(result)
```

```
SUMMARY OF ANOMINATE OBJECT
----------------------------


Number of Legislators:          108 (4 legislators deleted)
Number of Votes:          616 (80 votes deleted)
Number of Dimensions:          1
alpha Mean: 0.999
alpha Percentiles:
       2.5%    5%   50%  95% 97.5%
      0.996 0.997 0.999    1     1
```

```
par(mfrow=c(1,2))
traceplot(as.mcmc(result$alpha), ylim=c(.9,1), main="Trace of alpha")
densplot(as.mcmc(result$alpha), main="Density of alpha")
```

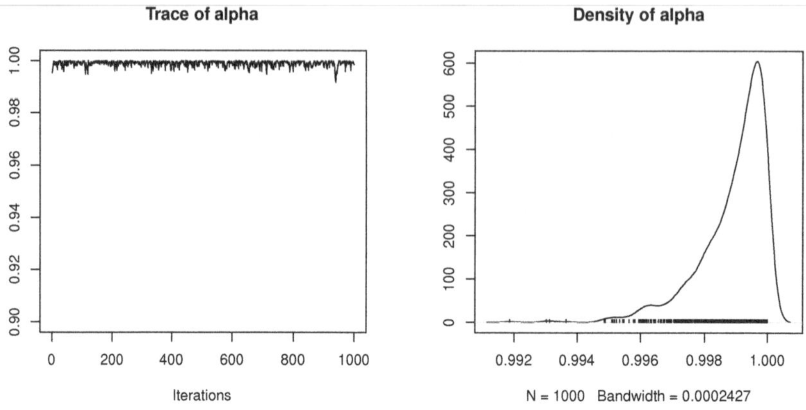

FIGURE 6.22: α-NOMINATE Trace and Density Plots for the α Parameter from the 111th US Senate

We might also wish to compare the legislator ideal points obtained from W-NOMINATE and α-NOMINATE. The `plot.legislators()` function included in the `anominate` package produces a scatterplot of the legislators' W-NOMINATE and α-NOMINATE coordinates on each dimension. The 95% credible intervals for the α-NOMINATE point estimates are also shown. Figure 6.23 shows that the W-NOMINATE and α-NOMINATE ideal point estimates for the 111th Senate are very similar (we calculate the correlation below as $r = 0.99$). This is not surprising as legislators' utility functions are estimated by α-NOMINATE to be approximately Gaussian as assumed by the W-NOMINATE model.

```
plot(result, main="")
```

6.6 Ordinal and Dynamic IRT Models

We introduced the IRT model earlier in this chapter and discussed how it can be adapted to estimate spatial voting models with the use of binary choice data, such as legislative roll call data. We now return to the IRT model

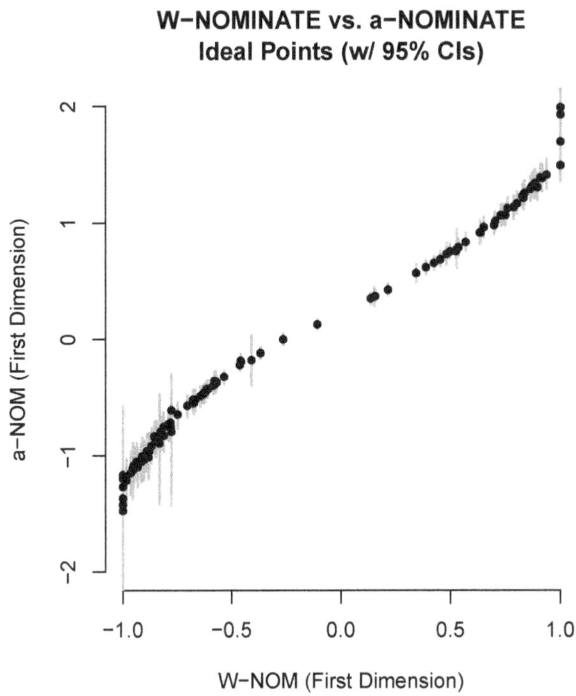

FIGURE 6.23: α-NOMINATE and W-NOMINATE Legislator Ideal Points

and discuss its application to the analysis of ordinal choice data and the measurement of latent ideology over time. Bayesian ordinal IRT has gained popularity as a method of estimating latent quantities in political science (e.g., Quinn 2004; Rosenthal and Voeten 2007; Treier and Jackman 2008; Treier and Hillygus 2009).* When latent variables are manifest in repeated observations over time, we might wish to estimate not only the latent variable itself but also its dynamics. Before moving to a dynamic specification, we first discuss the static Bayesian ordinal IRT model.

6.6.1 IRT with Ordinal Choice Data

The Bayesian mixed factor analysis model developed by Quinn (2004) and implemented in the `MCMCordfactanal()` function in the `MCMCpack` package (Martin, Quinn and Park, 2011) is equivalent to the standard two-parameter

*In this chapter, we limit our discussion to the estimation of the parametric ordinal IRT model using Bayesian methods. However, we note that a *nonparametric* ordinal IRT model is utilized in Mokken scale analysis. For more information, we refer readers to van Schuur (2003, 2011).

IRT model (with a probit link) when the data are polytomous or ordinal. Quinn's model allows for the inclusion of both continuous and ordinal responses in a standard factor model where i indexes respondents ($i = 1,...,n$), j indexes items ($j = 1,...,p$), k indexes factors or dimensions ($k = 1,...,s$), and c indexes categories in ordinal variables ($c = 1,...,C$). The latent variable x^*_{ij} is assumed to be continuous and underlie the observed responses x_{ij}:

$$x_{ij} = \begin{cases} x^*_{ij} \text{ for continuous variables } j \\ c \text{ if } x^*_{ij} \in (\gamma_{j(c-1)}, \gamma_{jc}) \text{ for ordinal variables } j \end{cases} \tag{6.37}$$

where γ_{jc} represents the cutpoint for category c in ordinal variable j.

For identification purposes, Quinn (2004) sets γ_{j0} to $-\infty$, γ_{j1} to 0, and γ_{jC_j} to ∞ for all items. The factor model for the n individuals is then:

$$x^*_i = \Lambda\phi_i + \varepsilon_i \tag{6.38}$$

where x^*_i is the p-length vector of individual i's latent responses, Λ is a $p \times s$ matrix of factor loadings for the s estimated factors, ϕ_i is an s-length vector of individual i's factor scores, and ε_i is the (normal) error term. Quinn (2004, p. 340) sets the first elements of $\phi_{1,...,n}$ to 1 so that the first element of $\Lambda_{1,...,p}$ serves as the item difficulty parameter. Hence, the first element of the ϕ_i terms are not reported by the `MCMCordfactanal()` function and ϕ_{i2} corresponds to individual i's factor score for the first factor, ϕ_{i3} is individual i's factor score for the second factor, and in general $\phi_{i(k+1)}$ is individual i's factor score for the kth factor.

When all of the items are ordinal, this model is equivalent to the normal ogive two-parameter IRT model in which the item difficulty and discrimination parameters (α_j, β_j) correspond to Λ_j and the individual ability parameters (x_i) to ϕ_i. As discussed, the first element of Λ_j is the item difficulty parameter for item j and subsequent elements are the item discrimination parameters on the first through s dimensions. For example, Λ_{j2} is item j's discrimination parameter on the first dimension, Λ_{j3} is item j's discrimination parameter on the second dimension, and more generally $\Lambda_{j(k+1)}$ is item j's discrimination parameter on the kth dimension. The cutpoints between the categories in each item j (γ_{jc}) are also estimated and reported by the `MCMCordfactanal()` function.

To demonstrate use of the `MCMCordfactanal()` function to perform ordinal IRT, we use data from the 2004 American National Election Study (ANES). The first 16 columns of the dataframe `ANES2004` are issue scales (shown in 6.2), the 17th column is the respondent's reported 2004 presidential vote (`Bush` or `Kerry`), and the final column is party identification (0 for Strong Democrat to 6 for Strong Republican). For all variables missing data are coded as `NA`.

```
data(ANES2004)
issues <- as.matrix(ANES2004[,1:16])
presvote <- ANES2004$presvote
partyid <- ANES2004$partyid
```

Table 6.2: Issue Scales in the 2004 ANES

Issue	Variable	Categories
Liberal - Conservative	`libcon`	1 (left) - 7 (right)
Diplomacy - Military Force	`diplomacy`	1 (left) - 7 (right)
Bush's Handling of Iraq War	`iraqwar`	1 (right) - 4 (left)
Government Spending/Services	`govtspend`	1 (right) - 7 (left)
Defense Spending	`defense`	1 (left) - 7 (right)
Bush Tax Cuts	`bushtaxcuts`	1 (right) - 4 (left)
Government Health Insurance	`healthinsurance`	1 (left) - 7 (right)
Guaranteed Jobs	`govtjobs`	1 (left) - 7 (right)
Government Aid to Blacks	`aidblacks`	1 (left) - 7 (right)
Government Abortion Funding	`govtfundsabortion`	1 (left) - 4 (right)
Partial-Birth Abortion Ban	`partialbirthabortion`	1 (right) - 4 (left)
Environment - Jobs	`environmentjobs`	1 (left) - 7 (right)
Death Penalty	`deathpenalty`	1 (right) - 4 (left)
Gun Regulations	`gunregulations`	1 (left) - 5 (right)
Women's Role	`womenrole`	1 (left) - 7 (right)
Gay Marriage	`gaymarriage`	1 (left) - 3 (right)

The `MCMCordfactanal()` function accepts several arguments; we discuss 10 of these. The first argument is the specification of the formula or matrix of ordinal responses. The number of latent factors or dimensions to be estimated is set with `factors`. We estimate a two-dimensional solution since previous work (notably Treier and Hillygus (2009)) has found a second ideological dimension that encompasses social/cultural issue attitudes in American public opinion.

As in Section 6.4.5, the solution is identified through constraints on the item discrimination parameters. In two dimensions, this requires setting the polarity of an item on each dimension and constraining one item to load only onto one of the two dimensions. We constrain the discrimination parameter of `healthinsurance` to be positive on the first dimension and 0 on the second dimension and the discrimination parameter of `partialbirthabortion` to be negative on the second dimension (meaning that to the extent social/cultural issue attitudes load onto the second dimension, conservative preferences will be associated with positive scores).

The details of the Gibbs sampler are specified with the arguments `burnin` (the burn-in period), `mcmc` (the number of iterations to be stored), and `thin` (retain every nth iteration from `mcmc`). We implement a burn-in period of 25,000 iterations and thin the 25,000 subsequent iterations by 25, producing 1,000 posterior samples for each of the parameters. The arguments `l0` and `L0` specify the means and precisions of the normal priors for Λ (the item parameters in the IRT context).

Finally, `store.lambda` and `store.scores` are logical arguments indicating whether the samples for Λ (the item parameters) and the factor scores ϕ (the individual ideal points) should be stored. When the number of individuals is large, it may be desirable to discard the ideal points to preserve memory resources.

```
library(MCMCpack)
result <- MCMCordfactanal(issues, factors=2,
    lambda.constraints=list(healthinsurance=list(2,"+"),
    healthinsurance=list(3,0), partialbirthabortion=list(3,"-")),
    burnin=25000, mcmc=25000, thin=25,
    l0=0, L0=0.1, store.lambda=TRUE, store.scores=TRUE)
```

We next store the posterior means and standard deviations of the parameters in the object `means.sds`. We then split up the individual ideal points and the discrimination and difficulty parameters using the `grepl()` command to only select those rows that include the specified character string ("phi" for the ideal points and "Lambda" for the item parameters).

```
means.sds <- summary(result)[[1]][,1:2]
ideal.points <- means.sds[grepl("phi",rownames(means.sds)),]
item.params <- means.sds[grepl("Lambda",rownames(means.sds)),]
```

To evaluate the dimensionality of the space and the substantive meaning of the dimensions, we first examine the item discrimination parameters of seven of the sixteen issues.[†] Recall that the first element of each issue (e.g., `lambda.libcon.1`) is the difficulty parameter for the liberal-conservative scale and the subsequent elements are the discrimination parameters on each dimension (e.g., `lambda.libcon.2` is the liberal-conservative scale's first dimension discrimination parameter).

Clearly, the first dimension represents the classic liberal-conservative divide; not only does the liberal-conservative self-identification item load strongly onto this dimension, but items tapping into attitudes about the role of government in the economy (e.g., the government spending and services and guaranteed jobs and standard of living scales) are nearly exclusively associated with the first dimension. This is not surprising, as we constrained the government-private health insurance item to load only onto the first dimension. Interestingly, the diplomacy-military force issue scale also loads strongly on the first dimension, suggesting that (at least in 2004) these attitudes conform to the classic liberal-conservative cleavage.

Three prominent social issues – abortion, gay marriage and gun regulations – all load about equally well on the first and second dimensions. Though based only on a few issues in a single year, respondents' social issue attitudes appear to be at least partly intertwined with the dominant liberal-conservative

[†]We choose only seven issues to preserve space, but the item parameters of the other nine issues support our analysis based on this subset.

ideological dimension. Finally, note that all of the discrimination parameters have the expected sign; that is, higher scores correspond to a greater likelihood of providing a conservative response to the corresponding survey item.

```
print(item.params[grep("libcon|diplomacy|govtspend|govtjobs|
    |partialbirthabortion|gunregulations|gaymarriage",
    rownames(item.params)),])
```

```
                                   Mean         SD
Lambdalibcon.1                3.05194083 0.08914799
Lambdalibcon.2                0.92930684 0.06382006
Lambdalibcon.3                0.67519113 0.07293708
Lambdadiplomacy.1              1.27217882 0.06485718
Lambdadiplomacy.2              0.51491472 0.05367758
Lambdadiplomacy.3              0.55317759 0.06531035
Lambdagovtspend.1              2.04886590 0.07473797
Lambdagovtspend.2             -0.68119705 0.05014929
Lambdagovtspend.3             -0.04416152 0.05241738
Lambdagovtjobs.1               2.13117126 0.18600591
Lambdagovtjobs.2               1.53833755 0.17540138
Lambdagovtjobs.3              -0.25501827 0.13205883
Lambdapartialbirthabortion.1 -0.14310743 0.04263040
Lambdapartialbirthabortion.2 -0.26495719 0.05048816
Lambdapartialbirthabortion.3 -0.42997822 0.05423703
Lambdagunregulations.1         0.16272277 0.04098471
Lambdagunregulations.2         0.37929965 0.04468674
Lambdagunregulations.3         0.30147653 0.05103918
Lambdagaymarriage.1           0.46914990 0.04821257
Lambdagaymarriage.2           0.43322807 0.06508491
Lambdagaymarriage.3           0.65153823 0.07434421
```

We next plot the respondent ideal points, first storing the posterior means of their first and second dimension coordinates in the objects `irt.means` and `irt2.means`. As with the item parameters, the results for each dimension are staggered, and so to retain the first and second dimension scores we select every other row of the matrix `ideal.points`.

```
irt1.means <- ideal.points[seq(1,nrow(ideal.points), by=2),1]
irt2.means <- ideal.points[seq(2,nrow(ideal.points), by=2),1]
```

Figure 6.24 shows the first and second dimension coordinates for the respondents, labeled "B" if they voted for Bush and "K" if they voted for Kerry. We find a major cleavage between the Bush and Kerry voters along the first dimension, but the two groups are more mixed along the second dimension.

```
oirt.df <- data.frame(
    D1=irt1.means, D2=irt2.means,
    presvote = presvote
)
```

```
oirt.df <- na.omit(oirt.df)
scatter <- ggplot(oirt.df,
    aes(x=D1, y=D2, colour=presvote, shape=presvote)) +
  geom_point(size=3) +
  scale_colour_manual(values=c("gray70", "gray30"),
    name="Vote", labels=c("Bush", "Kerry")) +
  scale_shape_manual(values=c("B", "K"), name="Vote",
    labels=c("Bush", "Kerry")) + theme_bw() +
    theme(legend.position="bottom") +
    labs(x="Liberal - Conservative", y="Second Dimension")
empty <- ggplot()+geom_point(aes(1,1), colour="white")+
theme(axis.ticks=element_blank(),
      panel.background=element_blank(),
      axis.text.x=element_blank(), axis.text.y=element_blank(),
      axis.title.x=element_blank(), axis.title.y=element_blank())
h1 <- ggplot(oirt.df, aes(x=D1)) +
    geom_histogram(data=subset(oirt.df, presvote == "Bush"),
      fill="gray70", alpha=.8) +
    geom_histogram(data=subset(oirt.df, presvote == "Kerry"),
      fill="gray30", alpha=.5) + theme_bw() + labs(x="", y="")
h2 <- ggplot(oirt.df, aes(x=D2)) +
    geom_histogram(data=subset(oirt.df, presvote == "Bush"),
      fill="gray70", alpha=.8) +
    geom_histogram(data=subset(oirt.df, presvote == "Kerry"),
      fill="gray30", alpha=.5) + theme_bw() + labs(x="", y="") +
    coord_flip()
library(gridExtra)
grid.arrange( h1, empty, scatter, h2,  ncol=2, nrow=2,
  widths=c(4,1), heights=c(1, 4))
```

Finally, we evaluate the influence of respondents' policy attitudes on their presidential vote choice in 2004 by running three separate probit models using the base `glm()` (for Generalized Linear Models) function in `R`. The probit link function is implemented using the argument `family=binomial(link=''probit'')` in the `glm()` call. Presidential vote (1 = Kerry, 0 = Bush) is regressed onto party identification (a 7-point scale) in the first model, the first and second dimension ideological scores from the ordinal IRT in the second model, and both sets of variables in the third model.

We use the `apsrtable()` function in the `apsrtable` package (Malecki, 2012) to generate the LaTeX code that creates Table 6.3: a side-by-side comparison of the estimates from models 1-3. The third (combined) model clearly outperforms the first two models. With this specification, both party identification and first dimension (liberal-conservative) scores are significant, in the expected direction, and substantively meaningful.

To convey the magnitude of the effect of respondents' policy positions on presidential vote choice and the moderating effect of partisanship, we generate predicted probabilities separately for Republicans, Democrats, and Indepen-

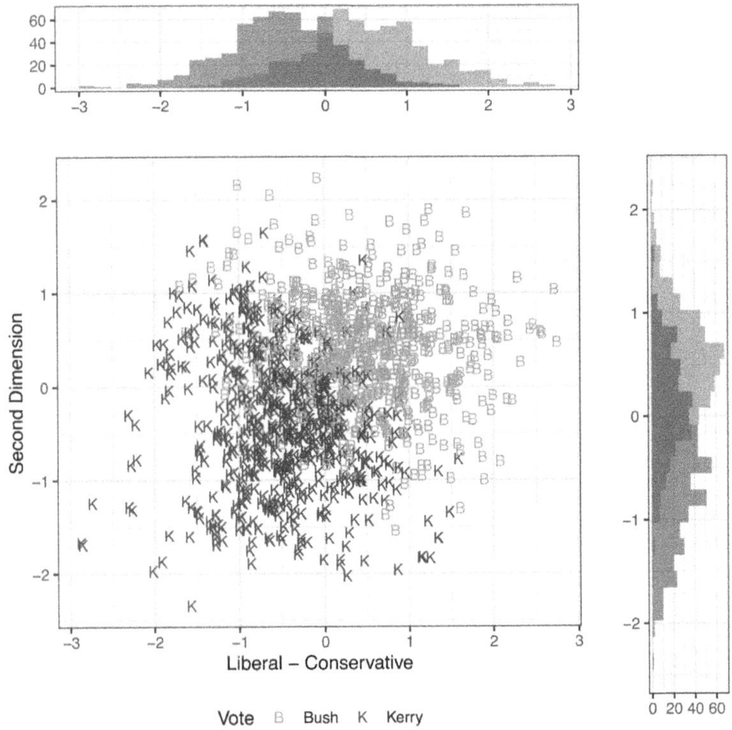

FIGURE 6.24: Ordinal IRT Scaling of Respondents

dents using interaction terms between partisan identification and first dimension ideological scores. First, we collapse the 7-point party identification scale into three categories (strong Democrats, weak Democrats and Democratic leaners; pure independents; and strong Republicans, weak Republicans, and Republican leaners) (Petrocik, 1974, 2009). Then, we use the `glm()` function to estimate a probit model with an interaction term for first dimension ideological score and partisanship.‡ This interaction term will provide three different coefficients of the effect of ideology on vote choice for each of the three partisan groups.§

We next use the `effects()` function in John Fox's `effects` package (Fox, 2003) to generate predicted probabilities (with 95% confidence intervals) of a vote for Democratic Senator John Kerry over Republican President George W. Bush from the interaction term `irt1.means*party`. We calculated predicted

‡The `glm()` function automatically includes the component terms of an interaction (in this case, `irt1.means` and `party`) as variables in the model.

§All coefficients are significant at $p < 0.05$.

```
#install.packages('apsrtable')
library(apsrtable)
mod1 <- glm(presvote ~ partyid, family=binomial(link="probit"))
mod2 <- glm(presvote ~ irt1.means + irt2.means,
    family=binomial(link="probit"))
mod3 <- glm(presvote ~ irt1.means + irt2.means + partyid,
    family=binomial(link="probit"))
apsrtable(mod1, mod2, mod3, Sweave=T)
```

	Model 1	Model 2	Model 3
(Intercept)	1.88*	0.03	1.28*
	(0.10)	(0.05)	(0.12)
partyid	−0.66*		−0.46*
	(0.03)		(0.04)
irt1.means		−1.51*	−1.16*
		(0.09)	(0.11)
irt2.means		−1.46*	−1.14*
		(0.09)	(0.11)
N	1132	1145	1132
AIC	789.12	751.23	557.26
BIC	829.38	811.75	637.77
$\log L$	−386.56	−363.62	−262.63

Standard errors in parentheses

* indicates significance at $p < 0.05$

Table 6.3: Probit Models of 2004 Presidential Vote

probabilities across 1,000 levels of the first dimension IRT score to create smooth probability curves. The estimates are then plotted in Figure 6.25.

What is striking about Figure 6.25 is the similarity between the probability curves for Democrats, Independents, and Republicans (see also Jessee (2009)). In each group, the most liberal respondents are predicted to almost certainly vote for Kerry, and as we move rightward across the ideological spectrum, respondents become increasingly likely to vote for Bush. What is different is the point on the x-axis at which respondents cross over from being more likely to vote for Kerry to being more likely to vote for Bush. For Democrats, this point is further rightward than for Independents, and for Republicans, this point is further leftward than for Independents. This result is easily interpretable: partisans are more likely to follow their party allegiances, even in the face of ideological incongruity. That is, holding policy distance equal, Republicans will be more likely to vote for Bush, and Democrats will be more likely to vote for Kerry. However, there comes a point when policy distance becomes so great that party identifiers are predicted to cross over to their rival party's candidate; in fact, this point is not so extreme for either Democrats or

Republicans, illustrating the importance of ideological factors in contemporary American voting behavior.

```
library(ggeffects)
party <- NA
party[partyid<=3] <- "Democratic"
party[partyid==4] <- "Independent"
party[partyid>=5] <- "Republican"
party <- as.factor(party)
mod <- glm(presvote ~ irt1.means*party,
           family=binomial(link="probit"))
ggp <- ggpredict(mod, c("irt1.means [n=25]", "party"))
ggp %>%
  ggplot(aes(x=x, y=predicted, ymin = conf.low, ymax=conf.high)) +
  geom_line() +
  geom_ribbon(alpha=.25) +
  facet_wrap(~group) +
  labs(x="First Dimension IRT Score (Liberal - Conservative)",
    y="Probability of Kerry Vote") +
  theme(aspect.ratio=1) +
  theme_bw()
```

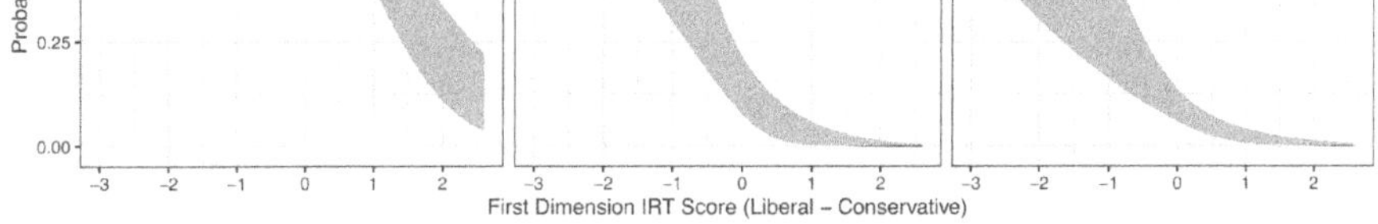

FIGURE 6.25: Effect of Ideology on 2004 Presidential Vote by Party Identification

6.6.2 Dynamic IRT

It is often the case that we want to estimate latent constructs over time. This provides the researcher both challenges and opportunities. One of the main opportunities is that the temporal structure of the data can be exploited to smooth changes over time. Generally, this has both statistical and substantive appeal. From a statistical point of view, the smoothing generally results in smaller posterior variances—that is, the latent point estimates are more tightly bound. From a substantive point of view, smoothing results in reducing temporal variability by setting the prior mean of the score to its value in the previous period. This indicates that in the absence of information about time t's latent variable score, our best guess would be centered at time t-1's score. Exploiting this temporal structure is accomplished easily by incorporating a random walk prior on the latent variable. We demonstrate this below using data from the Comparative Manifestos Project (Mikhaylov, Laver and Benoit, 2006). The project codes political party manifestos in every election across roughly 50 countries since 1945. Each statement in the manifesto is coded into one of a number of categories identifying its content. The main variables of interest here are the balanced or "quasi-balanced" items (i.e., those with clear left and right versions of the same question). Table 6.4 identifies the variables used in the analysis and their coding.

Table 6.4: Coding of Manifesto Variables Used in the Dynamic IRT Model

Left	Right
military - (per105)	military + (per104)
internat + (per107)	internat - (per109)
constitut - (per204)	constitut + (per203)
central + (per302)	decentral + (per301)
protectionism + (per406)	protectionism - (per407)
welfare + (per504)	welfare - (per505)
education + (per506)	education - (per507)
nat way life - (per602)	nat way life + (per601)
trad moral - (per604)	trad moral + (per603)
multicult + (per607)	multicult - (per608)
labour + (per701)	labour - (per702)
market regulation + (per403)	free enterprise + (per401)
econom planning + (per404)	incentives + (per402)
controlled econ + (per412)	econ orthodoxy + (per414)
nationalization + (per413)	

Note: The items below the horizontal line are items that were grouped together to form "quasi-balanced" economic items.

The data themselves provide the percentage of statements in each category. We turn them into proportions and then using the variable indicating the total number of statements in the manifesto, produce a number of statements in each category. The model treats the data as binomial with n equal to the sum of both left and right statements made (e.g., the sum of negative statements about the military and positive statements about the military). The number of successes in the binomial is the number of right statements made on the issue (e.g., the number of positive statements made about the military). When both left and right categories have zero statements, we treat the number of successes as unknown and the number of statements made as 10. This will allow us to impute the number of statements out of 10 that *would have been made* had the party chosen to make statements in that area.¶

The model looks as follows:

$$k_{itj} \sim \text{Binomial}(p_{itj}, N_{itj})$$
$$\text{logit}(p_{itj}) = \lambda_{0j} + \lambda_{1j}\phi_{it}$$

The λ parameters are given standard normal priors. For identification purposes, λ_{11} is set deterministically to 1 and the other λ_{1j} $j = \{2, \ldots, J\}$ are constrained to be positive. The latent variables at time 1 have the following prior: $\phi_{i1} \sim N(0, \tau_1)$ and in times $t = \{2, \ldots, T\}$ it is: $\phi_{it} \sim N(\phi_{i,t-1}, \tau_2)$. Note that τ_1 sets the scale of the space and τ_2 is the variance of the time-series innovations (i.e., the changes over time). Both precision parameters are given Gamma priors with shape equal to 1 and rate equal to 0.1. The model can be run in R with:

```
mod <- "model{
for(i in 1:npe){ #loop through party-elections
    for(j in 1:ncolx){ #loop through issues
        X[i,j] ~ dbin(p[i,j], n[i,j])
            logit(p[i,j]) <- alpha[j] + beta[j] * Z[party[i], elec[i]]
            }
        }
beta[1] <- 1
alpha[1] ~ dnorm(0,1)
for(i in 2:ncolx){
    beta[i] ~ dnorm(0, 1)T(0, ) # Here, betas are truncated to be
    alpha[i] ~ dnorm(0,1)   # positive, theoretically defensible
}
for(j in 1:nparty){
Z[j, 1] ~ dnorm(0, tau.Z[1])
for(i in 2:nelec){
Z[j, i] ~ dnorm(Z[j, i-1], tau.Z[2])
}
}
tau.Z[1] ~ dgamma(1, .1)
tau.Z[2] ~ dgamma(1, .1)
}"
```

¶This is a more interesting way of dealing with structural zeros than the method used by the CMP.

```
jags.data <- list(
    nelec = max(as.numeric(as.factor(uk$date))),
    nparty = max(as.numeric(as.factor(uk$partyname))),
    npe = nrow(uk),
    ncolx = ncol(X),
    X=as.matrix(X), n=as.matrix(n),
    party = as.numeric(as.factor(uk$partyname)),
    elec = as.numeric(as.factor(uk$date))
)
library(runjags)
out <- run.jags(mod, data=jags.data, monitor=c("beta", "alpha",
    "tau.Z", "Z"), adapt=5000, burnin=20000, sample=5000, thin=5,
    summarise=F)
```

Martin and Quinn (2002) suggest loosely informative priors for these models, as the likelihood functions in dynamic models are relatively flat.‖ Our model is slightly different from the one Martin and Quinn specify, as theirs has discrimination and difficulty parameters that also evolve over time. Ours are assumed static. This mitigates the problem relative to their model, but even they find that their models are not sensitive to prior specification. This will not always be the case and analysts should assess sensitivity to prior specifications, particularly in these models.

In terms of generating loosely informative priors, several things should be considered. First, observations that are presumed to be extreme should be placed in different places than those presumed to having middling values on the underlying dimension. For dynamic models like this, only the $t = 0$ initial values must be provided. The others are defined by the data and the priors. These more extreme observations could be ordered according to their presumed *a priori* ordering. In our example up above, we could put the Liberal Party at -1, the Liberal Democratic Party at 0 and the Conservative Party at 1. Obviously these values will have to be consistent with any other identifying constraints in the model (e.g., prior distribution censoring or truncation). Next, we could think about starting values for the other model parameters. In our model, we have used the regression parameterization of the IRT model, so α isn't a difficulty parameter; it is more of an "easiness" parameter, such that higher values indicate higher baseline probabilities of a correct answer. For the UK data, we could calculate $p_j = \frac{1}{NT}\sum_{i=1}^{N}\sum_{t=1}^{T}\frac{x_{itj}}{n_{itj}}$ - these are the probability of a right-wing response across all parties and times for each of the j issues. We could then put these back on the link scale by $\alpha_j^* = \Lambda^{-1}(p_j)$, where $\Lambda(\cdot)$ is the PDF of the logistic distribution. We could then use these α_j^* as our initial values. Next, we could make the coefficient initial values. Here, we want to calculate the standard deviation of the probabilities, rather than the mean - we call these s_j. The idea is that items with greater variance are likely

‖We thank one of the anonymous reviews of the book for suggesting the inclusion of this idea.

to be more discriminating. We then recognize that $\beta_1 = 1$ by definition in our model, so we enforce that by making $\beta_j^* = \frac{s_j}{s_1}$. It is also important to make sure that the magnitude of the starting values you choose for the model parameters and initial ideal points is consistent with the underlying distribution. For example, since $\Lambda_{-1}(0.01) = -4.59$ and by symmetry, $\Lambda^{-1}(.99) = 4.59$, you would want to make sure that the initial ideal points you choose are well inside these boundaries. You would want to do that with the coefficients as well, ensuring that none get too large.

```
probs <- jags.data$X/jags.data$n
probsm <- colMeans(probs, na.rm=TRUE)
alpha.init <- qlogis(probsm)
probsd <- apply(probs, 2, sd, na.rm=TRUE)
beta.init <- probsd/probsd[1]
beta.init[1] <- NA
inform.inits <- list(
    Z = matrix(c(1,-1,0, rep(NA, 17*3)), nrow=3),
    alpha=alpha.init,
    beta=beta.init,
    tau.Z = c(.1,.1))
out1 <- run.jags(mod, data=jags.data,
                 monitor=c("beta", "alpha", "tau.Z", "Z"),
    inits = list(inform.inits, inform.inits), adapt=5000,
    burnin=20000, sample=5000, thin=5, summarise=F)
om <- out$mcmc
om1 <- out1$mcmc
p <- om[,c(grep("alpha", colnames(om[[1]])),
           grep("beta", colnames(om[[1]])))]
p1 <- om1[,c(grep("alpha", colnames(om1[[1]])),
             grep("beta", colnames(om1[[1]])))]
sp <- summary(p)$statistics
sp1 <- summary(p1)$statistics
cor(sp[,1], sp1[,1])
mean(sp[,2]/sp1[,2], na.rm=TRUE)
```

We implemented all of these initial values in our model and provided these initial values to both chains, along with the initial values that $\tau = \{.1, .1\}$. The posterior means of this model versus the random initialization that JAGS does automatically correlate at 0.9999. The ratio of posterior variances for α and β for the two models was, on average, 1.002. This suggests that the modes were not sensitive to the initial conditions that were provided. If they were, then those with plausible or "loosely informative" values would be preferred.

There are a number of different model results that could be meaningful. While the coefficients themselves could be of interest, it is probably better to view their effect through plotting the item response functions. These show how the probability of success (i.e., making right-leaning statements) changes as a function of the latent dimension. Figure 6.26 shows the item response functions for the 12 variables in the model. It seems that issues of traditional morality, economy, national way of life, and the military are better discriminators of left-right sentiment than other issues such as welfare, education, internationalism, and multiculturalism.

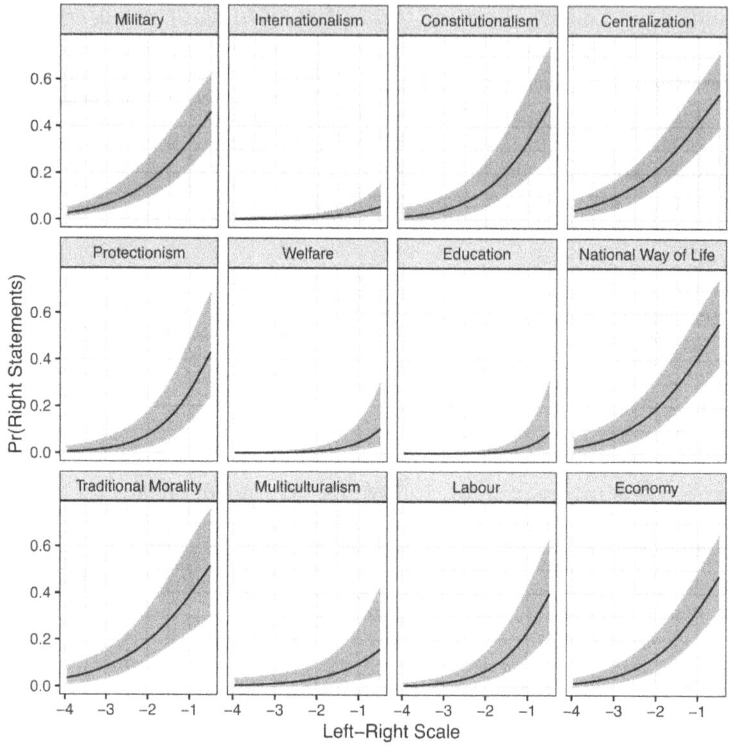

FIGURE 6.26: Item Response Functions for Dynamic IRT Model

Perhaps more importantly than the coefficients or even the item response functions is the distribution of the latent variable. That is to say, what are the left-right scores of the various parties over time? Figure 6.27 shows a dot plot of the four parties with estimated placements from the UK. This appears to identify what look to be important moves over time, for all parties. The start of Thatcher's reign as Prime Minister is visible as the Conservative Party moves considerably to the right in 1979, then moves back to the left in subsequent elections. In 1983, Michael Foot as Labour Party leader attempted to move the party significantly to the left, but those moves were later countered by Neil Kinnock—a series of events visible in the change in Labour's placement over time.

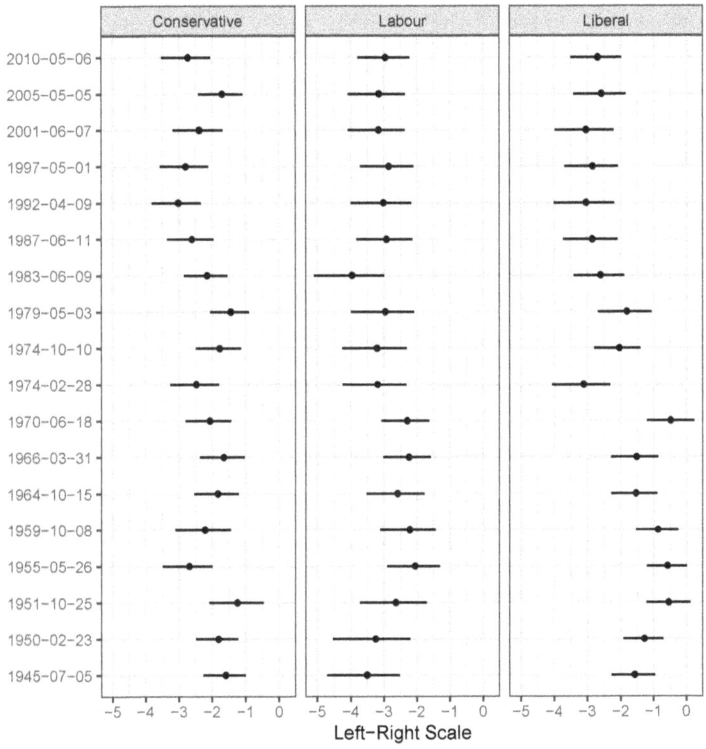

FIGURE 6.27: Party Placements and 95% Credible Intervals for the UK

While the overlap in credible intervals is interesting, it does not constitute a proper test of difference. If we wanted to know whether election-to-election changes are statistically interesting (i.e., the overlap in posteriors is sufficiently small), we can investigate a bit further. For example, consider the Conservative Party: ϕ_{1t}, a $n_{iter} \times T$ matrix of placements for the Conservative Party. An easy way to calculate each test is to create a matrix such that in each column, there are $T-2$ 0's, one 1 and one -1 such that the positive and minus one indicate the two columns being tested. Consider a simple example where there are three stimuli (e.g., party placements) being tested for equality. In this case, there are three pairwise differences to be tested. The matrix that will allow all tests to be performed is,

$$\Delta = \begin{bmatrix} 1 & 1 & 0 \\ -1 & 0 & 1 \\ 0 & -1 & -1 \end{bmatrix}$$

where Δ has dimensions $T \times T-1$. We can then calculate $D = \phi_{1t}\Delta$ and then figure out what proportion of the D values bigger is than zero with

$$p_t = \frac{\#\{D_t > 0\}}{n_{iter}} \tag{6.39}$$

This can be accomplished easily in R for a single set of estimates with:

```
## identify only conservative party latent variable scores
conspart <- lats[,grep("^Z\\[1", colnames(lats))]
## any changes (i.e., from any time to any other time)
combs <- combn(18,2)
d <- matrix(0, ncol=ncol(combs), nrow=ncol(conspart))
d[cbind(combs[1,], 1:ncol(combs))] <- -1
d[cbind(combs[2,], 1:ncol(combs))] <- 1
diff <- conspart %*%d
p <- apply(diff, 2, function(x)mean(x > 0))
p <- ifelse(p > .5, 1-p, p)
## only year-over-year changes
dy <- d[, which(combs[2,] - combs[1,] == 1)]
diff2 <- conspart %*% dy
p2 <- apply(diff2, 2, function(x)mean(x >0))
p2 <- ifelse(p2 > .5, 1-p2, p2)
pmat <- p
p2mat <- p2
```

This needs to be replicated for as many parties as there are in the data.

We transform p_t such that $p_t = 1 - p_t$ if $p_t > 0.5$. These values are presented in Table 6.5. While we do not present the table here, it would also be just as easy to test all temporal differences (e.g., the difference between ϕ_{1j} and ϕ_{ij} for $j \neq k$). Table 6.6 identifies the proportion of differences significant both for the election-over-election comparisons presented above and the all-times comparisons. All parties, save the SDP, see election-over-election changes, though none more often than not. The Labour Party seems considerably less likely to change over time than the Liberals or Conservatives. Not surprisingly, the same trend is visible when considering all changes rather than just the election-over-election changes.

Table 6.5: p-values for Election-Over-Election Changes in Placement

	Conservative	Labour	Liberal
1950-1945	0.20	0.30	0.14
1951-1950	0.05	0.11	0.01
1955-1951	0.00	0.10	0.45
1959-1955	0.09	0.35	0.16
1964-1959	0.14	0.18	0.01
1966-1964	0.28	0.18	0.48
1970-1966	0.07	0.43	0.00
1974.1-1970	0.08	0.01	0.00
1974.2-1974	0.01	0.48	0.00
1979-1974.2	0.10	0.28	0.26
1983-1979	0.00	0.01	0.02
1987-1983	0.02	0.00	0.22
1992-1987	0.04	0.40	0.31
1997-1992	0.18	0.33	0.30
2001-1997	0.08	0.18	0.28
2005-2001	0.01	0.48	0.09
2010-2005	0.00	0.28	0.36

Table 6.6: Proportion of Interesting Differences Across Time

	Conservative	Labour	Liberal
Election-over-Election	0.41	0.18	0.35
All Differences	0.52	0.37	0.67

6.7 EM IRT

The rise of "big data" has afforded scholars the opportunity to estimate ideological configurations from massive—and often novel—data sets. These data may come in the form of traditional roll call votes or survey responses that are bridged over time and/or space, or they may represent more exotic quantities such as word usage by political candidates and parties or the network of relationships between political actors on social media platforms. In either case, the ideal point estimation methods discussed so far do not typically scale to data that is exceptionally long (a large number of rows/individuals) and/or wide (a large number of columns/choices). Specifically, researchers will frequently encounter heavy (occasionally prohibitive) computational costs and convergence issues when analyzing large data sets with standard methods.**

To address this problem, Imai, Lo and Olmsted (2016) adapt the EM (Expectation-Maximization) algorithm to efficiently estimate a large class of Bayesian IRT models. These include the basic two-parameter model for binary choice data (discussed in Section 6.4) and its extension to ordinal choice data (discussed in Section 6.6.1). For these two ideal point models, they derive a closed-form version of the EM algorithm that exactly maximizes the posterior distributions of the unknown parameters. In cases where a closed-form derivation of the EM algorithm is not possible—as with dynamic, hierarchical, textual, and network IRT models—they develop a variational EM algorithm to approximately maximize the posterior distributions (see Grimmer, 2011).

The EM algorithm provides an efficient means of estimating likelihood-based models when missing data (often in the form of latent variables) prevents a direct maximization of the likelihood function (Dempster, Laird and Rubin, 1977).†† In cases with two sets of unknowns—the values of the model parameters $\boldsymbol{\theta}$ (or as in the two-parameter IRT model, $\boldsymbol{\alpha}$ and $\boldsymbol{\beta}$) and the values of the missing/latent data Z (or in our derivation of the two-parameter IRT model, x_i)—the EM algorithm iteratively switches between two steps. After initializing values for $\boldsymbol{\theta}$, the first (Expectation or E) step calculates an expectation of the "Q function" (the log likelihood function in frequentist models or the log joint posterior distribution in Bayesian models) holding $\boldsymbol{\theta}$ fixed. The second (Maximization or M) step then updates the estimates of $\boldsymbol{\theta}$ to maximize the Q function holding Z fixed. The process rotates between these two steps until some convergence criterion is satisfied (by default, `emIRT` uses the rule that the correlation between the estimates from the current and previous iterations exceed $1 - 10^{-6}$).

The basic logic behind the EM algorithm—flipping between multiple sets of unknown parameters to iteratively improve estimates of each—also forms the

**This is especially true when using MCMC-based methods to estimate Bayesian ideal point models.

††Of course, this naturally generalizes to Bayesian models, where we are instead concerned with maximizing the joint posterior distribution of a set of parameters (i.e., a weighted combination of the prior distributions and the likelihood function) rather than the likelihood function itself. These are sometimes referred to as the MAP (maximum a posteriori) estimates, which are equivalent to the maximum likelihood estimates if we assume uniform priors.

basis of alternating least-squares and alternating maximum likelihood procedures and underlies several methods discussed in this book. These include the MLSMU6 unfolding algorithm, NOMINATE, Optimal Classification, and the Gibbs sampler. In this case, though, Imai, Lo and Olmsted (2016) explicitly make use of the EM algorithm to dramatically speed up estimation for large-scale ideal point models.

As one example, recall from Section 6.4 the basic two-parameter IRT model with legislators indexed by i (i in $1,\ldots,n$) and roll call votes indexed by j (j in $1,\ldots,q$). In this model, the probability that legislator i votes Yea on the jth roll call is:

$$P_{ijy} = \Phi\left(\beta_j' x_i - \alpha_j\right) \tag{6.40}$$

Placing vague, conjugate (normal) priors on the legislator ideal points (x_i) and the vote-specific parameters (β_j and α_j) yields the posterior distribution:

$$p(\boldsymbol{B},\boldsymbol{\alpha},\boldsymbol{X} \mid \boldsymbol{Y}) \propto p(\boldsymbol{B},\boldsymbol{\alpha},\boldsymbol{X}) \times L(\boldsymbol{B},\boldsymbol{\alpha},\boldsymbol{X} \mid \boldsymbol{Y}) \tag{6.41}$$

where $\boldsymbol{Y}$ is the $n \times q$ matrix of observed roll call votes. This implies the full joint posterior distribution that also includes the $n \times q$ matrix of latent "Yea" vote probabilities $\boldsymbol{P}$:

$$p(\boldsymbol{P},\boldsymbol{B},\boldsymbol{\alpha},\boldsymbol{X} \mid \boldsymbol{Y}) \tag{6.42}$$

which we normally explore and summarize using Markov chain Monte Carlo (MCMC) methods. To speed up computation, Imai, Lo and Olmsted (2016) first perform the E step of the EM algorithm by calculating the Q function at iteration t using estimates from the previous iteration $(t-1)$:

$$Q(\boldsymbol{B},\boldsymbol{\alpha},\boldsymbol{X}) = \mathrm{E}\left[\log p(\boldsymbol{P},\boldsymbol{B},\boldsymbol{\alpha},\boldsymbol{X} \mid \boldsymbol{Y}) \mid \boldsymbol{Y},\boldsymbol{B}^{(t-1)},\boldsymbol{\alpha}^{(t-1)},\boldsymbol{X}^{(t-1)}\right] \tag{6.43}$$

where the entries of $\boldsymbol{P}$ are updated using:

$$P_{ijy}^{(t)} = \mathrm{E}\left(P_{ijy} \mid \boldsymbol{B}^{(t-1)},\boldsymbol{\alpha}^{(t-1)},\boldsymbol{X}^{(t-1)},Y_{ij}\right) \tag{6.44}$$

Then, the M step is performed by maximizing the Q function for the roll call and legislator parameters:

$$\boldsymbol{B}^{(t)} = \mathrm{E}\left(\boldsymbol{B} \mid \boldsymbol{\alpha}^{(t-1)},\boldsymbol{X}^{(t-1)},\boldsymbol{P}^{(t)}\right) \tag{6.45}$$

$$\boldsymbol{\alpha}^{(t)} = \mathrm{E}\left(\boldsymbol{\alpha} \mid \boldsymbol{B}^{(t-1)},\boldsymbol{X}^{(t-1)},\boldsymbol{P}^{(t)}\right) \tag{6.46}$$

$$\boldsymbol{X}^{(t)} = \mathrm{E}\left(\boldsymbol{X} \mid \boldsymbol{B}^{(t-1)},\boldsymbol{\alpha}^{(t-1)},\boldsymbol{P}^{(t)}\right) \tag{6.47}$$

The algorithm then switches back to the E step and repeats until convergence. Imai, Lo and Olmsted (2016) demonstrate that the emIRT algorithm produces estimates that are highly correlated with—often virtually identical to—the estimates from the original methods. The largest deviations occur in cases with a large amount of missing data. The improvement in computational efficiency is also, of course, contingent on the data. For large roll call matrices on the order of 10,000 legislators and 1,000 roll call votes, emIRT converges in less than 15 minutes while W-NOMINATE takes 2.5 hours.

To demonstrate the functionality of the `emIRT` package in `R` (Imai, Lo and Olmsted, 2017), we use the package's implementation of the Poisson IRT (or "Wordfish") model developed by Slapin and Proksch (2008) to analyze text data from presidential State of the Union (SOTU) addresses between 1985 and 2019.[‡‡] In the Wordfish model, the data are organized as word frequencies in a $J \times K$ matrix, with the j in $1,\ldots,J$ words on the rows and the k in $1,\ldots,K$ documents on the columns.[*] The word frequencies y_{jk} are assumed to be generated by a Poisson process that accounts for the general prevalence of a word j and the latent ideological position of document k. That is,

$$p(y_{jk} \mid \alpha_j, \beta_j, \psi_k, x_i) = \text{Poisson}(\lambda_{jk}) \tag{6.48}$$
$$\lambda_{jk} = \exp(\psi_k + \alpha_j + \beta_j x_i) \tag{6.49}$$

where ψ_k represents the verboseness of document k, the item parameters α_j and β_j represent the popularity and ideological discrimination, respectively, of word j, and x_i is actor i's ideal point.[†]

The `emIRT` implementation of the generalized Wordfish model is provided in the `poisIRT()` function. First, though, we need to pre-process the text data and organize it as a $J \times K$ document-feature matrix (with words on the rows, SOTU addresses on the columns, and entries representing word frequencies). We begin by loading the `emIRT` package alongside two text-processing packages (`quanteda` and `tm`) in `R` (Benoit et al., 2018; Feinerer, Hornik and Meyer, 2008).

```
library(emIRT)
library(tm)
library(quanteda)
```

[‡‡]The name "Wordfish" is a play on the French meaning of the word *poisson*. Additional details about the model are available from `wordfish.org`.

[*]The documents most regularly represent party/candidate manifestos, platforms, speeches, or other statements of a political actor's ideological or policy goals.

[†]Usually, the actors and documents are treated interchangeably, but the generalized Wordfish model also allows the user to nest documents within actors (i.e., to estimate $x_{i|k|}$) (Slapin and Proksch, 2008, p. 649).

We next load the file `SOTUcorpus` from the `asmcjr` package, which contains the text of each presidential State of the Union address since 1790.[‡] We examine the basic structure of the file with the `summary()` function below.

```
data(SOTUcorpus, package="asmcjr")
head(summary(SOTUcorpus))
```

```
             Text Types Tokens Sentences FirstName
1 Washington-1790   460   1167        24    George
2 Washington-1790b  591   1501        38    George
3 Washington-1791   814   2471        58    George
4 Washington-1792   769   2282        59    George
5 Washington-1793   800   2116        54    George
6 Washington-1794  1134   3192        77    George
   President       Date delivery type       party
1 Washington 1790-01-08   spoken SOTU Independent
2 Washington 1790-12-08   spoken SOTU Independent
3 Washington 1791-10-25   spoken SOTU Independent
4 Washington 1792-11-06   spoken SOTU Independent
5 Washington 1793-12-03   spoken SOTU Independent
6 Washington 1794-11-19   spoken SOTU Independent
```

The `quanteda` package includes several functions to examine and transform textual data. One of these, `kwic()`, can be used to locate keywords and their immediate context (i.e., the words that directly precede and follow the keyword). For example, the code below identifies the most 10 recent SOTU addresses that include the word "homeland," as well as the three words before and after the keyword "homeland" in the address. From this, it appears that the word "homeland" is usually used by Republican presidents in a national security-based context, and hence its frequency may be a useful indicator of latent right-wing ideology.

```
tail(kwic(SOTUcorpus, "homeland", 3), 10)
```

```
 [Bush-2005, 3275]        to make our | homeland |
 [Bush-2006, 2471]      military, and | homeland |
 [Bush-2007, 2518]       to guard the | homeland |
 [Bush-2008, 4783]          , and our | homeland |
[Obama-2010, 6017] investments in our | homeland |
[Trump-2017, 1087] the Departments of | Homeland |
[Trump-2017, 4129]  the Department of | Homeland |
[Trump-2018, 3558]       our country, | Homeland |
[Trump-2018, 5672]  soon threaten our | homeland |
[Trump-2019, 1763]       , protect our | homeland |
```

[‡]These data were collected and assembled by The American Presidency Project at the University of California, Santa Barbara and can be downloaded at `presidency.ucsb.edu/sou.php`.

```
safer, and
security. These
. We know
. The enemy
security and disrupted
Security and Justice
Security to create
Security Investigations Special
. We are
, and secure
```

Since we are estimating a single latent dimension of ideology, we next subset the data to analyze only SOTU addresses given since 1985 (by which time contemporary partisan polarization was well in progress). We also store each president's party affiliation in the object `party`, which will be useful when we generate starting values for the ideal points (x_i). Next, we create the $J \times K$ Document-Feature Matrix (DFM) using the `dfm()` function. This function also allows us to extract the stem of all words (removing suffixes such as "-ing", "-ed", and "-es") and remove punctuation marks and uninformative stop words (words such as "a", "an", and "the"). Finally, we use the `dfm_trim()` function to remove words that have been used less than three times across SOTU addresses (we could also use the `min_docfreq` argument to remove words that have appeared in less than some number of documents, but below we simply set this value to 1).

```
SOTUcorpus.subset <- corpus_subset(SOTUcorpus,
                                   Date > as.Date("1985-01-01"))
party <- docvars(SOTUcorpus.subset, "party")
dfmat <- dfm(SOTUcorpus.subset,
    stem = TRUE,
    remove_punct = TRUE,
    remove = stopwords("english"))
dfm.trimmed <- dfm_trim(dfmat, min_termfreq = 3, min_docfreq = 1)
```

When we examine the dimensions of the object `dfm.trimmed`, we see that the 35 SOTU addresses are on the rows and the 7025 remaining words are on the columns. We need to transpose this to create our $J \times K$ DFM, which we do and store in the object `dat`. Hence, in this example $J = 3341$ and $K = 35$.

```
dim(dfm.trimmed)
```

```
[1]   35 3342
```

```
dat <- t(as.matrix(dfm.trimmed))
J <- nrow(dat)
K <- ncol(dat)
```

The data is now in the required format to pass to `poisIRT()`, but as a final step we must create starting values and specify prior distributions for the unknown parameters. We create random starting values for α, β, and ψ; but for the ideal points x we use starting values of -0.5 and 0.5 for Democratic and Republican presidents, respectively. We then place diffuse normal priors ($\mu = 0$ and $\sigma^2 = 100$) on all of the parameters.

```
starts <- list(alpha = matrix(runif(K, -1, 1)),
    x = matrix(as.numeric(
              party=="Republican") - 0.5),
    psi = matrix(runif(J, 0, 1)),
    beta = matrix(runif(J, 0, 1)))
priors <- list(alpha = list(mu = 0, sigma2 = 100),
    x = list(mu = 0, sigma2 = 100),
    psi = list(mu = 0, sigma2 = 100),
    beta = list(mu = 0, sigma2 = 100))
```

We can now run the `emIRT` implementation of the Wordfish model. To do so, we need to specify six arguments: `.rc` (the DFM we just created), the lists `.starts` and `.priors`, `i` (a vector indicating which actor corresponds to each of the K documents; when the two are identical, we specify `0:(K-1)`), `NI` (the number of unique actors; in this case K), and `.control` (an optional list of control parameters specifying the convergence threshold, the maximum number of iterations, and the number of cores (`threads`) to use). We see that the algorithm completed in 559 iterations, which took less than 10 seconds.

```
set.seed(12943021)
lout <- poisIRT(.rc = dat,
    .starts = starts,
    .priors = priors,
    i = 0:(K-1),
    NI = K,
    .control = {list(
        threads = 1,
        thresh = 1e-6,
        maxit=1000)}
    )
```

The estimated means and variances of the parameters are stored in the objects `lout$means` and `lout$vars`. Below, we extract the ideal points for each of the 35 SOTU addresses and store them in the object `df.plot`. We then use `ggplot` to create a dot plot of the scaling results in Figure 6.28. The estimates indicate perfect ideological divergence between Democratic and Republican SOTU addresses, although President George H.W. Bush's 1992 address and President Barack Obama's 2016 address nearly overlap. According to the Wordfish model, President Bill Clinton's 1993 address (which was technically an Address to a Joint Session of Congress rather than an official

SOTU address, and focused on his economic and health care agenda) is the most liberal of the last 35 years, while President George W. Bush's 2003 address (which focused on terrorism and the impending Iraq War) is the most conservative.

```
df.plot <- data.frame(
  idealpoint=lout$means$x,
  sd = sqrt(c(lout$vars$x)),
  name = colnames(dat))
df.plot <- df.plot %>%
              mutate(
                lower = idealpoint - 1.96*sd,
                upper = idealpoint + 1.96*sd,
                year = str_extract(
                  as.character(df.plot$name), "\\d{4}")) %>%
              mutate(year = as.numeric(year))
df.eras <- data.frame(
  x=c(1984.5, 1988.5, 1988.5, 1984.5, 1984.5,
      1992.5, 2000.5, 2000.5, 1992.5, 1992.5,
      2008.5, 2016.5, 2016.5, 2008.5, 2008.5),
  y=c(-.1, -.1, .1,.1,-.1,
      -.1, -.1, .1,.1,-.1,
      -.1, -.1, .1,.1,-.1),
  group=as.factor(rep(1:3, each=5)))
ggplot(df.plot) +
    geom_point(aes(x=year, y=idealpoint), size=2, alpha=0.7) +
  geom_segment(aes(x=year, y=lower, xend=year, yend=upper)) +
  geom_polygon(data=df.eras, aes(x=x, y=y, group=group),
               fill="gray75", alpha=.25) +
    xlab("\nIdeal Point") +
    ylab("") +
  coord_cartesian(ylim=c(-.08, .08)) +
  scale_x_continuous(sec.axis=sec_axis(trans=~.,
         breaks=c(1986.5, 1990.5, 1996.5, 2004.5, 2012.5, 2018),
         labels=c("Reagan", " GHW Bush", "Clinton",
                  "GW Bush", "Obama","Trump"))) +
  theme_bw()
```

6.8 Conclusion

Throughout this chapter, we have focused on the Bayesian inferential framework as a means for estimating many of the models we talked about in previous chapters. The versatility of Bayesian models and the natural means for

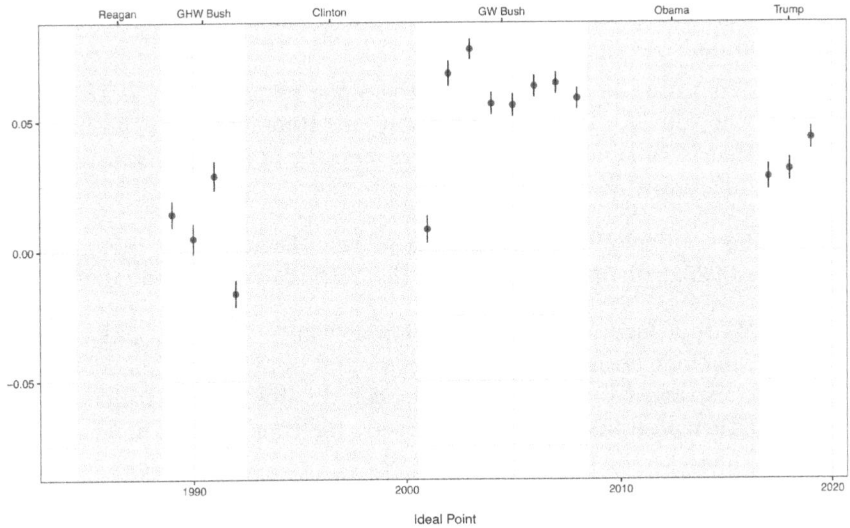

FIGURE 6.28: Estimated Ideological Positions of Presidential State of the Union Addresses, 1985-2019.

capturing uncertainty in latent point estimates are the main reasons that we devote an entire chapter to this topic.

6.9 Exercises

1. Data from the 2012 American National Election Study (ANES) is stored in the data set `ANES2012.Rda`. `ANES2012` is a list that stores respondents' placements of themselves and four stimuli (Barack Obama, Mitt Romney, and the Democratic and Republican Parties) on seven-point liberal-conservative. (`ANES2012$libcon.placements`) gives the 0-100 feeling thermometer ratings of eight stimuli (Barack Obama, Mitt Romney, Joe Biden, Paul Ryan, Hillary Clinton, George W. Bush, and the Democratic and Republican Parties) (`ANES2012$thermometers`) and presidential vote choice (`ANES2012$presvote`). Missing values are coded as 999. Use Bayesian Aldrich-McKelvey scaling to analyze respondents' placements of themselves and Barack Obama, Mitt Romney, and the Democratic and Republican Parties on the liberal-conservative scale in the matrix `ANES2012$libcon.placements` using 25,000 iterations of two chains with a burn-in period of 20,000 iterations.

(a) Plot the posterior means and 95% credible intervals of the four stimuli locations in a dot plot.

(b) What are the mean values of the alpha parameter for respondents who voted for Barack Obama and Mitt Romney? Does this indicate that Obama and Romney voters used the liberal-conservative scale differently? Explain.

2. Use Bakker-Poole Bayesian MDS to analyze the agreement score matrix in the `CCES2010.GA` data set using the `CCES2010.GA.bug` model file.

   (a) Write a function to generate initial values for the `z` and `tau` parameters, noting that three coordinates in the BUGS model file are constrained to be zero and four coordinates are constrained to be either positive or negative. Provide the code used to program this function.

   (b) Perform Bayesian MDS using the `jags.model()` function with a burn-in period of 50,000. Then take 10,000 samples of `z` with the `coda.samples()` function.

   (c) Store the means and 95% credible intervals for the respondent coordinates in a single matrix. Print the matrix.

   (d) Plot the respondent coordinates and 95% credible intervals, clearly labeling the letter identifiers for the respondents.

3. Use the Bakker-Poole Bayesian unfolding procedure to analyze the 2012 ANES feeling thermometers data in two dimensions. Make sure that the thermometers matrix includes only respondents who provided at least five valid thermometer ratings and values have been transformed to distances between 0 and 2. These steps were taken in Exercise 1.

   (a) Convert the missing values (now coded as 999) to `NA`.

   (b) Run the Bayesian unfolding procedure on the data using the same parameters (e.g., `nburn = 500` and `nslice = 1500`).

   (c) Plot the configuration of stimuli and respondents estimated by the L-BFGS optimization procedure.

   (d) Plot the configuration (the posterior means) of stimuli and respondents estimated by Bayesian unfolding.

4. Assemble the posterior means and 95% credible intervals of the stimuli locations from Bayesian unfolding of the 2012 ANES feeling thermometers data.

   (a) Neatly format and print this table.

   (b) Plot the stimuli point estimates (the means) with cross-hairs for the 95% credible intervals on each dimension.

i. Generally, is there greater uncertainty associated with the first or second dimension locations of the stimuli? Does this make sense? Why?

5. From Chapter 5, run `wnominate()` on the Fourth European Parliament data (stored in the `rollcall` object `ep4` from Exercise 2) in two dimensions.

   (a) Plot the W-NOMINATE coordinates of members from the following European Party Groups: E (European People's Party), S (Socialists), G (European Democrats/Union for Europe), L (Liberal Democrats) and V (Greens).

   (b) Perform IRT on the `ep4` data using the `ideal()` function in two dimensions with the default setting of the other parameters. Use the W-NOMINATE coordinates of three members to identify the solution.

      i. Provide the code used to run `ideal()` and identify the solution.

   (c) Perform a Procrustes rotation of the IRT coordinates of members of the Fourth European Parliament using the W-NOMINATE coordinates as the target configuration.

   (d) Do a side-by-side plot of the W-NOMINATE coordinates of members of the Fourth European Parliament in the left plot and the rotated IRT coordinates in the right plot.

      i. Compare and contrast the two configurations.

6. Use the `anominate()` function to estimate the α parameters for the `sen104kh`, `ep4`, and `CASEN.19931994` roll call data sets.

   (a) Print the means and 95% credible intervals for the α parameters for each legislature.

   (b) What do these results indicate about the shape of these legislators' utility functions? Explain.

7. Estimate a unidimensional IRT model of the`sen104kh` data using:

   (a) The `ideal()` function.

   (b) The `binIRT()` function in the `emIRT()` package.

   (c) Plot and compare the two sets of legislator ideal point estimates.

References

Abramowitz, Alan I. 2010. *The Disappearing Center: Engaged Citizens, Polarization, and American Democracy.* New Haven: Yale University Press.

Abramowitz, Alan I., John McGlennon, Ronald B. Rapoport and Walter J. Stone. 2001. "Activists in the United States Presidential Nomination Process, 1980-1996." Ann Arbor, MI: Inter-university Consortium for Political and Social Research (ICPSR). ICPSR06143-v2.
URL: *http://dx.doi.org/10.3886/ICPSR06143.v2*

Aitchison, John and James Alan Calvert Brown. 1957. *The Lognormal Distribution.* Cambridge: Cambridge University Press.

Aldrich, John H. 1983. "A Downsian Spatial Model with Party Activism." *The American Political Science Review* 77(4):974–990.

Aldrich, John H. and Richard D. McKelvey. 1977. "A Method of Scaling with Applications to the 1968 and 1972 Presidential Elections." *American Political Science Review* 71(1):111–130.

Alesina, Alberto. 1988. "Credibility and Policy Convergence in a Two-Party System with Rational Voters." *American Economic Review* 78(4):796–805.

Alesina, Alberto and Alex Cukierman. 1990. "The Politics of Ambiguity." *Quarterly Journal of Economics* 105(4):829–850.

Alesina, Alberto and Howard Rosenthal. 1995. *Partisan Politics, Divided Government and the Economy.* New York: Cambridge University Press.

Alvarez, R. Michael. 1997. *Information and Elections.* Ann Arbor: University of Michigan Press.

Ansolabehere, Stephen. 2010. *CCES Common Content, 2010.* V2 [Version].
URL: *http://projects.iq.harvard.edu/cces/*

Ansolabehere, Stephen, Jonathan Rodden and James M. Snyder, Jr. 2006. "Purple America." *The Journal of Economic Perspectives* 20(2):97–118.

Arabie, Phipps, J. Douglas Carroll and Wayne S. DeSarbo. 1987. *Three Way Scaling: A Guide to Multidimensional Scaling and Clustering.* Newbury Park, CA: Sage.

Aranson, Peter H. and Peter C. Ordeshook. 1972. "Spatial Strategies for

Sequential Elections." In *Probability Models of Collective Decision Making*, ed. Richard G. Niemi and Herbert F. Weisberg. Columbus, Ohio: Charles E. Merrill, pp. 298–331.

Arrow, Kenneth J. 1951. *Social Choice and Individual Values*. New York: Wiley.

Bafumi, Joseph and Michael C. Herron. 2010. "Leapfrog Representation and Extremism: A Study of American Voters and Their Members in Congress." *American Political Science Review* 104(3):519–542.

Bailey, Michael A. 2007. "Comparable Preference Estimates across Time and Institutions for the Court, Congress, and Presidency." *American Journal of Political Science* 51(3):433–448.

Bakker, Ryan, Catherine de Vries, Erica Edwards, Liesbet Hooghe, Seth Jolly, Gary Marks, Jonathan Polk, Jan Rovny, Marco Steenbergen and Milada Anna Vachudova. 2015. "Measuring Party Positions in Europe: The Chapel Hill Expert Survey Trend File, 1999–2010." *Party Politics* 21(1):143–152.

Bakker, Ryan and Keith T. Poole. 2013. "Bayesian Metric Multidimensional Scaling." *Political Analysis* 21(1):125–140.

Bakker, Ryan, Seth Jolly, Jon Polk and Keith T. Poole. 2013. "The European Common Space: Using Anchoring Vignettes to Scale Party Positions across Europe." Working paper.

Baron, David P. 1993. "Government Formation and Endogenous Parties." *American Political Science Review* 87(1):34–47.

Bartholomew, David J., Fiona Steele, Irini Moustaki and Jane I. Galbraith. 2008. *Analysis of Multivariate Social Science Data*. 2nd ed. Boca Raton, FL: Chapman & Hall/CRC.

Benoit, Kenneth, Kohei Watanabe, Haiyan Wang, Paul Nulty, Adam Obeng, Stefan Müller and Akitaka Matsuo. 2018. "quanteda: An R Package for the Quantitative Analysis of Textual Data." *Journal of Open Source Software* 3(30):774.

Benoit, Kenneth and Michael Laver. 2006. *Party Policy in Modern Democracies*. London: Routledge.

Berinsky, Adam J. and Jeffrey B. Lewis. 2007. "An Estimate of Risk Aversion in the U.S. Electorate." *Quarterly Journal of Political Science* 2(2):139–154.

Berndt, E.K., Bronwyn H. Hall, Robert E. Hall and Jerry Hausman. 1974. "Estimation and Inference in Nonlinear Structural Models." *Annals of Economic and Social Measurement* 3/4:653–666.

Black, Duncan. 1948. "On the Rationale of Group Decision-Making." *Journal of Political Economy* 56(1):23–34.

Black, Duncan. 1958. *The Theory of Committees and Elections*. Cambridge: Cambridge University Press.

Bolt, Daniel M. and Venessa F. Lall. 2003. "Estimation of Compensatory and Noncompensatory Multidimensional Item Response Models Using Markov Chain Monte Carlo." *Applied Psychological Measurement* 27(6):395–414.

Bonica, Adam. 2013. "Ideology and Interests in the Political Marketplace." *American Journal of Political Science* 57(2):294–311.

Borg, Ingwer and Patrick J.F. Groenen. 2010. *Modern Multidimensional Scaling: Theory and Applications*. 2nd ed. New York: Springer.

Brady, Henry E. 1985. "The Perils of Survey Research: Inter-Personally Incomparable Responses." *Political Methodology* 11(3–4):269–291.

Brady, Henry E. 1990. "Dimensional Analysis of Ranking Data." *American Journal of Political Science* 34(4):1017–1048.

Brady, Henry E. and Stephen Ansolabehere. 1989. "The Nature of Utility Functions in Mass Publics." *American Political Science Review* 83(1):143–163.

Brooks, S. P. and B. J. T. Morgan. 1995. "Optimization Using Simulated Annealing." *The Statistician* 44(2):241–257.

Butters, Ross and Christopher Hare. 2020. "Polarized Networks? New Evidence on American Voters' Political Discussion Networks." Forthcoming, *Political Behavior*.

Cahoon, Lawrence S., Melvin J. Hinich and Peter C. Ordeshook. 1976. "A Multidimensional Statistical Procedure for Spatial Analysis." Manuscript, Carnegie-Mellon University.

Carroll, J. Douglas and Jih-Jie Chang. 1970. "Analysis of Individual Differences in Multidimensional Scaling via an N-Way Generalization of 'Eckart-Young' Decomposition." *Psychometrika* 35(3):283–319.

Carroll, J. Douglas and Jih-Jie Chang. 1972. "IDIOSCAL (Individual Differences in Orientation SCALing): A Generalization of INSCAL Allowing Idiosyncratic Reference Systems as well as an Analytic Approximation to INDSCAL." Presented at the Spring Meeting of the Psychometric Society, Princeton, NJ.

Carroll, Royce and Hiroki Kubo. 2018. "Explaining Citizen Perceptions of Party Ideological Positions: The Mediating Role of Political Contexts." *Electoral Studies* 51:14–23.

Carroll, Royce, Jeffrey B. Lewis, James Lo, Keith T. Poole and Howard Rosenthal. 2009. "Comparing NOMINATE and IDEAL: Points of Difference and Monte Carlo Tests." *Legislative Studies Quarterly* 34(4):555–591.

Carroll, Royce, Jeffrey B. Lewis, James Lo, Keith T. Poole and Howard Rosenthal. 2013. "The Structure of Utility in Spatial Models of Voting." *American Journal of Political Science* 57(4):1008–1028.

Carroll, Royce and Keith T. Poole. 2014. "Roll Call Analysis and the Study of Legislatures." In *The Oxford Handbook of Legislative Studies*, ed. Shane Martin, Thomas Saaleld and Kaare Strøm. Oxford: Oxford University Press.

Carsey, Thomas M. 2000. *Campaign Dynamics: The Race for Governor.* Ann Arbor: University of Michigan Press.

Carson, Jamie L., Michael H. Crespin, Jeffery A. Jenkins and Ryan J. Vander Wielen. 2004. "Shirking in the Contemporary Congress: A Reappraisal." *Political Analysis* 12(2):176–179.

Carter, Dan T. 1995. *The Politics of Rage: George Wallace, the Origins of the New Conservatism, and the Transformation of American Politics.* Baton Rouge: Louisiana State University Press.

Chang, Jih-Jie and J. Douglas Carroll. 1969. "How to Use MDPREF, a Computer Program for Multidimensional Analysis of Preference Data." In *Multidimensional Scaling Program Package of Bell Laboratories.* Murray Hill, NJ: Bell Laboratories.

Cleveland, William S. 1981. "LOWESS: A Program for Smoothing Scatterplots by Robust Locally Weighted Regression." *The American Statistician* 35(1):54.

Clinton, Joshua D. and Simon Jackman. 2009. "To Simulate or NOMINATE?" *Legislative Studies Quarterly* 34(4):593–621.

Clinton, Joshua, Simon Jackman and Douglas Rivers. 2004. "The Statistical Analysis of Roll Call Data." *American Political Science Review* 98(2):355–370.

Converse, Philip E. 1964. "The Nature of Belief Systems in Mass Publics." In *Ideology and Discontent*, ed. David E. Apter. New York: Free Press, pp. 206–261.

Coombs, Clyde H. 1950. "Psychological Scaling without a Unit of Measurement." *Psychological Review* 57(3):145–158.

Coombs, Clyde H. 1952. "A Theory of Psychological Scaling." In *Engineering Research Bulletin Number 34.* Ann Arbor: University of Michigan Press.

Coombs, Clyde H. 1958. "On the Use of Inconsistency of Preferences in Psychological Measurement." *Journal of Experimental Psychology* 55(1):1–7.

Coombs, Clyde H. 1964. *A Theory of Data.* New York: Wiley.

Cox, Trevor F. and Michael A.A. Cox. 2001. *Multidimensional Scaing.* 2nd ed. Boca Raton, FL: Chapman & Hall/CRC.

Davis, Otto A., Melvin J. Hinich and Peter C. Ordeshook. 1970. "An Expository Development of a Mathematical Model of the Electoral Process." *The American Political Science Review* 64(2):426–448.

de Leeuw, Jan. 1977. "Applications of Convex Analysis to Multidimensional Scaling." In *Recent Developments in Statistics*, ed. J.R. Barra, F. Brodeau, G. Romer and B. van Custem. Amsterdam: North Holland Publishing Company, pp. 133–145.

de Leeuw, Jan. 1988. "Convergence of the Majorization Method for Multidimensional Scaling." *Journal of Classification* 5(2):163–180.

de Leeuw, Jan and Patrick Mair. 2009. "Multidimensional Scaling Using Majorization: SMACOF in R." *Journal of Statistical Software* 31(3):1–30. **URL:** *http://www.jstatsoft.org/v31/i03/*

de Leeuw, Jan and Willem J. Heiser. 1977. "Convergence of Correction Matrix Algorithms for Multidimensional Scaling." In *Geometric Representations of Relational Data*, ed. James C. Lingoes. Ann Arbor: Mathesis Press, pp. 735–752.

Delli Carpini, Michael X. and Scott Keeter. 1996. *What Americans Know about Politics and Why It Matters.* New Haven, CT: Yale University Press.

Dempster, Arthur P., Nan M. Laird and Donald B. Rubin. 1977. "Maximum Likelihood from Incomplete Data Via the EM Algorithm." *Journal of the Royal Statistical Society: Series B (Methodological)* 39(1):1–22.

Desposato, Scott W. 2006. "Parties for Rent? Ambition, Ideology, and Party Switching in Brazil's Chamber of Deputies." *American Journal of Political Science* 50(1):62–80.

Dougherty, Keith L. and Jac C. Heckelman. 2006. "A Pivotal Voter from a Pivotal State: Roger Sherman at the Constitutional Convention." *American Political Science Review* 100(2):297–302.

Downs, Anthony. 1957. *An Economic Theory of Democracy.* New York: Harper & Row.

Eckart, Carl H. and Gale Young. 1936. "The Approximation of One Matrix by Another of a Lower Rank." *Psychometrika* 1:211–218.

Efron, Bradley and Robert J. Tibshirani. 1993. *An Introduction to the Bootstrap.* New York: Chapman & Hall/CRC.

Ekman, Gosta. 1954. "Dimensions of Color Vision." *The Journal of Psychology*

38(2):467–474.

Enelow, James M. and Melvin J. Hinich. 1984. *The Spatial Theory of Voting.* New York: Cambridge University Press.

Epstein, Lee, Andrew D. Martin, Jeffrey A. Segal and Chad Westerland. 2007. "The Judicial Common Space." *Journal of Law, Economics, and Organization* 23(2):303–325.

Feinerer, Ingo, Kurt Hornik and David Meyer. 2008. "Text Mining Infrastructure in R." *Journal of Statistical Software* 25(5):1–54.

Finocchiaro, Charles J. and Jeffery A. Jenkins. 2008. "In Search of Killer Amendments in the Modern U.S. House." *Legislative Studies Quarterly* 33(2):263–294.

Fletcher, Roger. 1987. *Practical Methods of Optimization.* 2nd ed. New York: Wiley.

Fox, John. 2003. "Effect Displays in R for Generalised Linear Models." *Journal of Statistical Software* 8(15):1–27.
URL: *http://www.jstatsoft.org/v08/i15/*

Gelfand, Alan E. and Adrian F.M. Smith. 1990. "Sampling-Based Approaches to Calculating Marginal Densities." *Journal of the American Statistical Association* 85(410):398–409.

Gelman, Andrew. 1992. "Iterative and Non-iterative Simulation Algorithms." *Computing Science and Statistics* 24:433–438.

Gelman, Andrew and Donald B. Rubin. 1992. "Inference from Iterative Simulation Using Multiple Sequences." *Statistical Science* 7(4):457–472.

Gelman, Andrew and Donald B. Rubin. 1996. "Markov Chain Monte Carlo Methods in Biostatistics." *Statistical Methods in Medical Research* 5(4):339–355.

Gelman, Andrew, John B Carlin, Hal S Stern, David B Dunson, Aki Vehtari and Donald B Rubin. 2014. *Bayesian Data Analysis.* Vol. 2, Boca Raton, FL: CRC Press.

Geman, Donald and Stuart Geman. 1984. "Stochastic Relaxation, Gibbs Distributions, and the Bayesian Restoration of Images." *IEEE Transactions on Pattern Analysis and Machine Intelligence* 6(721–741).

Gerber, Elisabeth R. and Jeffrey B. Lewis. 2004. "Beyond the Median: Voter Preferences, District Heterogeneity, and Political Representation." *Journal of Political Economy* 112(6):1364–1383.

Geweke, John. 1992. "Evaluating the Accuracy of Sampling-Based Approaches to the Calculation of Posterior Moments." In *Bayesian Statistics 4*, ed.

J.M. Bernardo, J. Berger, A.P. Dawid and A.F.M. Smith. Oxford: Oxford University Press, pp. 169–193.

Geweke, John. 1993. "Bayesian Treatment of the Independent Student-t Linear Model." *Journal of Applied Econometrics* 8(S1):S19–S40.

Gill, Jeff. 2008. *Bayesian Methods: A Social and Behavioral Sciences Approach.* 2nd ed. Boca Raton, FL: Chapman & Hall/CRC.

Goffe, William L., Gary D. Ferrier and John Rogers. 1994. "Global Optimization of Statistical Functions with Simulated Annealing." *Journal of Econometrics* 60(1–2):65–99.

Gordon, Alex. 2004. "The Partial-Birth Abortion Ban Act of 2003." *Harvard Journal on Legislation* 41(2):501–516.

Greene, Kenneth F. 2007. *Why Dominant Parties Lose: Mexico's Democratization in Comparative Perspective.* New York: Cambridge University Press.

Grimmer, Justin. 2011. "An Introduction to Bayesian Inference via Variational Approximations." *Political Analysis* 19(1):32–47.

Guttman, Louis L. 1944. "A Basis for Scaling Qualitative Data." *American Sociological Review* 9:139–150.

Guttman, Louis L. 1950. "The Basis for Scalogram Analysis." In *Measurement and Prediction: The American Soldier Vol. IV.* New York: Wiley.

Hainmueller, Jens and Chad Hazlett. 2014. "Kernel Regularized Least Squares: Reducing MisspecificationBias with a Flexible and Interpretable Machine-Learning Approach." *Political Analysis* 22:143–168.

Hare, Christopher, David A Armstrong, Ryan Bakker, Royce Carroll and Keith T Poole. 2015. "Using Bayesian Aldrich-McKelvey Scaling to Study Citizens' Ideological Preferences and Perceptions." *American Journal of Political Science* 59(3):759–774.

Hare, Christopher and Keith T. Poole. 2014*a*. "Psychometric Methods in Political Science." In *The Wiley-Blackwell Handbook of Psychometric Testing*, ed. Paul Irwing, Tom Booth and David Hughes. New York: Wiley-Blackwell.

Hare, Christopher and Keith T. Poole. 2014*b*. "Psychometric Methods in Political Science." In *The Wiley-Blackwell Handbook of Psychometric Testing*, ed. Paul Irwing, Tom Booth and David Hughes. New York: Wiley-Blackwell.

Hare, Christopher, Tzu-Ping Liu and Robert N. Lupton. 2018. "What Ordered Optimal Classification Reveals about Ideological Structure, Cleavages, and Polarization in the American Mass Public." *Public Choice* 176(1-2):57–78.

Hastings, W.K. 1970. "Monte Carlo Sampling Methods Using Markov Chains and Their Applications." *Biometrika* 57(1):97–109.

Hinich, Melvin J. and Michael C. Munger. 1994. *Ideology and the Theory of Political Choice*. Ann Arbor: University of Michigan Press.

Hinich, Melvin J. and Michael C. Munger. 1997. *Analytical Politics*. Cambridge: Cambridge University Press.

Hinich, Melvin J. and Walker Pollard. 1981. "A New Approach to the Spatial Theory of Electoral Competition." *American Journal of Political Science* 25(2):323–341.

Hix, Simon, Abdul Noury and Gérard Roland. 2006. "Dimensions of Politics in the European Parliament." *American Journal of Political Science* 50(2):494–520.

Hix, Simon, Abdul G. Noury and Gérard Roland. 2007. *Democratic Politics in the European Parliament*. Cambridge: Cambridge University Press.

Horan, C. B. 1969. "Multidimensional Scaling: Combining Observations When Individuals Have Different Perceptual Structures." *Psychometrika* 34(2):139–165.

Hotelling, Harold. 1929. "Stability in Competition." *The Economic Journal* 39:41–57.

Imai, Kosuke, James Lo and Jonathan Olmsted. 2016. "Fast Estimation of Ideal Points with Massive Data." *American Political Science Review* 110(4):631–656.

Imai, Kosuke, James Lo and Jonathan Olmsted. 2017. *emIRT: EM Algorithms for Estimating Item Response Theory Models*. R package version 0.0.8. **URL:** *http://cran.r-project.org/package=emIRT*

Jackman, Simon. 2000*a*. "Estimation and Inference Are Missing Data Problems: Unifying Social Science Statistics via Bayesian Simulation." *Political Analysis* 8(4):307–332.

Jackman, Simon. 2000*b*. "Estimation and Inference via Bayesian Simulation: An Introduction to Markov Chain Monte Carlo." *American Journal of Political Science* 44(2):375–404.

Jackman, Simon. 2001. "Multidimensional Analysis of Roll Call Data via Bayesian Simulation: Identification, Estimation, Inference, and Model Checking." *Political Analysis* 9(3):227–241.

Jackman, Simon. 2009. *Bayesian Analysis for the Social Sciences*. New York: Wiley.

Jackman, Simon. 2012. *pscl: Classes and Methods for R Developed in the Political Science Computational Laboratory, Stanford University*. R package

version 1.04.1.
URL: *http://pscl.stanford.edu/*

Jacoby, William G. 1986. "Levels of Conceptualization and Reliance on the Liberal-Conservative Continuum." *Journal of Politics* 48(2):423–432.

Jacoby, William G. 1991. *Data Theory and Dimensional Analysis.* Thousand Oaks, CA: Sage.

Jacoby, William G. 2006. "Value Choices and American Public Opinion." *American Journal of Political Science* 50(3):706–723.

Jacoby, William G. 2009. "Public Opinion during a Presidential Campaign: Distinguishing the Effects of Environmental Evolution and Attitude Change." *Electoral Studies* 28(3):422–436.

Jacoby, William G. 2013. "Individual Value Structures and Personal Political Orientations: Determining the Direction of Influence." Presented at the Annual Meeting of the Midwest Political Science Association, Chicago, IL.

Jacoby, William G. and David A. Armstrong, II. 2014. "Bootstrap Confidence Regions for Multidimensional Scaling Solutions." *American Journal of Political Science* 58(1):264–278.

Jenkins, Jeffery A. and Brian R. Sala. 1998. "The Spatial Theory of Voting and the Presidential Election of 1824." *American Journal of Political Science* 42(4):1157–1179.

Jeong, Gyung-Ho, Gary J. Miller, Camilla Schofield and Itai Sened. 2011. "Cracks in the Opposition: Immigration as a Wedge Issue for the Reagan Coalition." *American Journal of Political Science* 55(3):511–525.

Jessee, Stephen A. 2009. "Spatial Voting in the 2004 Presidential Election." *American Political Science Review* 103(1):59–81.

Johnson, Norman L., Samuel Kotz and Narayanaswamy Balakrishnan. 1994. *Continuous Univariate Distributions.* Vol. 1 New York: Wiley.

Jost, John T., Christopher M. Federico and Jaime L. Napier. 2009. "Political Ideology: Its Structure, Functions, and Elective Affinities." *Annual Review of Psychology* 60(1):307–337.

King, Gary. 1989. *Unifying Political Methodology: The Likelihood Theory of Statistical Inference.* Cambridge: Cambridge University Press.

King, Gary, Christopher J.L. Murray, Joshua A. Salomon and Ajay Tandon. 2004. "Enhancing the Validity and Cross-Cultural Comparability of Measurement in Survey Research." *American Political Science Review* 98(1):191–207.

King, Gary and Jonathan Wand. 2007. "Comparing Incomparable Survey Responses: Evaluating and Selecting Anchoring Vignettes." *Political Analysis* 15(1):46–66.

Kruskal, Joseph B. 1964*a*. "Multidimensional Scaling by Optimizing a Goodness of Fit to a Nonmetric Hypothesis." *Psychometrika* 29(1):1–27.

Kruskal, Joseph B. 1964*b*. "Nonmetric Multidimensional Scaling: A Numerical Method." *Psychometrika* 29(2):115–129.

Kruskal, Joseph B. and Myron Wish. 1978. *Multidimensional Scaling*. Quantiative Applications in the Social Sciences Beverly Hills, CA: Sage.

Ladha, Krishna K. 1991. "A Spatial Model of Legislative Voting with Perceptual Error." *Public Choice* 68(1-3):151–174.

Lauderdale, Benjamin E. 2010. "Unpredictable Voters in Ideal Point Estimation." *Political Analysis* 18(2):151–171.
URL: *https://www.cambridge.org/core/journals/political-analysis/article/div-classtitleunpredictable-voters-in-ideal-point-estimationdiv/F128B5910CE7229E28A2C954E199855F*

Lauderdale, Benjamin E. 2013. "Does Inattention to Political Debate Explain the Polarization Gap between the U.S. Congress and Public?" *Public Opinion Quarterly* 77(S1):2–23.

Layman, Geoffrey C. 2001. *The Great Divide: Religious and Cultural Conflict in American Party Politics*. New York: Columbia University Press.

Layman, Geoffrey C., Thomas M. Carsey, John C. Green, Richard Herrera and Rosalyn Cooperman. 2010. "Activists and Conflict Extension in American Party Politics." *American Political Science Review* 104(2):324–346.

Lewis, Jeffrey B. and Gary King. 1999. "No Evidence on Directional vs. Proximity Voting." *Political Analysis* 8(1):21–33.

Lewis, Jeffrey B. and Keith T. Poole. 2004. "Measuring Bias and Uncertainty in Ideal Point Estimates via the Parametric Bootstrap." *Political Analysis* 12(2):105–127.

Likert, Rensis. 1932. "A Technique for the Measurement of Attitudes." *Archives of Psychology* 22(140):44–53.

Limpert, Eckhard, Werner A. Stahel and Markus Abbt. 2001. "Log-normal Distributions across the Sciences: Keys and Clues." *BioScience* 51(5):341–352.

Lo, James, Keith Poole, Jeff Lewis and Christopher Hare. 2013. *anominate: alpha-NOMINATE Ideal Point Estimator*. R package version 0.01.
URL: *http://www.voteview.com/alphanominate.asp*

Mackie, Gerry. 2003. *Democracy Defended*. Cambridge: Cambridge University

Press.

MacRae, Jr., Duncan. 1958. *Dimensions of Congressional Voting.* Berkeley: University of California Press.

MacRae, Jr., Duncan. 1967. *Parliament, Parties, and Society in France 1946-1958.* New York: St. Martin's Press.

MacRae, Jr., Duncan. 1970. *Issues and Parties in Legislative Voting.* New York: Harper & Row.

Malecki, Michael. 2012. *apsrtable: apsrtable Model-Output Formatter for Social Science.* R package version 0.8-8.
URL: *http://CRAN.R-project.org/package=apsrtable*

Martin, Andrew D. 2003. "Bayesian Inference for Heterogeneous Event Counts." *Sociological Methods and Research* 32:30–63.

Martin, Andrew D. and Kevin M. Quinn. 2002. "Dynamic Ideal Point Estimation via Markov Chain Monte Carlo for the U.S. Supreme Court, 1953-1999." *Political Analysis* 10(2):134–153.

Martin, Andrew D. and Kevin M. Quinn. 2007. "Assessing Preference Change on the US Supreme Court." *Journal of Law, Economics, and Organization* 23(2):365–385.

Martin, Andrew D., Kevin M. Quinn and Jong Hee Park. 2011. "MCMCpack: Markov chain Monte Carlo in R." *Journal of Statistical Software* 42(9):1–21.

Martin, Andrew D., Kevin M. Quinn and Lee Epstein. 2005. "The Median Justice on the United States Supreme Court." *North Carolina Law Review* 83:1275–1322.

Martin, Olivier C. and Steve W. Otto. 1996. "Combining Simulated Annealing with Local Search Heuristics." *Annals of Operations Research* 63(1):57–75.

Matsumoto, Makoto and Takuji Nishimura. 1998. "Mersenne Twister: A 623-Dimensionally Equidistributed Uniform Pseudo-Random Number Generator." *ACM Transactions on Modeling and Computer Simulation* 8(1):3–30.

May, William, Keith Poole and Nolan McCarty. 2019. *dwnominate: An interface to the DW-NOMINATE roll call scaling program.* R package version 1.1.2.

McCann, James A. 2012. "Changing Dimensions of National Elections in Mexico." In *The Oxford Handbook of Mexican Politics*, ed. Roderic A. Camp. New York: Oxford University Press, pp. 497–522.

McCarty, Nolan, Keith T. Poole and Howard Rosenthal. 2006. *Polarized America: The Dance of Ideology and Unequal Riches.* Cambridge, MA: MIT Press.

McCarty, Nolan, Keith T. Poole and Howard Rosenthal. 2013. *Political Bubbles: Financial Crises and the Failure of American Democracy.* Princeton, NJ: Princeton University Press.

McCarty, Nolan M. and Keith T. Poole. 1995. "Veto Power and Legislation: An Empirical Analysis of Executive and Legislative Bargaining from 1961 to 1986." *Journal of Law, Economics, & Organization* 11(2):282–312.

McCarty, Nolan M., Keith T. Poole and Howard Rosenthal. 1997. *Income Redistribution and the Realignment of American Politics.* AEI Studies on Understanding Economic Inequality Washington, DC: AEI Press.

McFadden, Daniel L. 1976. "Quantal Choice Analaysis: A Survey." *Annals of Economic and Social Measurement* 5(4):363–390.

McKelvey, Richard D. 1976. "Intransitivities in Multidimensional Voting Models and Some Implications for Agenda Control." *Journal of Economic Theory* 12(3):472–482.

McKelvey, Richard D. 1979. "General Conditions for Global Intransitivities in Formal Voting Models." *Econometrica* 47(5):1085–1112.

Mebane, Jr., Walter R. and Jasjeet S. Sekhon. 2011. "Genetic Optimization Using Derivatives: The rgenoud for R." *Journal of Statistical Software* 42(11):1–26.
URL: *http://www.jstatsoft.org/v42/i11/*

Merrill, Samuel, III and Bernard Grofman. 1999. *A Unified Theory of Voting: Directional and Proximity Spatial Models.* Cambridge: Cambridge University Press.

Metropolis, Nicholas and Stanislaw Ulam. 1949. "The Monte Carlo Method." *Journal of the American Statistical Association* 44(247):335–341.

Mikhaylov, Slava, Michael Laver and Kenneth Benoit. 2006. "Coder Reliability and Misclassification in Comparative Manifesto Project Codings." Presented at the Annual Meeting of the Midwest Political Science Association, Chicago, IL.

Moreno, Alejandro. 2007. "The 2006 Mexican Presidential Election: The Economy, Oil Revenues, and Ideology." *PS: Political Science and Politics* 40(1):15–19.

Mulaik, Stanley A. 2009. *Foundations of Factor Analysis.* 2nd ed. Boca Raton, FL: Chapman & Hall/CRC.

Neal, Radford M. 2003. "Slice Sampling." *The Annals of Statistics* 31(3):705–741.

Nelder, John A. and Roger Mead. 1965. "A Simplex Method for Function Minimization." *Computer Journal* 7(4):308–313.

Nokken, Timothy P. and Keith T. Poole. 2004. "Congressional Party Defection in American History." *Legislative Studies Quarterly* 29:545–568.

Oh, Man-Suk and Adrian E. Raftery. 2001. "Bayesian Multidimensional Scaling and Choice of Dimension." *Journal of the American Statistical Association* 96(455):1031–1044.

Okada, Kensuke and Kazuo Shigemasu. 2010. "Bayesian Multidimensional Scaling for the Estimation of a Minkowski Exponent." *Behavior Research Methods* 42(4):899–905.

Palfrey, Thomas R. 1984. "Spatial Equilibrium with Entry." *The Review of Economic Studies* 51(1):139–156.

Palfrey, Thomas R. and Keith T. Poole. 1987. "The Relationship between Information, Ideology, and Voting Behavior." *American Journal of Political Science* 31(3):511–530.

Peress, Michael. 2012. "Identification of a Semiparametric Item Response Model." *Psychometrika* 77(2):223–243.

Petrocik, John R. 1974. "An Analysis of Intransitivities in the Index of Party Identification." *Political Methodology* 1:31–47.

Petrocik, John R. 2009. "Measuring Party Support: Leaners are not Independents." *Electoral Studies* 28(4):562–572.

Plott, Charles R. 1967. "A Notion of Equilibrium and Its Possibility under Majority Rule." *American Economic Review* 57(4):787–806.

Plummer, Martyn. 2003*a*. JAGS: A Program for Analysis of Bayesian Graphical Models Using Gibbs Sampling. In *Proceedings of the 3rd International Workshop on Distributed Statistical Computing, Vienna, Austria*, ed. Kurt Hornik, Friedrich Leisch and Achim Zeileis.

Plummer, Martyn. 2003*b*. *JAGS: A Program for Analysis of Bayesian Graphical Models Using Gibbs Sampling.*
URL: *http://mcmc-jags.sourceforge.net/*

Plummer, Martyn. 2013. *rjags: Bayesian Graphical Models using MCMC.* R package version 3-10.
URL: *http://CRAN.R-project.org/package=rjags*

Plummer, Martyn, Nicky Best, Kate Cowles and Karen Vines. 2006. "CODA: Convergence Diagnosis and Output Analysis for MCMC." *R News* 6(1):7–11.
URL: *http://CRAN.R-project.org/doc/Rnews/Rnews_2006-1.pdf*

Poole, Keith, Howard Rosenthal, Jeffrey Lewis, James Lo and Royce Carroll. 2013. *basicspace: A Package to Recover a Basic Space from Issue Scales.* R package version 0.07.

URL: *http://CRAN.R-project.org/package=basicspace*

Poole, Keith, Jeffrey Lewis, James Lo and Royce Carroll. 2011. "Scaling Roll Call Votes with wnominate in R." *Journal of Statistical Software* 42(14):1–21.
URL: *http://www.jstatsoft.org/v42/i14/*

Poole, Keith, Jeffrey Lewis, James Lo and Royce Carroll. 2012. *oc: OC Roll Call Analysis Software.* R package version 0.93.
URL: *http://cran.r-project.org/web/packages/oc/index.html*

Poole, Keith T. 1981. "Dimensions of Interest Group Evaluation of the U.S. Senate, 1969-1978." *American Journal of Political Science* 25(1):49–67.

Poole, Keith T. 1984. "Least Squares Metric, Unidimensional Unfolding." *Psychometrika* 49(3):311–323.

Poole, Keith T. 1990. "Least Squares Metric, Unidimensional Scaling of Multivariate Linear Models." *Psychometrika* 55(1):123–149.

Poole, Keith T. 1998*a*. "How to Use the Black Box." Manuscript, University of Georgia.
URL: *http://voteview.com/blackuse.pdf*

Poole, Keith T. 1998*b*. "Recovering a Basic Space from a Set of Issue Scales." *American Journal of Political Science* 42(3):954–993.

Poole, Keith T. 2000. "Nonparametric Unfolding of Binary Choice Data." *Political Analysis* 8(3):211–237.

Poole, Keith T. 2001. "The Geometry of Multidimensional Quadratic Utility in Models of Parliamentary Roll Call Voting." *Political Analysis* 9(3):211–226.

Poole, Keith T. 2005. *Spatial Models of Parliamentary Voting.* New York: Cambridge University Press.

Poole, Keith T. 2007. "Changing Minds? Not in Congress!" *Public Choice* 131(3/4):435–451.

Poole, Keith T. 2017. "The Scientific Status of Geometric Models of Choice and Similarities Judgment." *Public Choice* 171(3/4):245–256.

Poole, Keith T. and Howard Rosenthal. 1983. "A Spatial Model for Legislative Roll Call Analysis." GSIA Working Paper No. 5-83-84.

Poole, Keith T. and Howard Rosenthal. 1984. "U.S. Presidential Elections 1968–80: A Spatial Analysis." *American Journal of Political Science* 28(2):282–312.

Poole, Keith T. and Howard Rosenthal. 1985. "A Spatial Model for Legislative Roll Call Analysis." *American Journal of Political Science* 29(2):357–384.

Poole, Keith T. and Howard Rosenthal. 1991. "Patterns of Congressional Voting." *American Journal of Political Science* 35(1):228–278.

Poole, Keith T. and Howard Rosenthal. 1997. *Congress: A Political-Economic History of Roll Call Voting*. New York: Oxford University Press.

Poole, Keith T. and Howard Rosenthal. 2007. *Ideology and Congress*. New Brunswick, NJ: Transaction.

Poole, Keith T. and R. Steven Daniels. 1985. "Ideology, Party, and Voting in the U.S. Congress, 1959–1980." *American Political Science Review* 79(2):373–399.

Quinn, Kevin M. 2004. "Bayesian Factor Analysis for Mixed Ordinal and Continuous Responses." *Political Analysis* 12(4):338–353.

Quinn, Kevin M. and Andrew D. Martin. 2002. "An Integrated Computational Model of Multiparty Electoral Competition." *Statistical Science* 17:405–419.

Quinn, Kevin M., Andrew D. Martin and Andrew B. Whitford. 1999. "Voter Choice in Multi-party Democracies: A Test of Competing Theories and Models." *American Journal of Political Science* 43(4):1231–1247.

Rabinowitz, George. 1975. "An Introduction to Nonmetric Multidimensional Scaling." *American Journal of Political Science* 19(2):343–390.

Rabinowitz, George. 1978. "On the Nature of Political Issues: Insights from a Spatial Analysis." *American Journal of Political Science* 22(4):793–817.

Rabinowitz, George and Stuart Elaine Macdonald. 1989. "A Directional Theory of Issue Voting." *American Political Science Review* 83(1):93–121.

Ramsay, James O. 1977. "Maximum Likelihood Estimation in Multidimensional Scaling." *Psychometrika* 42(2):241–266.

Reckase, Mark D. 1985. "The Difficulty of Test Items That Measure More Than One Ability." *Applied Psychological Measurement* 9(4):401–412.

Reckase, Mark D. 2010. A Linear Logistic Multidimensional Model for Dichotomous Item Response Data. In *Handbook of Modern Item Response Theory*, ed. Wim J. van der Linden and Ronald K. Hambleton. New York: Springer pp. 271–286.

Rice, Stuart A. 1928. *Quantitative Methods in Politics*. New York: Knopf.

Riker, William H. 1980. "Implications from the Disequilibrium of Majority Rule for the Study of Institutions." *American Political Science Review* 74(2):432–446.

Riker, William H. 1982. *Liberalism against Populism: A Confrontation Between the Theory of Democracy and the Theory of Social Choice*. San Francisco: W.H. Freeman.

Riker, William H. 1986. *The Art of Political Manipulation.* New Haven, CT: Yale University Press.

Riker, William H. 1990*a*. "Heresthetic and Rhetoric in the Spatial Model." In *Advances in the Spatial Theory of Voting*, ed. James M. Enelow and Melvin J. Hinich. New York: Cambridge University Press, pp. 46–65.

Riker, William H. 1990*b*. "Political Science and Rational Choice." In *Perspectives on Positive Political Economy*, ed. James E. Alt and Kenneth A. Shepsle. New York: Cambridge University Press, pp. 163–181.

Riker, William H. 1996. *The Strategy of Rhetoric: Campaigning for the American Constitution.* New Haven, CT: Yale University Press.

Riker, William H. and Peter C. Ordeshook. 1973. *An Introduction to Positive Political Theory.* Englewood Cliffs, NJ: Prentice-Hall.

Rivers, Douglas. 2003. "Identification of Multidimensional Spatial Voting Models." Manuscript, Stanford University.

Robert, Christian P. and George Casella. 2010. *Introducing Monte Carlo Methods with R.* New York: Springer.

Rosenthal, Howard and Erik Voeten. 2004. "Analyzing Roll Calls with Perfect Spatial Voting: France 1946–1958." *American Journal of Political Science* 48(3):620–632.

Rosenthal, Howard and Erik Voeten. 2007. "Measuring Legal Systems." *Journal of Comparative Economics* 35(4):711–728.

Saad, Lydia. 2003. "Americans Agree with Banning 'Partial-Birth Abortion'." November 6, 2003. Gallup News Service `http://www.gallup.com/poll/9658/americans-agree-banning-partialbirth-abortion.aspx`.

Sammon, Jr., John W. 1969. "A Nonlinear Mapping for Data Structure Analysis." *IEEE Transactions on Computers* C-18(5):401–409.

Schofield, Norman. 1978. "Instability of Simple Dynamic Games." *Review of Economic Studies* 45(3):575–594.

Schofield, Norman, Andrew D. Martin, Kevin M. Quinn and Andrew B. Whitford. 1998. "Multiparty Electoral Competition in the Netherlands and Germany: A Model Based on Multinomial Probit." *Public Choice* 97(3):257–293.

Schönemann, Peter. 1970. "On Metric Multidimensional Unfolding." *Psychometrika* 35(3):349–366.

Sekhon, Jasjeet S. and Walter R. Mebane, Jr. 1998. "Genetic Optimization Using Derivatives." *Political Analysis* 7(1):187–210.

Shapiro, Robert Y. and Benjamin I. Page. 1988. "Foreign Policy and the

Rational Public." *The Journal of Conflict Resolution* 32(2):211–247.

Shepard, Roger N. 1987. "Toward a Universal Law of Generalization for Psychological Science." *Science* 237(4820):1317–1323.

Shepsle, Kenneth A. 2010. *Analyzing Politics: Rationality, Behavior, and Institutions.* 2nd ed. New York: W.W. Norton.

Shor, Boris and Nolan McCarty. 2011. "The Ideological Mapping of American Legislatures." *American Political Science Review* 105(3):530–551.

Slapin, Jonathan B. and Sven-Oliver Proksch. 2008. "A Scaling Model for Estimating Time-Series Party Positions from Texts." *American Journal of Political Science* 52(3):705–722.

Smithies, Arthur. 1941. "Optimum Location in Spatial Competition." *Journal of Political Economy* 49(3):423–439.

Spearman, Charles E. 1904. "'General Intelligence' Objectively Determined and Measured." *American Journal of Psychology* 15:201–293.

Stimson, James A. 2004. *Tides of Consent: How Public Opinion Shapes American Politics.* Cambridge: Cambridge University Press.

Stone, Walter J. and Alan I. Abramowitz. 1983. "Winning May Not Be Everything, but It's More Than We Thought: Presidential Party Activists in 1980." *American Political Science Review* 77(4):945–956.

Struthers, Cory L., Christopher Hare and Ryan Bakker. 2020. "Bridging the Pond: Measuring Policy Positions in the United States and Europe." *Political Science Research and Methods* p. 1–15.

Takane, Yoshio. 1981. "Multidimensional Successive Categories Scaling: A Maximum Likelihood Method." *Psychometrika* 46(1):9–28.

Takane, Yoshio, Forrest Young and Jan de Leeuw. 1977. "Nonmetric Individual Differences Multidimensional Scaling: An Alternating Least Squares Method with Optimal Scaling Features." *Psychometrika* 42(1):7–67.

Tausanovitch, Chris and Christopher Warshaw. 2013. "Measuring Constituent Policy Preferences in Congress, State Legislatures, and Cities." *Journal of Politics* 75(2):330–342.

Team, R Core. 2018. "R: A language and environment for statistical computing [Internet]. Vienna, Austria: R Foundation for Statistical Computing; 2018.".

Thurstone, L.L. 1932. "Isolation of Blocs in a Legislative Body by the Voting Records of its Members." *Journal of Social Psychology* 3(4):425–433.

Tomz, Michael and Robert P. Van Houweling. 2008. "Candidate Positioning and Voter Choice." *American Political Science Review* 102(3):303–318.

Torgerson, Warren S. 1952. "Multidimensional Scaling: I. Theory and Method." *Psychometrika* 17:401–419.

Torgerson, Warren S. 1958. *Theory and Methods of Scaling*. New York: Wiley.

Treier, Shawn and D. Sunshine Hillygus. 2009. "The Nature of Political Ideology in the Contemporary Electorate." *Public Opinion Quarterly* 73(4):679–703.

Treier, Shawn and Simon Jackman. 2008. "Democracy as a Latent Variable." *American Journal of Political Science* 52(1):201–217.

Tufte, Edward R. 1983. *The Visual Display of Quantitative Information*. Cheshire, CT: Graphics Press.

van Schuur, Wijbrandt H. 1992. "Nonparametric Unidimensional Unfolding for Multicategory Data." *Political Analysis* 4(1):41–74.

van Schuur, Wijbrandt H. 2003. "Mokken Scale Analysis: Between the Guttman Scale and Parametric Item Response Theory." *Political Analysis* 11(2):139–163.

van Schuur, Wijbrandt H. 2011. *Ordinal Item Response Theory: Mokken Scale Analysis*. Thousand Oaks, CA: Sage.

Vavreck, Lynn. 2009. *The Message Matters: The Economy and Presidential Campaigns*. Princeton, NJ: Princeton Universty Press.

Venables, W. N. and B. D. Ripley. 2002. *Modern Applied Statistics with S*. Fourth ed. New York: Springer.

Voeten, Erik. 2000. "Clashes in the Assembly." *International Organization* 54(2):185–215.

Weisberg, Herbert F. 1968. "Dimensional Analysis of Legislative Roll Calls." Doctoral dissertation, University of Michigan.

Weisberg, Herbert F. and Jerrold G. Rusk. 1970. "Dimensions of Candidate Evaluation." *American Political Science Review* 64(4):1167–1185.

Weller, Susan C. and A Kimball Romney. 1990. *Metric Scaling: Correspondence Analysis*. Newbury Park, CA: Sage.

Wilcox, Clyde, Lee Sigelman and Elizabeth Cook. 1989. "Some Like It Hot: Individual Differences in Responses to Group Feeling Thermometers." *The Public Opinion Quarterly* 53(2):246–257.

Wink, Kenneth A., C. Don Livingston and James C. Garand. 1996. "Dispositions, Constituencies, and Cross-Pressures: Modeling Roll-Call Voting on the North American Free Trade Agreement in the U.S. House." *Political Research Quarterly* 49(4):749–770.

Wish, Myron. 1971. "Individual Differences in Perceptions and Preferences Among Nations." In *Attitude Research Reaches New Heights*, ed. Charles W. King and Douglas J. Tigert. Chicago: American Marketing Association, pp. 312–328.

Young, Forrest, Jan de Leeuw and Yoshio Takane. 1976. "Regression with Qualitative and Quantitative Variables: An Alternating Least Squares Method with Optimal Scaling Features." *Psychometrika* 41(4):505–529.

Young, Gale and Alston S. Householder. 1938. "Discussion of a Set of Points in Terms of their Mutual Distances." *Psychometrika* 3:19–22.

Zakharova, Maria and Paul V. Warwick. 2014. "The Sources of Valence Judgments: The Role of Policy Distance and the Structure of the Left–Right Spectrum." *Comparative Political Studies* 47(14):2000–2025.

Index